Springer Series in Reliability Engineering

Today's modern systems have become increasingly complex to design and build, while the demand for reliability and cost effective development continues. Reliability is one of the most important attributes in all these systems, including aerospace applications, real-time control, medical applications, defense systems, human decision-making, and home-security products. Growing international competition has increased the need for all designers, managers, practitioners, scientists and engineers to ensure a level of reliability of their product before release at the lowest cost. The interest in reliability has been growing in recent years and this trend will continue during the next decade and beyond.

The Springer Series in Reliability Engineering publishes books, monographs and edited volumes in important areas of current theoretical research development in reliability and in areas that attempt to bridge the gap between theory and application in areas of interest to practitioners in industry, laboratories, business, and government.

Now with 100 volumes!

Indexed in Scopus and EI Compendex

Interested authors should contact the series editor, Hoang Pham, Department of Industrial and Systems Engineering, Rutgers University, Piscataway, NJ 08854, USA. Email: hopham@rci.rutgers.edu, or Anthony Doyle, Executive Editor, Springer, London. Email: anthony.doyle@springer.com.

Vijay Kumar · Hoang Pham
Editors

Predictive Analytics in System Reliability

Springer

Editors
Vijay Kumar
Department of Mathematics
Amity Institute of Applied Sciences
Amity University Uttar Pradesh
Noida, Uttar Pradesh, India

Hoang Pham
Department of Industrial and Systems Engineering
Rutgers University
Piscataway, NJ, USA

ISSN 1614-7839 ISSN 2196-999X (electronic)
Springer Series in Reliability Engineering
ISBN 978-3-031-05346-7 ISBN 978-3-031-05347-4 (eBook)
https://doi.org/10.1007/978-3-031-05347-4

This Springer imprint is published by the registered company Springer Nature Switzerland AG
The registered company address is: Gewerbestrasse 11, 6330 Cham, Switzerland

Preface

The development of modern methodologies allows for efficient updating the system when information changes. Automatic model calibration with existing latest techniques from Machine Learning (ML), Artificial Intelligence (AI), Big Data, Genetic Algorithm (GA), Information Theory, Multi-criteria decision-making (MCDM) provides disruptive solutions for business insight. Predictive analytics refers to making predictions about the future based on different parameters which are from historical data. The historical data is fed into a mathematical model that considers key trends and patterns in the data. The model obtained is then applied to current data to predict the future trends. Machine Learning, Artificial Intelligence, Big Data, Genetic Algorithm (GA), Information Theory, Multi-criteria decision-making, and other techniques play a vital role to analyze the historical data and forecasts about what may happen in the future with an acceptable level of reliability. Overall, ML and other techniques are capable of providing novel insights and opportunities to solve important challenges in reliability and safety applications. Reliability analysis is one of the most multidimensional area in systems reliability engineering. The recent developments in system reliability has also created many opportunities and challenges for both industrialists and academicians. It has also revolutionized and completely transformed the systems engineering environment. Most of the modeling tasks can now be undertaken within a simulated environment using latest simulation and virtual reality technologies.

The main aim of the book is to publish the well-written original research studies and articles that describe the latest research and developments in the area of system reliability engineering with the application of Machine Learning, Artificial Intelligence, Big data, Genetic Algorithm, Information Theory, Multi-criteria decision-making, and other techniques. The book "*Predictive Analytics in System Reliability*" consists of 17 chapters featuring a broad range of topics including the latest applications of predictive analytics in comprehensive range of systems reliability engineering. Each chapter is written by well-known researchers and IT practitioners to present recent trends and research opportunities in the area of reliability engineering. More specifically, Chapter "Deep Learning Approach Based on Fault Correction Time for Reliability Assessment of Cloud and Edge Open Source Software" discussed a machine learning approach considering the characteristics reliability trends of edge open source project. Authors proposed a method to extract the characteristics data in order to comprehend the trend of failure big data recorded on the bug tracking system in edge open source project. Chapter "System Reliability Models with Random Shocks and Uncertainty: A State-of-the-Art Review" focused on the review of system reliability models with random shocks and the uncertainty of the degradation process. The system reliability models based on five random shock models that are commonly used in Reliability Engineering, cumulative shock model, extreme shock model, run shock model, δ-shock model, and mixed shock model were reviewed. Chapter "A Hybrid Approach for Evaluation and Prioritization of Software Vulnerabilities" has sub-grouped software vulnerability into code execution vulnerabilities and improper authentication vulnerabilities and focused on assessing the vulnerabilities which are most prone to attacks. This chapter describes a hybrid methodology comprising of the fuzzy Best Worst Method to prioritize the identified software vulnerabilities, followed by a two-way analysis to integrate the opinion of decision-makers.

Chapter "Investigating Bad Smells with Feature Selection and Machine Learning Approaches" discussed optimisation features of Android code smells in terms of software metrics using feature selection technique based on Correlation on 2896 instances of open source projects which are extracted from GitHub. Further, authors have examined the performance measures like accuracy, precision, F-measure and execution time, etc., with the reduced features data set of Android code smells and discussed about implementation of correlation-based feature selection algorithm to reduce the features of code smells. Chapter "SDE Based SRGM Considering Irregular Fluctuation in Fault Introduction Rate" proposed a software reliability growth model considering the irregular fluctuation of fault introduction rate over time with non-constant fault detection rate and assuming that fault introduction changes non-linearly over time and the fault introduction rate fluctuates irregularly. Chapter "Ant Colony Optimization Algorithm with Three Types of Pheromones for the Component Assignment Problem in Linear Consecutive-k-out-of-n:F Systems" presented an ACO algorithm with three types of pheromones for solving the component assignment problem of the linear consecutive--out-of-:F system. This configuration can be used to represent a real system in which consecutive failed components cause system failures.

Chapter "Reliability Assessment and Profit Analysis of Automated Teller Machine System Under Copular Repair Policy" presented the structure for analyzing Automated Teller Machines (ATMs) failures that allows for the identification of the most appropriate methods for removing them and formulated expressions for system availability, reliability, mean time to failure (MTTF), and cost function related to performance measures. Chapter "An Efficient Regression Test Cases Selection & Optimization Using Mayfly Optimization Algorithm" adopted Mayfly Optimization Algorithm to solve the regression test case selection problem to minimize the maintenance cost with the aim to optimize the number of test cases to re-execute to reduce the execution time and cost. Chapter "Development of Reliability Block Diagram (RBD) Model for Reliability Analysis of a Steam Boiler System" used reliability block diagrams (RBD) to estimate the reliability of boiler systems used in Indian textile industries considering their physical arrangement in the system and analyzed the impact of item failures on system availability.

Chapter "Computation Signature Reliability of Computer Numerical Control System Using Universal Generating Function" aimed to deal with a complex manufacturing system using the Computer Numerical Control as the bottom case manufacturing system, where the arrangement of various complex subsystems is in series, parallel, or in both the configurations. Chapter "Evaluate and Measure Agile Software Efficiency by the Integrated Strategy of Fuzzy MOORA and AHP" used an approach or proposed an integrated strategy of Fuzzified Multi-Objective Optimization on the bases of Ratio Analysis (MOORA) and Analytic Hierarchy Process (AHP) to determine the efficiency of Agile software. Chapter "Software Reliability Models and Multi-attribute Utility Function Based Strategic Decision for Release Time Optimization" discussed a realistic approach for determining when to stop software testing that considers reliability and cost. A multi-attribute utility theory-based proposed decision model is analyzed on various separate weighted combinations of utility functions.

Chapter "Reliability Analysis of Centerless Grinding Machine Using Fault Tree Analysis" addresses the Fault Tree Analysis (FTA) of the Centerless Grinding Machine (CGM) for safety purposes. FTA is one of the reliability evaluation technique that plays a crucial role in the design process. Fault Tree is a graphical representation of major faults or critical failures associated with a system. It uses Boolean logic and low-level event methods to analyze the possible mechanisms of failures and evaluate the expected frequency of their occurrences by describing undesired states of the system. Chapter "Machine Learning Based Software Defect Categorization Using Crowd Labeling" proposed a learning model which learns effectively to predict the impact category of software defects using the expectation maximization algorithm and shows the better performance according to the various types of metrics by improving the existing technique by 8% and 11% accuracy for Compendium and Mozilla datasets respectively. Chapter "Development of an Algorithm Using the Vikor Method to Increase Software Reliability" used the VIKOR (VIsekriterijumska optimizacija i KOmpromisno Resenje) method for the development of an algorithm to increase software reliability. Chapter "Mathematical Modeling for Evaluation Reliability of a Bleaching System" deals with the various reliability measures analysis for a complex bleaching system. The system has a complex structure with three subsystems associated with each other in series arrangement. Chapter "An Effort Allocation Model for a Three Stage Software Reliability Growth Model" analyzes the optimal efforts allocation plan for minimizing overall cost during the testing phase of the software development life cycle using three stages of fault detection, isolation, and removal in a dynamic environment.

The book is enriched by figures, examples, and case studies. The main benefit of the book is to look at recent methods and algorithms, techniques, ML with AI related areas and their applications in system reliability. The book is a timely publication and will be a valuable source of reference for graduate, post graduate, research students, and for professionals in research groups of large companies involved in reliability engineering. We hope our readers will enjoy the book and will find it both interesting and useful. As Editors of this book, we very much thank the authors for accepting to contribute with their invaluable research, their efforts, and time. Our special thanks to Dr. Anthony Doyle, the Executive Editor and Kavitha Sathish, the Project coordinator, for their extensive support and cooperation in bringing out this book. We acknowledge Springer for this opportunity and professional support extended to us in the project.

Noida, India
Piscataway, USA
March 2022

Vijay Kumar
Hoang Pham

Contents

About the Editors

Dr. Vijay Kumar received his M.Sc. in Applied Mathematics and M.Phil. in Mathematics from Indian Institute of Technology (IIT), Roorkee, India in 1998 and 2000, respectively. He has completed his PhD from the Department of Operational Research, University of Delhi. Currently, he is an Associate Professor in the Department of Mathematics, Amity Institute of Applied Sciences, Amity University Uttar Pradesh, Noida, India. He is the coauthor of one book and has published more than 55 research papers in the areas of software reliability, mathematical modeling and optimization in international journals and conferences of high repute. His current research interests include software reliability growth modeling, optimal control theory, and marketing models in the context of innovation diffusion theory. He has reviewed many papers for Soft Computing (Springer), IJRQSE, IJQRM, IJSAEM, and other reputed journals. He has edited special issues of IJAMS, IJCMESM (Taylor & Francis), and RIO journal. He is an editorial board member of IJMEMS. He is a life member of Society for Reliability Engineering, Quality and Operations Management (SREQOM).

Dr. Hoang Pham is a Distinguished Professor and former Chairman (2007–2013) of the Department of Industrial and Systems Engineering at Rutgers University, New Jersey. Before joining Rutgers, he was a Senior Engineering Specialist with the Boeing Company and the Idaho National Engineering Laboratory. He has been served as Editor-in-Chief, Editor, Associate Editor, Guest Editor, and board member of many journals. He is the author or coauthor of 7 books and has published over 190 journal articles and edited 15 books including *Springer Handbook in Engineering Statistics* and *Handbook in Reliability Engineering*. He has delivered over 40 invited keynote and plenary speeches at many international conferences. His numerous awards include the 2009 IEEE Reliability Society *Engineer of the Year Award*. He is a Fellow of the Institute of Electrical and Electronics Engineers (IEEE) and the Institute of Industrial Engineers (IIE).

Deep Learning Approach Based on Fault Correction Time for Reliability Assessment of Cloud and Edge Open Source Software

Hironobu Sone, Shoichiro Miyamoto, Yukinobu Kashihara, Yoshinobu Tamura, and Shigeru Yamada

Abstract We discuss a method of machine learning in order to consider the characteristics reliability trends of edge open source project. Then, we focus on the method based on deep learning analysis. Thereby, the proposed method will be able to extract the characteristics data in order to comprehend the trend of fault big data recorded on the bug tracking system in edge open source project. Moreover, several numerical examples are shown by using actual fault big data in the edge open source project. Then, the illustrative results based on the deep learning are shown by using our methods discussed in this chapter. We discuss that our method by deep learning and prediction model are useful to assess the quality and reliability of the edge open source project.

H. Sone
IBM Japan, Ltd., Tokyo, Japan

S. Miyamoto · Y. Tamura (✉)
Yamaguchi University, Yamaguchi, Japan
e-mail: tamuray@yamaguchi-u.ac.jp

S. Miyamoto
e-mail: i083fe@yamaguchi-u.ac.jp

Y. Kashihara
Tokyo City University, Tokyo, Japan
e-mail: g2181424@tcu.ac.jp

S. Yamada
Tottori University, Tottori, Japan
e-mail: yamada@tottori-u.ac.jp

V. Kumar and H. Pham (eds.), *Predictive Analytics in System Reliability*,
Springer Series in Reliability Engineering,
https://doi.org/10.1007/978-3-031-05347-4_1

1 Introduction

In the future, software services based on the edge computing will be in widespread use all over the world. For example, *StarlingX* is known as the cloud infrastructure software stack for the edge. *StarlingX* is one of the component in the cloud service software *OpenStack*.

At present, almost of commercial software are included some open source software. Considering the edge open source software, it will be used in various area. We discuss the method of machine learning in order to consider the characteristics reliability trends of edge open source project. Then, this chapter focuses on the prediction method of mean time between software failures based on deep learning. Thereby, the proposed method will be able to extract the characteristics data in order to comprehend the trend of fault big data recorded on the bug tracking system in edge open source project. Then, we focus on the major fault level in fault severity in terms of the reliability.

Moreover, several numerical examples are shown by using actual large scale fault data in the edge open source project. Then, numerical illustrations focused on the major fault level are shown by using our methods proposed. Finally, we discuss that our method by deep learning and prediction model are useful to control the quality and reliability of the edge open source project.

Traditionally, many of open source software (OSS) are developed under various OSS projects. Several researchers discuss the reliability assessment methods of OSS [1]. Historically, many methods of software reliability management based on the stochastic process models have been proposed by several researchers [2–5]. Moreover, our research group has been proposed the method of reliability assessment for the OSS [1]. However, the plausible researches focused on the reliability management by deep learning for OSS fault big data have not been proposed in the past. On the other hand, it is important for the OSS managers to control the OSS project in terms of the quality management. The proper control of the OSS faults will directly relate to the quality, reliability, and cost. Therefore, it will be successful in achieving reliability improvement and the reduction of development cost in OSS if the OSS project is properly managed.

This chapter discusses the reliability index of the correction time of software faults for OSS project considering various situation in performance resulting from OSS operation. In particular, this chapter discusses the learning situation of deep learning for the actual data sets. Then, we propose the suitable learning conditions by the prediction method of the correction time of software faults based on deep learning. Then, we offer the estimation examples by applying the proposed method. Especially, we discuss the learning simulation for several conditions of data dividing in the learning data sets. Moreover, several numerical examples based on the proposed method by using the actual fault big data are shown. Then, the actual data sets of three pattern for the learning phase in deep learning are used.

2 Estimation of Correction Time of Software Faults Based on Deep Learning

There are several approaches for the software reliability management by using the machine learning [6, 7]. Traditionally, the comparison results between the software reliability models and the method of neural network have been discussed in the past. Especially, the past research papers based on the neural network have been used the fault data only. On the other hand, we use several different data type depended on the software reliability in the proposed method. The unique features of our research is to use several kinds of explanatory variables as the input data sets. Moreover, we focus on the fault level in this chapter. We show the fault levels recorded on the bug tracking system in the actual data set as follows:

- High
- Low
- Medium
- Unspecified
- Urgent

We focus on the high level fault, because the number of detected high level faults is larger than the other level. On the other hand, the number of detected low level faults is smaller than the other level.

Moreover, the algorithm of the deep learning is shown in Fig. 1. Several algorithms of deep learning have been discussed by some researchers [8–13]. In this chapter, we apply the deep feedforward neural network to learn the fault big data on bug tracking systems of OSS projects. We apply the following amount of information to estimate the parameters of deep learning. Then, the objective variable is given as the correction time of software faults in the cases of high level faults only, because the correction time of high level faults will become long necessarily. Moreover, the high level faults deeply depend on the reliability of OSS.

As mentioned above, we apply the correction time of software failures as the output values. Therefore, the following 12 kinds of explanatory variables are set to the amount of input layer:

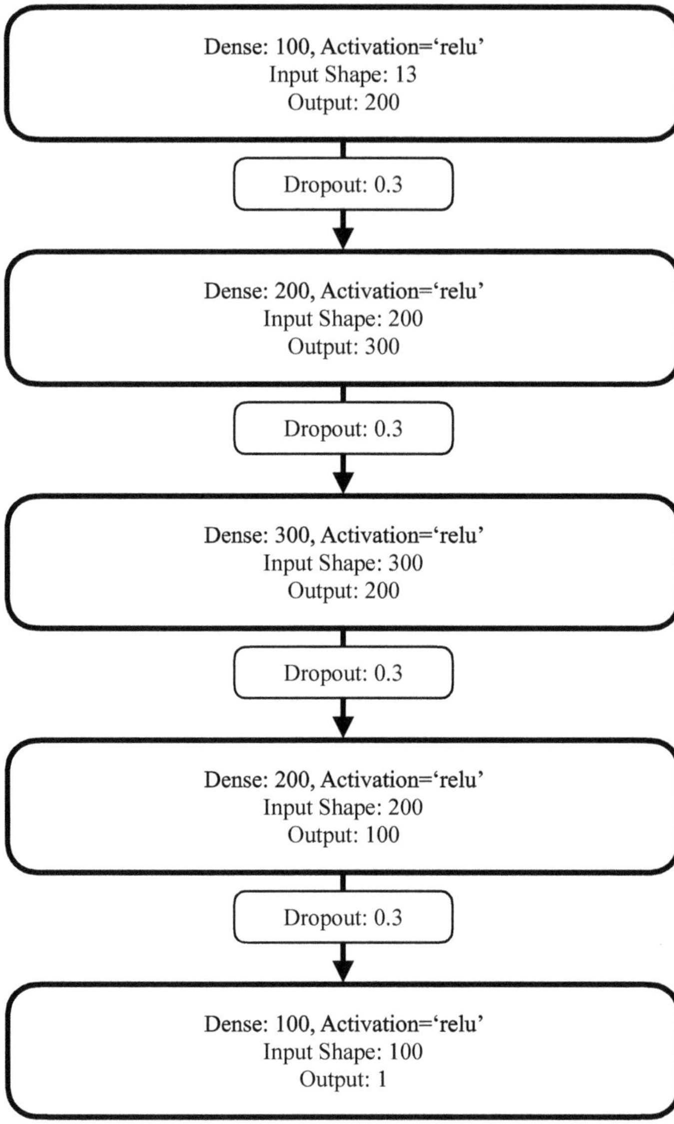

Fig. 1 The structure of deep feedforward neural network in this chapter

- Opened
- Changed
- Reporter
- Product
- Component
- Status

- Resolution
- Hardware
- OS
- Severity
- Version
- Summary

The large scale data sets on several explanatory variables are converted from the text data to the numerical one such as fault occurrence rate by using the frequency encoding. The correction time of OSS faults will be useful to understand the property of correction difficulty in the OSS fault.

3 Data for Numerical Illustration Based on Deep Feedforward Neural Network

In the historical deep learning, several algorithms based on deep learning have been discussed by some researchers [8–13]. We use the Adam (Adaptive moment estimation) optimizer [13] known as the optimization algorithms of deep learning. The Adam is the algorithms improved the AdaGrad and RMSProp. The algorithm of Adam is shown in the reference [13] in detail. Figure 1 shows the detailed parameters based on Adam optimizer in the proposed method.

We show the data used in the proposed method in the Figs. 2 and 3. In particular, we use mainly the count encoding and frequency encoding methods converting from the raw data to the numeric values. Then, the data will be converted from Figs. 2 to 3. There is no way to know that any variables out of 12 kinds of explanatory variables depend on the reliability. Therefore, we will be able to automatically identify several variables in terms of the reliability by using the deep learning.

4 Comparison Results Based on the Amount of Learning Data

We analyze the fault big data in terms of the time between software failures (faults) in OpenStack [14] including the edge component such as "StarlingX ". This fault big data set includes the following items such as "2 Estimation of Correction Time of Software Failures Based on Deep Learning".

Bug ID	Opened	Changed	Reporter	Product	Component	Status	Resolution	Hardware	OS	Severity	Version	Summary
985361	2013/7/17 10:56	2014/1/9 19:40	Jaroslav Henner	Red Hat OpenStack	openstack-packstack	CLOSED	ERRATA	x86_64	Linux	medium	3	packstack doesn't acc
1146938	2014/9/26 11:50	2016/4/26 13:26	Sunil Thaha	Red Hat OpenStack	openstack-glance	CLOSED	CURRENTRELEASE	Unspecified	All	unspecified	5.0 (RHEL 7)	Unittests fail when ru
1155592	2014/10/22 12:52	2016/4/26 13:50	Marian Krcmarik	Red Hat OpenStack	openstack-cinder	CLOSED	ERRATA	Unspecified	Unspecified	unspecified	5.0 (RHEL 7)	"openstack-service lis
1157619	2014/10/27 11:31	2016/4/26 13:35	Marko Myllynen	Red Hat OpenStack	openstack-cinder	CLOSED	CURRENTRELEASE	Unspecified	Unspecified	unspecified	5.0 (RHEL 7)	cinder provides obsole
1163421	2014/11/12 16:42	2015/9/10 11:45	Lon Hohberger	Red Hat OpenStack	openstack-trove	CLOSED	ERRATA	Unspecified	Unspecified	unspecified	5.0 (RHEL 7)	Adjust log permissions
1169145	2014/11/30 17:56	2016/4/27 2:17	bkopilov	Red Hat OpenStack	openstack-glance	CLOSED	CURRENTRELEASE	x86_64	Linux	high	6.0 (Juno)	glance -failed on Valu
1170343	2014/12/3 21:01	2016/4/26 22:06	Lon Hohberger	Red Hat OpenStack	openstack-glance	CLOSED	ERRATA	Unspecified	Unspecified	unspecified	6.0 (Juno)	Rebase openstack-gla
1174760	2014/12/16 12:37	2016/4/26 22:58	Dafna Ron	Red Hat OpenStack	openstack-cinder	CLOSED	DUPLICATE	x86_64	Linux	urgent	6.0 (Juno)	[upgrade]: cinder-api s
1184349	2015/1/21 7:35	2016/4/26 18:02	bkopilov	Red Hat OpenStack	openstack-cinder	CLOSED	NOTABUG	x86_64	Linux	high	6.0 (Juno)	glusterfs - volume upl
1186395	2015/1/27 15:36	2016/4/26 15:16		Red Hat OpenStack	openstack-cinder	CLOSED	WONTFIX	Unspecified	Unspecified	unspecified	5.0 (RHEL 7)	Submission from Mell
1209584	2015/4/7 17:13	2015/5/12 4:04	Luigi Toscano	Red Hat OpenStack	openstack-trove	CLOSED	ERRATA	Unspecified	Unspecified	unspecified	6.0 (Juno)	trove-* services do no
1254711	2015/8/18 17:29	2016/4/26 17:48	Lon Hohberger	Red Hat OpenStack	openstack-glance	CLOSED	ERRATA	Unspecified	Unspecified	unspecified	5.0 (RHEL 7)	Rebase openstack-gla
1254718	2015/8/18 17:31	2016/4/26 14:59	Lon Hohberger	Red Hat OpenStack	openstack-glance	CLOSED	ERRATA	Unspecified	Unspecified	unspecified	5.0 (RHEL 6)	Rebase openstack-gla
1304111	2016/2/2 22:17	2016/12/14 15:22	Dustin Schoenbrun	Red Hat OpenStack	openstack-manila-ui	CLOSED	ERRATA	Unspecified	Unspecified	high	8.0 (Liberty)	RHEL-OSP Does not H
1314821	2016/3/4 15:45	2016/4/26 15:10	Flavio Percoco	Red Hat OpenStack	openstack-glance	CLOSED	ERRATA	Unspecified	Unspecified	unspecified	8.0 (Liberty)	Drop dependency on e
1333884	2016/5/6 14:37	2016/8/11 12:19	Harry Rybacki	Red Hat OpenStack	openstack-tempest	CLOSED	ERRATA	Unspecified	Unspecified	unspecified	9.0 (Mitaka)	config_tempest.py rais
1346749	2016/6/15 9:54	2016/8/11 12:25	Anshul Behl	Red Hat OpenStack	openstack-ironic	CLOSED	ERRATA	Unspecified	Unspecified	unspecified	9.0 (Mitaka)	RHOS9 tempest run re
1441796	2017/4/12 18:12	2017/4/12 18:13		Red Hat OpenStack	openstack-ironic-inspe	CLOSED	NOTABUG	Unspecified	Unspecified	unspecified	7.0 (Kilo)	[RFE][HA
1455490	2017/5/25 10:24	2020/7/16 9:39	Eduard Barrera	Red Hat OpenStack	openstack-tripleo	CLOSED	DUPLICATE	Unspecified	Unspecified	unspecified	10.0 (Newton	during introspection vi
1544713	2018/2/13 11:14	2019/9/9 16:48	Punit Kundal	Red Hat OpenStack	openstack-nova	CLOSED	INSUFFICIENT_DATA	Unspecified	Unspecified	medium	10.0 (Newton	Incorrect results being
1545722	2018/2/15 14:26	2021/3/11 17:11	David Vallee Delisle	Red Hat OpenStack	openstack-nova	CLOSED	WONTFIX	Unspecified	Unspecified	unspecified	10.0 (Newton	When spawning too m
1545809	2018/2/15 15:38	2021/3/11 17:12	KOSAL RAJ I	Red Hat OpenStack	openstack-nova	CLOSED	INSUFFICIENT_DATA	Unspecified	Unspecified	high	10.0 (Newton	Live migration failing a
1547969	2018/2/22 12:38	2019/9/9 13:10	Madhur Gupta	Red Hat OpenStack	openstack-nova	CLOSED	DUPLICATE	x86_64	Linux	medium	10.0 (Newton	[RFE] libvirt: add supp
1551911	2018/3/6 7:09	2019/9/9 15:56	Petersingh Anburaj	Red Hat OpenStack	openstack-nova	CLOSED	WONTFIX	Unspecified	Unspecified	high	11.0 (Ocata)	Wrong value for opens

Fig. 2 A portion of the raw data for the input in the deep learning

Opened	Changed	Reporter	Product	Component	Status	Resolution	Hardware	OS	Severity	Version
		0.00571607	1	0.0463	1	0.4148	0.1595	0.2304	0.1164	0.0204
436.037732	837.740266	0.00010393	1	0.0249	1	0.0892	0.7671	0.0292	0.2721	0.0567
26.0428935	0.01643519	0.00197464	1	0.0806	1	0.4148	0.7671	0.7401	0.2721	0.0567
4.9437963	0	0.00187071	1	0.0806	1	0.0892	0.7671	0.7401	0.2721	0.0567
16.2161227	0	0.01444606	1	0.0024	1	0.4148	0.7671	0.7401	0.2721	0.0567
18.0512847	229.605544	0.00602785	1	0.0249	1	0.0892	0.1595	0.2304	0.3469	0.0469
3.12871528	0	0.01444606	1	0.0249	1	0.4148	0.7671	0.7401	0.2721	0.0469
12.6499421	0.03587963	0.01060071	1	0.0806	1	0.134	0.1595	0.2304	0.2311	0.0469
35.7897454	0	0.00602785	1	0.0806	1	0.1728	0.1595	0.2304	0.3469	0.0469
6.33430556	0	0	1	0.0806	1	0.0752	0.7671	0.7401	0.2721	0.0567
70.0674769	0	0.00332571	1	0.0024	1	0.4148	0.7671	0.7401	0.2721	0.0469
133.011319	350.572558	0.01444606	1	0.0249	1	0.4148	0.7671	0.7401	0.2721	0.0567
0.00123843	0	0.01444606	1	0.0249	1	0.4148	0.7671	0.7401	0.2721	0.0174
168.198183	232.015822	0.00166286	1	0.0009	1	0.4148	0.7671	0.7401	0.3469	0.0685
30.7283796	0	0.00187071	1	0.0249	1	0.4148	0.7671	0.7401	0.2721	0.0685
62.9521412	106.880776	0.00270214	1	0.0097	1	0.4148	0.7671	0.7401	0.2721	0.0399
39.8039699	0.00443287	0.00083143	1	0.0263	1	0.4148	0.7671	0.7401	0.2721	0.0399
301.345833	244.241597	0	1	0.0062	1	0.1728	0.7671	0.7401	0.2721	0.0824
42.6751157	1190.64323	0.01278321	1	0.0397	1	0.134	0.7671	0.7401	0.2721	0.1183
264.034572	0	0.00187071	1	0.1365	1	0.0409	0.7671	0.7401	0.1164	0.1183
2.13289352	549.016458	0.00374143	1	0.1365	1	0.0752	0.7671	0.7401	0.2721	0.1183
0.05056713	6.94E-05	0.00228643	1	0.1365	1	0.0409	0.7671	0.7401	0.3469	0.1183
6.87478009	0	0.0007275	1	0.1365	1	0.134	0.1595	0.2304	0.1164	0.1183
11.7718171	0.11493056	0.00020786	1	0.1365	1	0.0752	0.7671	0.7401	0.3469	0.0419

Fig. 3 A portion of the actual data for the input in the deep learning

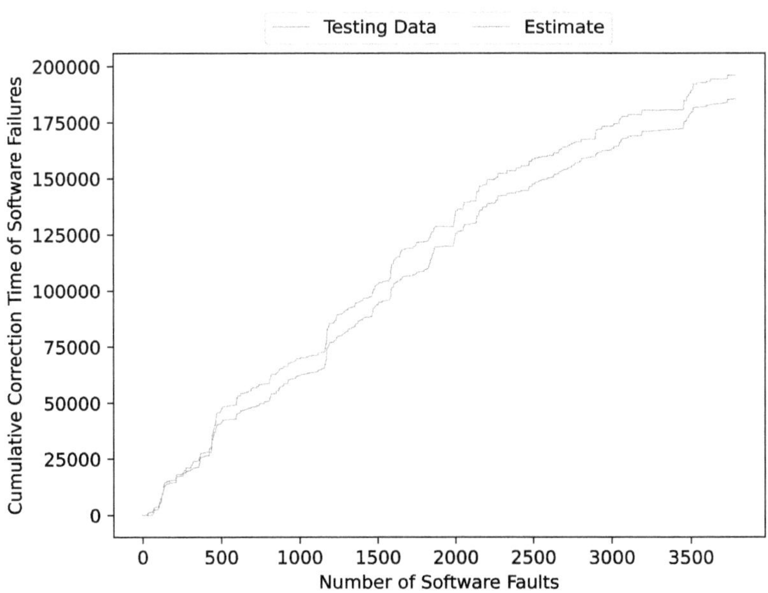

Fig. 4 The estimated correction time of software failures for 20% testing data

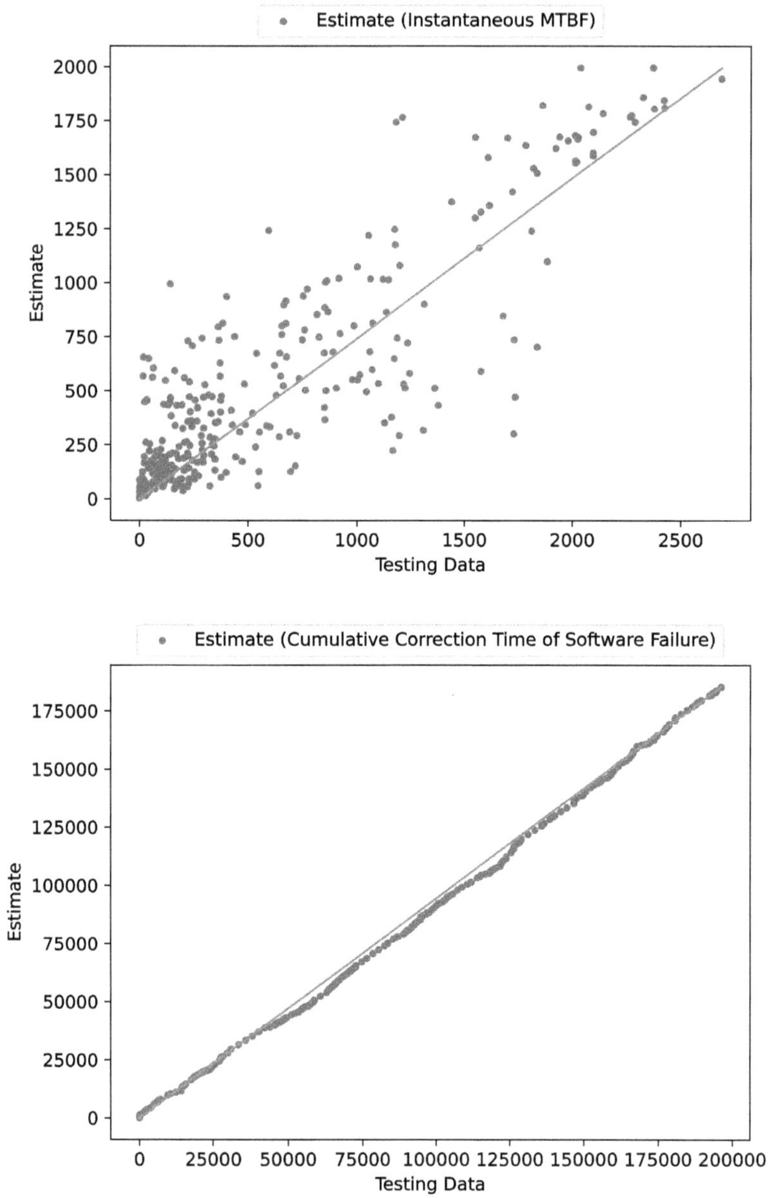

Fig. 5 The comparison results between the estimates and 20% testing data

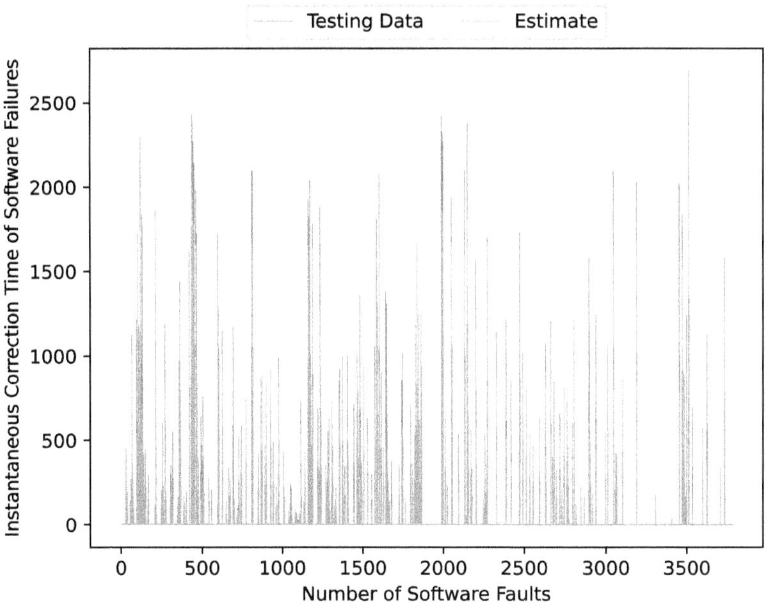

Fig. 6 The estimated instantaneous correction time of software failures for 20% testing data

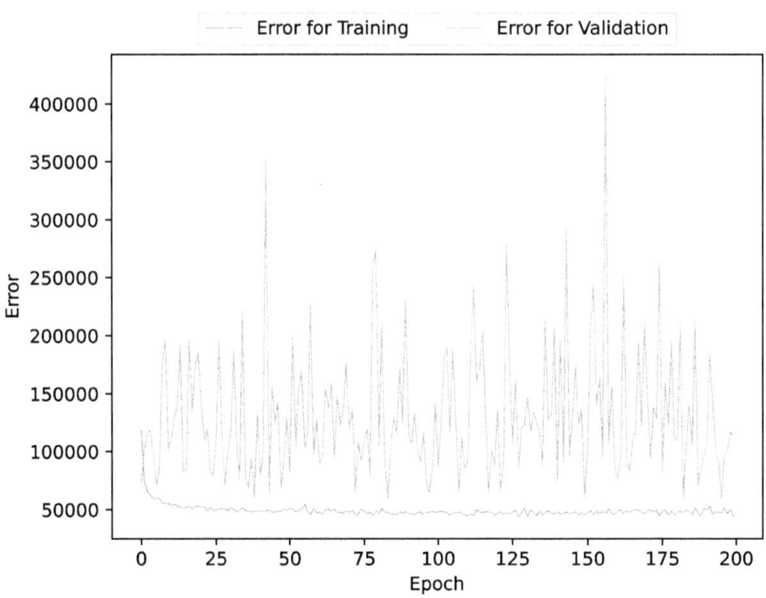

Fig. 7 The estimation error between learning data and validation in case of 20% testing data

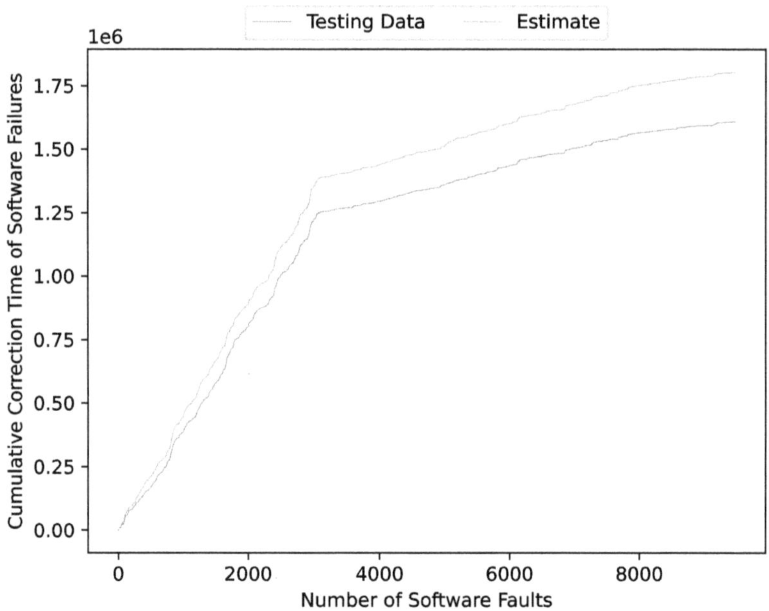

Fig. 8 The estimated cumulative mean time between software failures for 50% testing data

- Opened
- Changed
- Reporter
- Product
- Component
- Status
- Resolution
- Hardware
- OS
- Severity
- Version
- Summary

In this chapter, we discuss the cumulative correction time of software faults and the instantaneous correction time of software faults. Then, we define as follows:

The cumulative correction time of software faults:
the cumulative time from the detection of each new fault to the end of time for its fault correction.

The instantaneous correction time of software faults:
the time period from the detection of new fault to the correction of its fault.

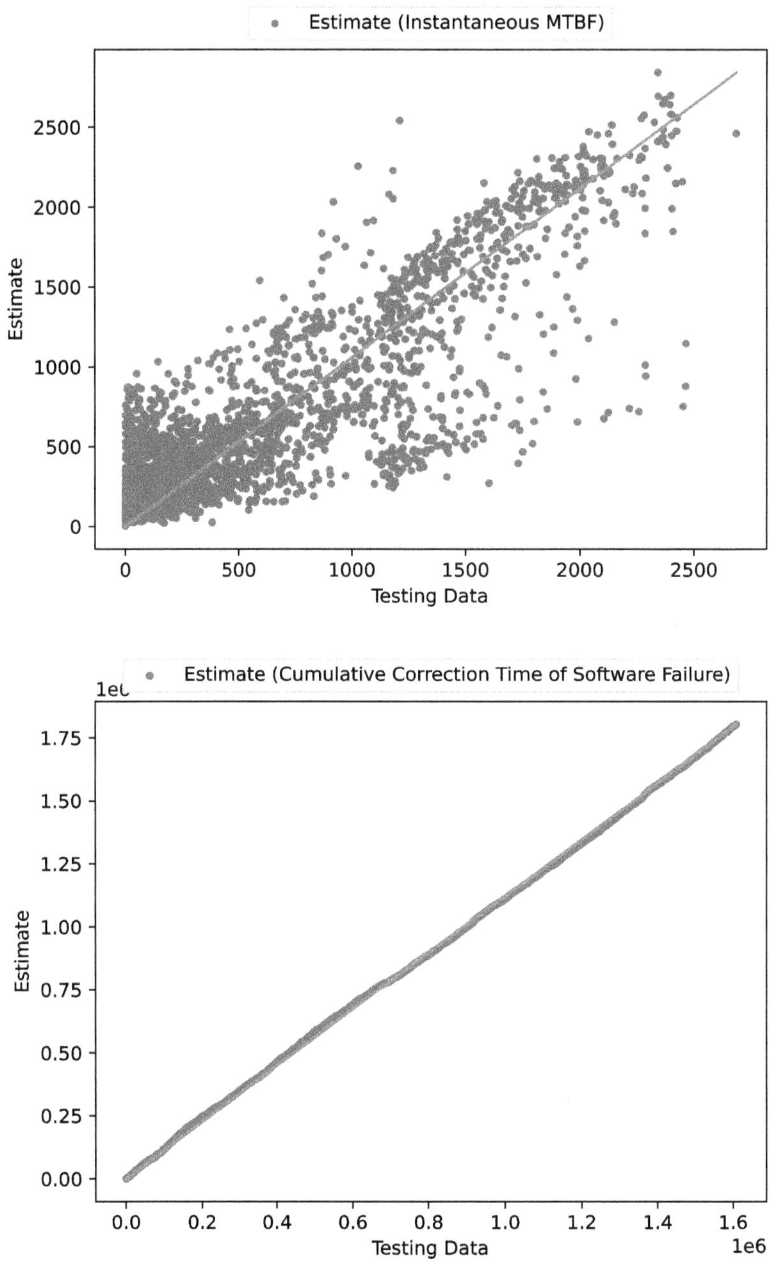

Fig. 9 The comparison results between the estimates and 50% testing data

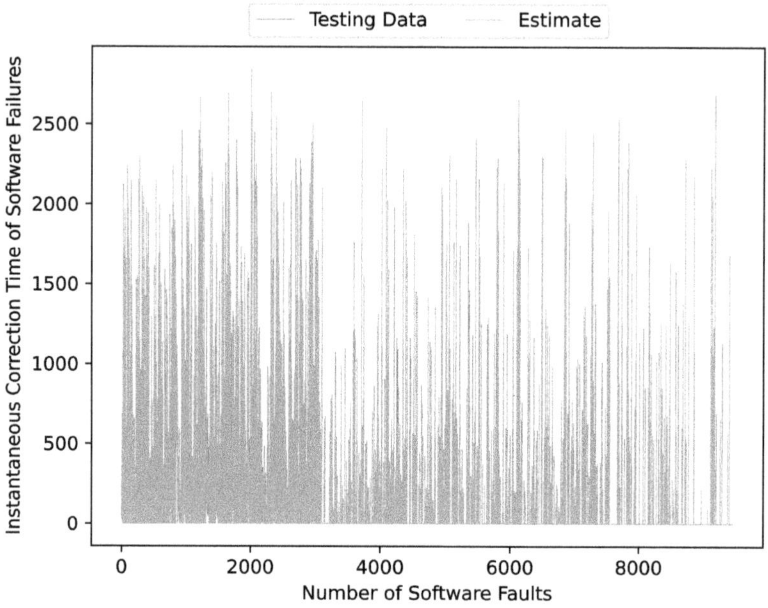

Fig. 10 The estimated instantaneous mean time between software failures for 50% testing data

Moreover, we compare the estimation results according to the rate of learning.

First, in case of 20% testing data, Figs. 4, 5, 6 and 7 show the estimated cumulative correction time of software failures (faults), the comparison results between the estimates, the estimated instantaneous correction time of software failures, and the absolute error between learning data and validation, respectively.

Second, in case of 50% testing data, Figs. 8, 9, 10 and 11 show the estimated cumulative correction time of software failures, the comparison results between the estimates, the estimated instantaneous correction time of software failures, and the error between learning data and validation, respectively.

Similarly, in case of 80% testing data, Figs. 12, 13, 14 and 15 show the estimated cumulative correction time of software failures, the comparison results between the estimates, the estimated instantaneous correction time of software failures, and the error between learning data and validation, respectively.

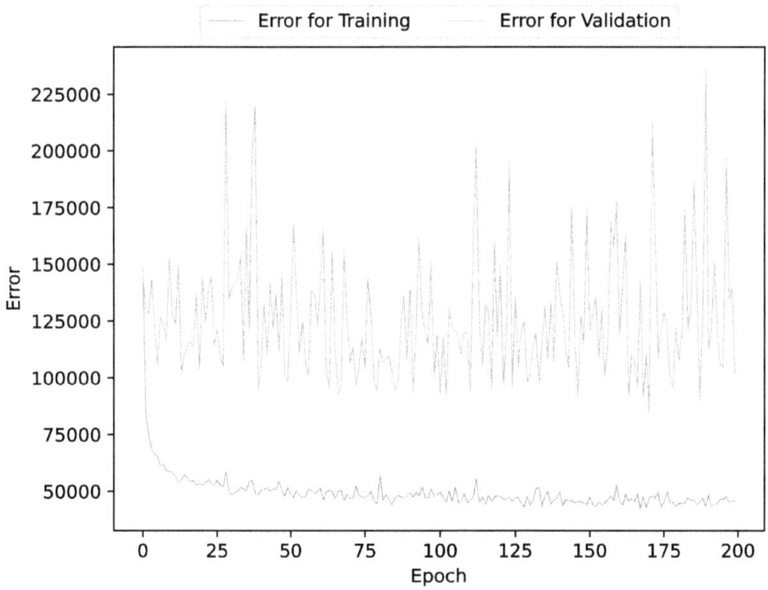

Fig. 11 The estimation error between learning data and validation in case of 50% testing data

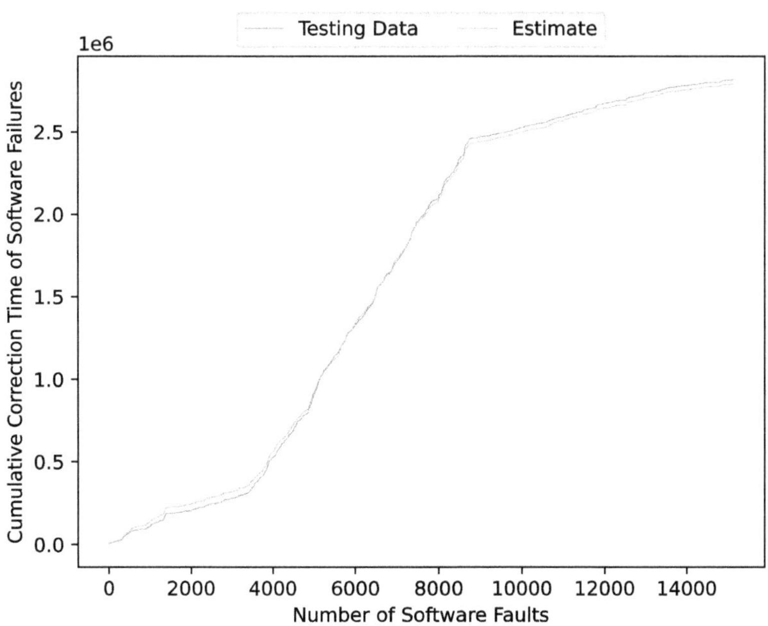

Fig. 12 The estimated cumulative mean time between software failures for 80% testing data

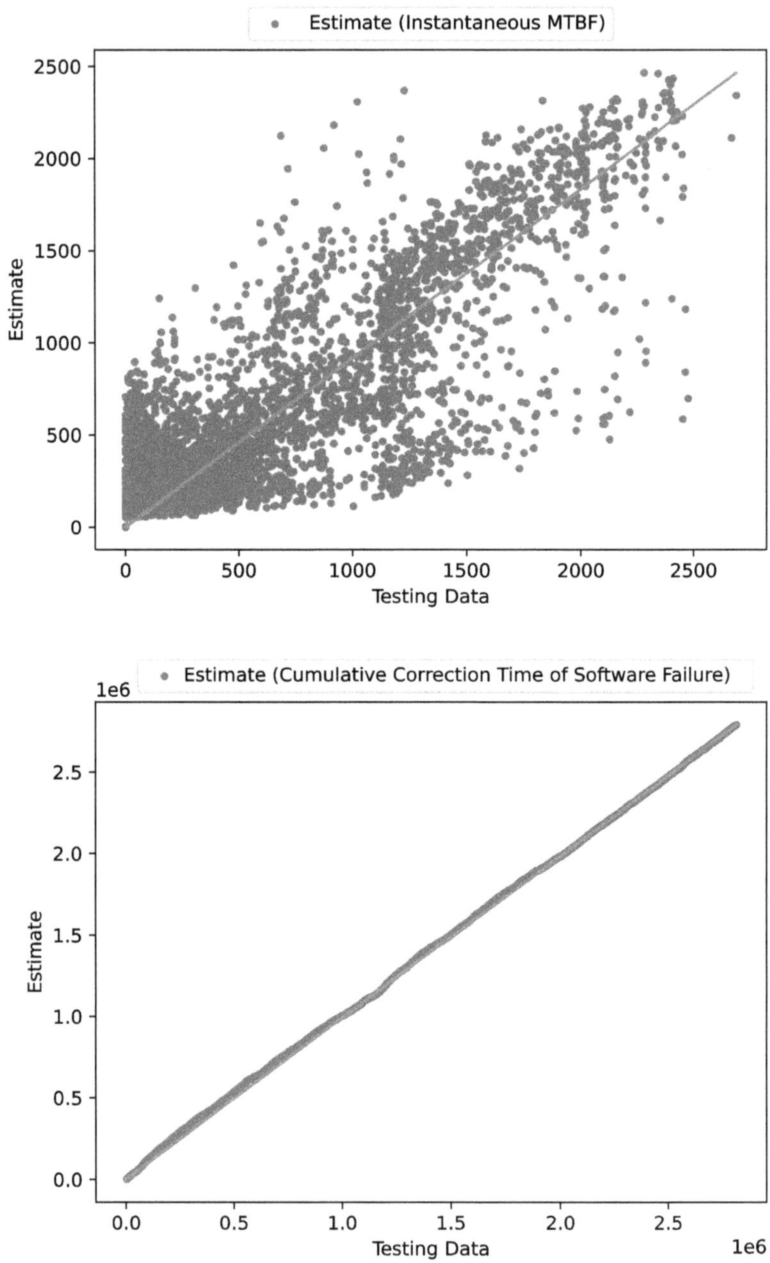

Fig. 13 The comparison results between the estimates and 80% testing data

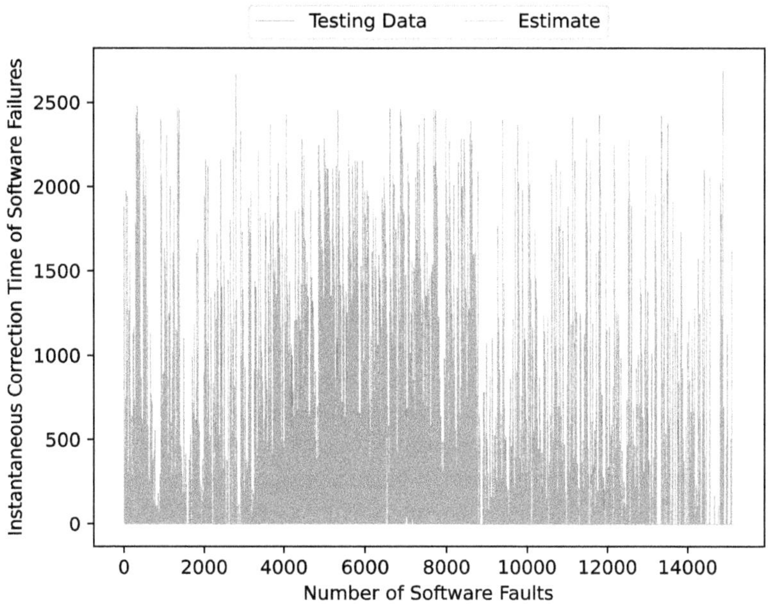

Fig. 14 The estimated instantaneous mean time between software failures for 80% testing data

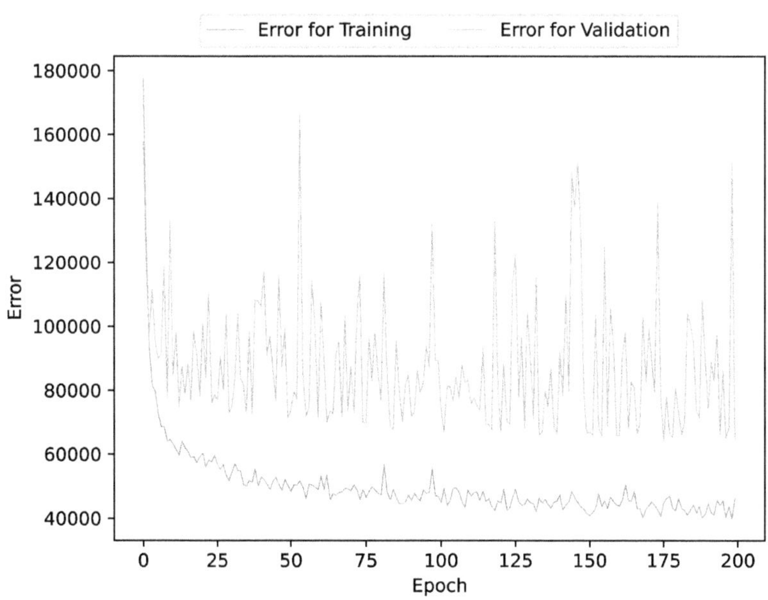

Fig. 15 The estimation error between learning data and validation in case of 80% testing data

5 Concluding Remarks

Many OSS are used for various situation such as the embedded system, server system, cloud computing, edge computing, and the component based system. Several OSS projects have been controlled by using the bug tracking system. The huge data sets in terms of many faults have been recorded on the bug tracking system. It will be helpful for OSS project managers to assess the reliability of OSS, if the huge data on bug tracking system are effectively utilized by using the automated machine learning such as the deep learning.

In this chapter, we have discussed the estimation method of correction time of software failures (faults) for the cloud and edge computing by OSS. It will be able to control the manpower in terms of the OSS reliability of edge computing, if the OSS managers can estimate the correction time of software failures. Also, the estimation method of the correction time of software failures based on the deep learning considering the complexity of OSS project has been developed in this chapter. In particular, the huge data sets on OSS bug tracking system is useful for the OSS managers to control the manpower of OSS project. Furthermore, we have compared the proposed method according to several cases of learning data.

In the future study, it is necessary to compare the estimation results in cases of the other fault levels such as Medium and Urgent.

Funding This study was funded by the JSPS KAKENHI Grant No. 20K11799 in Japan.
Conflict of Interest The authors declare that they have no conflict of interest.

Acknowledgements This work was supported in part by the JSPS KAKENHI Grant No. 20K11799 in Japan.

References

1. Yamada S, Tamura Y (2016) OSS reliability measurement and assessment. Springer International Publishing Switzerland
2. Lyu MR (ed) (1996) Handbook of software reliability engineering. IEEE Computer Society Press, Los Alamitos, CA
3. Yamada S (2014) Software reliability modeling: fundamentals and applications. Springer, Tokyo/Heidelberg
4. Kapur PK, Pham H, Gupta A, Jha PC (2011) Software reliability assessment with OR applications. Springer, London
5. Kingma DP, Rezende DJ, Mohamed S, Welling M (2014) Semi-supervised learning with deep generative models. In: Proceedings of the 27th international conference on neural information processing systems, pp 1–9
6. Karunanithi N, Whitley D, Malaiya YK (1992) Using neural networks in reliability prediction. IEEE Softw 9(4):53–59
7. Dohi T, Nishio Y, Osaki S (1999) Optimal software release scheduling based on artificial neural networks. Ann Softw Eng 8(1):167–185

8. Blum A, Lafferty J, Rwebangira MR, Reddy R (2004) Semi-supervised learning using randomized mincuts. In: Proceedings of the international conference on machine learning, pp 1–8
9. George ED, Dong Y, Li D, Alex A (2012) Context-dependent pre-trained deep neural networks for large-vocabulary speech recognition. IEEE Trans Audio Speech Lang Process 20(1):30–42
10. Vincent P, Larochelle H, Lajoie I, Bengio Y, Manzagol PA (2010) Stacked denoising autoencoders: Learning useful representations in a deep network with a local denoising criterion. J Mach Learn Res 11(2):3371–3408
11. Martinez HP, Bengio Y, Yannakakis GN (2013) Learning deep physiological models of affect. IEEE Comput Intell Mag 8(2):20–33
12. Hutchinson B, Deng L, Yu D (2013) Tensor deep stacking networks. IEEE Trans Pattern Anal Mach Intell 35(8):1944–1957
13. Kingma DP, Ba JL (2015) Adam: a method for stochastic optimizations. In: Proceedings of the international conference on learning representations, pp 1–15
14. The OpenStack project, OpenStack. http://www.openstack.org/

System Reliability Models with Random Shocks and Uncertainty: A State-of-the-Art Review

Yuhan Hu and Mengmeng Zhu

Abstract Reliability evaluation is an important task in safety–critical applications. The failure of a system is generally caused by random shocks resulting from adverse events or internal degradations. This chapter thus mainly focuses on the review of system reliability models with random shocks and the uncertainty of the degradation process. In the category of system reliability models with random shocks, we review system reliability models based on five random shock models that are commonly used in Reliability Engineering, cumulative shock model, extreme shock model, run shock model, δ-shock model, and mixed shock model. In addition, three sources of variabilities, commonly discussed in the literature, can result in the uncertainty of the degradation process, which are temporal variability in the degradation process, unit-to-unit variability, and measurement error caused by imperfect instruments or imperfect inspection. In the category of system reliability model with uncertainty, we review system reliability models using stochastic degradation models in terms of three stochastic processes, Wiener process, gamma process, and inverse Gaussian process.

Keywords Reliability model · Random shocks · Uncertainty · Stochastic process

1 Introduction

Reliability is defined as the probability that a product can function properly without failure during its designed life under the designed operating conditions [1]. The failure of a system has a wide-ranging societal impact. For example, a plane from Sudan lost control on the runway while landing due to the bad weather on June

Y. Hu · M. Zhu (✉)
Department of Textile Engineering, Chemistry and Science, North Carolina State University, Raleigh, NC 27606, USA
e-mail: mzhu7@ncsu.edu

M. Zhu
Operations Research Graduate Program, North Carolina State University, Raleigh, NC 27606, USA

V. Kumar and H. Pham (eds.), *Predictive Analytics in System Reliability*,
Springer Series in Reliability Engineering,
https://doi.org/10.1007/978-3-031-05347-4_2

10, 2008, which caused the death of 1 crew member and 29 passengers. In addition, random incidents in our daily life, such as collisions with vehicles, are also very common, which can influence the lifespan of the product, and even influence human life. Thus, considering the random incidents into the reliability modeling can effectively improve the accuracy of the reliability evaluation in safety–critical applications. These random incidents can exert random stresses in the system, which can be modeled by random shocks received by a system in the field of Reliability Engineering.

Generally, random shock models are classified into five groups, cumulative shock model, extreme shock model, run shock model, δ-shock model, and mixed shock model. The reliability model with random shocks is first proposed by Esary and Marshall [2], in which the shock loadings are assumed to be independently distributed. Based on this, the development of the random shock-based reliability model is further investigated in many studies [1–16]. Gut [3] proposed a cumulative shock model, where the system fails when the cumulative shock damage is larger than a preset threshold. Later, Che et al. [4] developed a reliability model with a mutually dependent degradation process and shock process. Dong et al. [5] developed a multi-component system reliability model with generalized cumulative shocks and a stochastic degradation process. The extreme shock model is first developed by Shanthikumar and Sumita [6], in which the system failure is determined by the magnitude of the shock and further studied by [4, 8]. Based on the studies [3, 6], Gut [9] investigated a mixed shock model that considers the cumulative shock model and extreme shock model, which assumed that the system fails when the magnitude of the shock is larger than a threshold or the accumulative shock damage is larger than another critical threshold. Later, other types of shock models are introduced, which are run shock model [10] and δ-shock model [11]. Their applications in the reliability estimation can be referred to studies [12–16].

With the increasing complexity of the system engineering problems, the uncertainty caused by the internal product properties or external factors has also become significant. Generally, there are three sources of variabilities [17, 18]. The first source is the temporal variability representing the inherent uncertainty in the degradation path [17]. The second source is the unit-to-unit variability. Take the battery management system in electric vehicles as an example. One battery pack consists of numerous battery cells. The degradation of one battery cell may be different from the other cells even they are manufactured from the same production line. When considering the degradation of a battery cell, it is necessary to consider the degradation differences between cells. The third source is the measurement error caused by an imperfect instrument or imperfect inspection which is the distinction between the true value and the measured value. In many cases, regardless of the precision of the instrument, the experimental data is always contaminated in the experiment, which will influence the reliability prediction accuracy of the target system. In literature, many studies have considered the three sources of uncertainty in the system reliability modeling [19–34]. Three classic stochastic processes that are commonly used to address the uncertainty are Wiener process [19, 20], gamma process [21–23], and inverse Gaussian process [24].

The rest of this chapter is organized as follows. In Sect. 2, we first introduce the definitions and applications of random shock models and then review the related work on system reliability models considering different types of random shocks. Section 3 reviews the literature on system reliability models with uncertainty in terms of different stochastic processes. Section 4 concludes this chapter.

2 System Reliability Models with Random Shocks

Typically, five random shock models are widely used in the field of Reliability Engineering, cumulative shock model, extreme shock model, run shock model, δ-shock model, and mixed shock model. These models are generally defined by the inter-arrival time between consecutive shocks and/or the damage from shocks. Section 2 is divided into two parts. Section 2.1 reviews the definitions and applications of random shock models. Section 2.2 reviews the literature on system reliability models incorporating different types of random shocks.

The following notations are defined for Sect. 2. $R(t)$ is the system reliability function by time t. $M(t)$ is the system degradation function by time t. $X(t)$ is the internal degradation function by time t. $S(t)$ is the cumulative shock damage function by time t. $N(t)$ is the total number of random shocks by time t.

2.1 *Shock Model Categorization*

Shock models, cumulative and extreme shock models, are initially proposed in the 1970s, to apply for predicting the system reliability in a random environment [2]. Based on these two classic models, other shocks models, such as run shock model and δ-shock model, have been developed in the early 2000s. Meanwhile, the mixed shock model is proposed, which combines two types of random shock models, such as the combination of the extreme shock model and δ-shock model. Nowadays, the mixed shock model is not limited to the combination of two types of random shocks but extends to integrating three types of random shocks in order to predicate the complex engineering system reliability.

The cumulative shock model commonly use the following equation:

$$S(t) = \sum_{k=1}^{N(t)} Y_k \tag{1}$$

where Y_k is the damage caused by k^{th} shock. The system fails when the accumulative damage $S(t)$ exceeds a pre-specified threshold. This model is applied in the situation where the system is subject to a series of random shocks. Take an example in practice. A car accident can be regarded as a random shock for the engine. For a vehicle, it is likely to have more than one accident. To assess the damage from all accidents on the engine, it is indeed to summarize these damages. When the total damage on the engine is larger than a threshold, the engine will fail. Other applications in literature, for example, Che et al. [4] regarded the contamination lock in the jet pipe servo value as one random shock that can cause wear debris on the value. The cumulative wear debris will increase with time and finally results in failure when the total wear exceeds a threshold. Dong et al. [5] predicted the reliability of micro electro-mechanical systems that withstand three different kinds of shocks, mechanical vibration, piezoelectric stimuli, and magnetic stimuli. The shock damage from different kinds of shocks can be summarized. The micro electro-mechanical systems will fail when the damage exceeds a threshold.

The extreme shock model is commonly defined as the system fails when the magnitude of any shock exceeds the given level. In other words, the system lifetime is determined by the magnitude of individual random shock. The applications of the extreme shock model are presented as follows. In Che et al. [4], because the contamination lock can fail suddenly when there is sufficient friction generated to withstand the normal actuating force, the extreme shock model is utilized to model this application. Wang et al. [7] applied the extreme shock model to model the impact load in the microelectromechanical system because the load can cause the system failure directly. Hao and Yang [8] regarded the vessel collision on the bridge as one random shock and classified the shocks as fatal and nonfatal according to their magnitude based on the extreme shock model.

The run shock model is first proposed by Mallor and Omey [10], which defined the system breaks down when there are consecutive shocks whose magnitudes are above a threshold. This model is usually applied in mechanical and electronic systems, which generally suffered from fatigue damage. Specifically, fatigue damage refers to the situation that the system is under repeated shocks above a critical threshold. It is noted that the run shock model measures the magnitude of consecutive shocks instead of the magnitude of an individual shock.

The δ-shock model is first proposed by Li and Kong [11], which defined the system failure when the inter-arrival time between two consecutive shocks less than a pre-specified threshold δ. Compared with the traditional shock models, cumulative shock, and extreme shock models, there are some phenomena that are more suitable to use the inter-arrival time to define system failure. For example, when the damage caused by random shocks is hard to be determined, it is more suitable to use the δ-shock model since it pays more attention to the shock occurrence rate instead of the individual or cumulative damage of shocks.

The mixed shock model defines the system failure caused by two or more random shock models. For example, if cumulative shock and extreme shock are considered, the system will fail when the cumulative shock loadings are larger than one threshold or the magnitude of an individual shock is larger than another threshold, whichever occurs first. Parvardeh and Balakrishnan [35] proposed two mixed shock models based on the δ-shock model. One is the combination of the extreme shock model and δ-shock model, and the other is the combination of the cumulative shock model and δ-shock model. Lorvand et al. [16] combined the extreme shock model and run shock model as a mixed shock model. The mixed shock model is not limited to the combination of two shock models. Rafiee et al. [36] developed a mixed shock model that employed extreme shock model, δ-shock model, and run shock model.

2.2 System Reliability Models with Shock Models

2.2.1 System Reliability Models with Cumulative Shock Model

Systems are generally subject to two competing risks, degradation, and random shocks. Given many research efforts have been focused on modeling the dependent relationship between degradation and random shocks. We review the literature on system reliability models with cumulative shock model, considering the dependent relationship between degradation and random shocks. In general, random shocks are commonly assumed to have two types of impacts on the system, sudden incremental jump on the system degradation, and degradation rate acceleration [4, 37–40, 43]. The cumulative shock model is usually used to describe these impacts and further employed to calculate the accumulated system degradation. To represent the sudden incremental jumps, many studies [4, 38–40, 43] described the individual shock damage as an independent and identically distributed random variable. The cumulative shock damage is the summation of individual shock damages, which is shown in Eq. (1). Other studies [37, 41, 42] applied the shock magnitude instead of the individual shock damage to $S(t)$[37]. Assumed each shock damage is linear dependent with its shock magnitude, namely, $S(t) = \sum_{k=1}^{N(t)} (\alpha W_k)$, $N(t) > 0$, where W_k is the magnitude of kth shock and α is the coefficient. In general, system degradation, $M(t)$, consists of the internal degradation $X(t)$ and cumulative shock damage $S(t)$:

$$M(t) = X(t) + S(t) \tag{2}$$

The failure happens when the system degradation exceeds a critical threshold. Generally, system reliability, $R(t)$, can be modeled as:

$$R(t) = P(X(t) + S(t) < H) \tag{3}$$

where H is the failure threshold.

The system may become more susceptible because of undertaking shocks, which makes degradation increase faster. Therefore, the degradation rate will not be ideally constant and will be accelerated by random shocks. Wang and Pham [38] developed a system reliability model with multiple degradation processes and random shocks. The arrival of random shocks follows a homogeneous Poisson process (HPP). The impacts of random shocks are classified into sudden incremental jump and degradation rate acceleration. The sudden incremental jump is adopted from Eq. (1). The degradation rate acceleration is illustrated by incorporating a time-scaled factor $G(t, \gamma^{(i)})$ into the ith degradation process, in which $G(t, \gamma^{(i)}) = \gamma_1^{(i)} N_2(t) + \gamma_2^{(i)} S(t)$, where $\gamma_1^{(i)}$ and $\gamma_2^{(i)}$ denote the impact magnitude on the degradation rate of the ith degradation process and $N_2(t)$ is the total number of nonfatal shocks by time t. The internal degradation for the ith degradation process is modeled by a basic multiplicative path function, which is $X_i \cdot \eta_i(t; \theta_i)$, where X_i is the random variable representing the unit-to-unit variability, $\eta_i(t; \theta_i)$ is the ith mean degradation path function with a parameter vectorθ_i. The ith degradation function is thus modeled as $X_i \cdot \eta_i\left(te^{G(t,\gamma^{(i)})}; \theta_i\right) + \sum_{k=1}^{N(t)} w_{ij}$, where w_{ij} is the loading from shock j in the ith degradation process. The marginal reliability function $R_i(t)$ of the ith degradation process with no fatal shock is: $R_i(t) = P\left(X_i \cdot \eta_i\left(te^{G(t,\gamma^{(i)})}; \theta_i\right) + \sum_{k=1}^{N_2(t)} w_{ij} < H_i\right)$, in which H_i is the corresponding failure threshold of the ith degradation process. The system reliability model is [38]:

$$R(t) = C(R_1(t), R_2(t), \ldots, R_m(t)\Pr(N_1(t) = 0) \tag{4}$$

where C is the joint copula of the marginal reliability function and $N_1(t)$ is the total number of fatal shocks by time t.

The degradation impact on random shocks can be reflected by shock arrival frequency. Fan et al. [41] assumed the number of random shocks follow the nonhomogeneous Poisson process (NHPP) with rate $\lambda(t)$. $\lambda(t)$ is assumed to be linearly related with the current internal degradation $X(t)$:

$$\lambda(t) = \lambda_0 + \beta \cdot X(t) \tag{5}$$

where λ_0 is the initial intensity of NHPP and β is the dependence factor.

Because the shocks with different magnitude have different impacts on the system, they are classified into three zones, safety zone, damage zone, and fatal zone [41]. Noting that only the shocks in the damage zone can generate damage on the system, while the fatal shock will fail the system directly, and the shocks in the safety zone have no effect on the system. The system can function if the internal degradation does not reach one threshold, the cumulative shock damage in the damage zone does not exceed another threshold, and there is no shock in the fatal zone. Thus, the reliability function is [41]:

$$R(t) = \sum_{k=0}^{\infty} P(X(t) < H_1, S(t) < H_2, N_2(t) = 0 | N_1(t) = k) P(N_1(t) = k) \tag{6}$$

where k denotes random shock, H_1 is the threshold for the internal degradation, H_2 is the threshold for cumulative shock damage, $N_1(t)$ is the number of shocks in the damage zone, and $N_2(t)$ is the number of shocks in the fatal zone.

Based on the study [41], Che et al. [4] carried out that the shock intensity $\lambda(t)$ is not only affected by the current internal degradation but also by the shock occurrences by time t. The intensity after k random shocks is defined as $\lambda_k(t) = (1 + \eta k)\lambda_0(t)$, where $\lambda_0(t)$ is the initial intensity influenced by the current degradation level and η is the facilitation factor. The formulation of system reliability in reference [4] is based on Eq. (4).

Moreover, a few studies did not follow the common assumption of cumulative shock models, which is the shocks follow a distribution. In reality, such an assumption may not be practical because the frequency of the shock occurrence can be determined by many factors. For example, Gong et al. [12] developed a system reliability model incorporating the influence of shocks from different sources under the cumulative shock model. The system is subject to random shocks, which come from m sources, and the probability of each source is π_i. The magnitude of shocks from each source follows a phase-type (PH) distribution. The continuous PH distribution is a probability distribution constructed by the convolution of exponential distributions. According to the property of PH distribution, the summation of independent PH random variables still follows PH distribution, which is further utilized in modeling system reliability.

Ranjkesh et al. [44] proposed a new cumulative shock model considering the dependency between shock damage and inter-arrival time, and utilized this model to predict the system reliability of civil structures, such as bridges. A parameter δ, which represents the system recover time, is set to determine the shock damage level. When the inter-arrival time between two consecutive shocks, X_k, is larger than δ, the damage level, Y_k, is defined as mild since the system may recover itself from the previous shock. When the shock time-lapse is less than δ, the damage level is defined as severe because the system does not have enough time to recover from the shock:

$$Y_k = \begin{cases} Y_{k1}, \ X_k \le \delta \\ Y_{k2}, \ X_k > \delta \end{cases} \tag{7}$$

where Y_{k1} is the severe damage of the kth shock and Y_{k2} is the mild damage of the kth shock. The system fails when the cumulative load and severe shock damage is larger than a certain threshold. Hence, the system reliability function [44] is as follows:

$$\begin{aligned} R(t) = {} & \sum_{m=0}^{\infty}\sum_{m_1=0}^{m}\binom{m}{m_1}P\Big(\sum_{k=0}^{m_1}Y_{k1} + \sum_{k=m_1+1}^{m}Y_{k2} < H\Big) \\ & P\big(X_1, \ldots, X_{m_1} \le \delta, X_{m_1+1}, \ldots, X_m > \delta, N(t) = m\big) \end{aligned} \tag{8}$$

Recently, Wang and Zhu [43] proposed a shock-loading based degradation model based on the magnitude of impacts caused by random shocks on degradation processes. Random shock are grouped into fatal shocks and nonfatal shocks. They incorporated a threshold H' to measure the temporal loading level. H' is a time-dependent critical ratio, calculated by $H' = S(t)/S$, where S is the cumulative shock loading that can cause the system failure. If $H^{'} < H_0$, where H_0 is the preset critical threshold, nonfatal shocks can only cause the degradation rate acceleration. If $H^{'} > H_0$, nonfatal shocks can cause both accelerate degradation rate and sudden incremental jump. The method to model system reliability with multiple dependent degradation process incorporating the proposed shock-loading based degradation model is based on Eq. (4).

2.2.2 System Reliability Models with Extreme Shock Model

The extreme shock model defined the system failure when the magnitude of any shock exceeds the given level [4, 37–41]. From this definition, many studies classified shocks based on the magnitude of shocks and further impacts on the system. For example, Wang and Pham [38] considered two types of shocks in the model, fatal and nonfatal shocks. Fatal shocks can fail the system directly, while nonfatal shocks can accelerate the degradation processes. Fan et al. [41] modeled the random shocks into three zones according to their magnitude, fatal, damage, and safety zones. Song et al. [40] classified random shocks into different sets according to their function, size, and affected components. Each component in the system has its own shock set, which indicates that only when the shock belongs to the shock set of that component, the damage will exist. Consider the magnitude of the kth shock that belongs to the jth shock set impacting component l, $W_{l,j,k}$, follows a normal distribution $W_{l,j,k} \sim N\left(\mu_{W_{l,j}}, \sigma^2_{W_{l,j}}\right)$, the reliability function of component l, $R_l(t)$, considering an extreme shock that belongs to the jth shock set is:

$$R_l(t) = P\left(W_{l,j,k} < H_l\right) = \phi\left(\frac{H_l - \mu_{W_{l,j}}}{\sigma_{W_{l,j}}}\right) \; for \; l = 1, 2, \ldots, n, \; j\epsilon\phi_l \tag{9}$$

where H_l is the failure threshold of the component l, $\phi(\cdot)$ is the cumulative density function (CDF) of a standard normally distributed function, and ϕ_l is the shock set for component l.

Some studies assumed that random shocks can be classified into fatal shocks and nonfatal shocks [38, 43, 45]. In this case, fatal shocks are extreme shocks. These studies [38, 43, 45] assumed that random shocks follow HPP with rate λ. The probability that a shock that could be fatal to the system at time t is $p(t)$. Thus, fatal shocks follow NHPP with rate $\lambda p(t)$. Rafiee et al. [39] proposed a system reliability model considering degradation and random shocks. The degradation rate is assumed to be changed by shocks because the system may become vulnerable. The first shock that leads to the degradation rate change is defined as a trigger shock, denoted as the kth shock. The overall degradation is represented as a linear path function: $X(t) = \beta t + \varphi + \varepsilon$, where φ is the initial degradation, β is the degradation rate, and ε is the measurement error. Considering the impact of the trigger shock, $X(t)$ is modeled as:

$$X(t) = \begin{cases} \beta_1 t + \varphi + \varepsilon, k > N(t) \\ \beta_1 T_k + \beta_2(t - T_k) + \varphi + \varepsilon, k \leq N(t) \end{cases} \quad (10)$$

where k is a random variable, T_k is the arrival time of the kth shock, β_1 is the initial degradation rate, and β_2 is the changed degradation rate. The reliability function with extreme shock model in reference [39] is developed based on two conditions, no shock occurs by time t and at least one shock occurs by time t.

Eryilmaz and Kan [46, 47] considered there are changes of distributions of the shock magnitude to propose a system reliability model. These models are preferable to use in the conditions that there is an urgent or a dramatic change in environments, which can cause a larger shock in the system. The change point is assumed to follow a certain distribution, for example, a geometric distribution with a given probability mass function. The reliability function can be derived based on the proposed extreme shock model.

2.2.3 System Reliability Models with Run Shock Model

System reliability models with run shock model is discussed in a few studies. For example, Gong et al. [48] assessed the reliability of the system under a run shock model with two thresholds H_1 and H_2, where $H_1 < H_2$. There are two cases that cause system failure: (1) more than k_1 successive shocks with the magnitude above H_1; (2) more than k_2 successive shocks with the magnitude above H_2. The inter-arrival time and the magnitude of shocks are modeled by PH distribution. Compared with the classic run shock model, adding one more threshold helps determine the severity of a shock. Ozkut and Eryilmaz [13] proposed a Marshall-Olkin run shock model to predict system reliability. The system is assumed to have two components subject to three sources of shocks. In this run shock model, the system failure occurs when k critical shocks arrive in succession and these shocks should come from the same source. Later, Wu et al. [14] proposed an N-critical shock model based on Markov renewal process. The run shock model is a special case of the developed model when N shocks occur consequently.

2.2.4 System Reliability Models with δ-shock Model

Wang and Peng [15] studied a generalized δ-shock model with two types of shocks, type 1 and type 2, with the recovery times are δ_1 and δ_2, respectively. Assume the arrival of shocks follow a HPP with rate λ, and the probability of being type 1 is p and type 2 is $q = 1 - p$. They also assume: (1) if either type of shock arrives during the recovery time, the system will fail; (2) if no shock occurs during the recovery time, the system will be recovered from the damage and shown as good as new. The reliability function of the δ-shock model is shown as [15]:

$$R(t) = \sum_{n_1=0}^{\infty}\sum_{n_2=0}^{\infty} P(T > t, N_1(t) = n_1, N_2(t) = n_2) \tag{11}$$

where $N_i(t)$ is the number of type i shock by time t[15]. Discussed several cases to compute Eq. (11). First, there is no shocks occurred by time t. Second, there is one type of shock occurred by time t. Third, there are two types of shocks occurred by time t, in which the authors proposed a reliability function with the generalized δ-shock model.

Poursaeed [49] developed a new δ-shock model with two thresholds δ_1 and δ_2, where $1 \leq \delta_1 < \delta_2$. When the time interval is smaller than δ_1, the system fails. When the time interval falls between δ_1 and δ_2, the probability of system failure is θ. When the time interval is larger than δ_2, the shocks do not cause damage to the system. Thus, the reliability function based on the proposed δ-shock model is:

$$R(t) = P\left(T_{\delta_1,\delta_2,\theta} > t\right) = P\left(\sum_{i=0}^{L_1} Y_i + \sum_{i=1}^{L_2} Z_i + W > t\right) \tag{12}$$

where $T_{\delta_1,\delta_2,\theta}$ is the system failure time, L_1 is the number of intervals in $[\delta_2, \infty)$, and L_2 is the number of intervals in (δ_1, δ_2). X_i is the time intervals between the ith and $(i+1)$th shock. $Y_i \sim X|X > \delta_2$,$Z_i \sim X|\delta_1 < X < \delta_2$ for $i = 1, 2, \ldots,$ and $W \sim X|\delta_1 < X \leq \delta_2$ or $W \sim X|X \leq \delta_1$, where $X \sim Y$ indicates that X and Y follow the same distribution.

Typically, the arrival of random shocks is modeled by HPP, in other words, the inter-arrival time between two consecutive shocks follows an exponential distribution, which is commonly adopted in many studies. HPP has the advantages of the simplicity of mathematical expressions; however, the limitation also exists. For example, Liu [50] pointed out that HPP can only fit the data which is equal-dispersion, that is, the mean should be equal to the variance. However, the mean and the variance of the shock inter-arrival time are not equal in most cases. Also, HPP can only represent the situation when the hazard rate is constant, while the rate can be either increasing or decreasing in the real life. Thus, a reliability model subject to degradation and random shocks is developed under the assumption that the inter-arrival time of shocks follows Weibull distribution. There are two advantages over HPP. First, Weibull distribution can model the under-dispersion data and over-dispersion data besides the equal-dispersion data. Second, Weibull distribution can specifically simulate the impact caused by related system failures. Under this assumption, the probability that n shocks occur is: $P(N(t) = n) = \sum_{j=n}^{\infty}\left[(-1)^{j+n}\left(\frac{t}{\lambda}\right)^{cj}\alpha_j^n\right]/\Gamma(cj+1), n = 0, 1, 2, \ldots,$ where $\Gamma(\cdot)$ is the Gamma function λ is the scale parameter of Weibull distribution, c is the shape parameter of Weibull distribution, $\alpha_j^0 = \Gamma(cj+1)/\Gamma(j+1)$, and $\alpha_j^{n+1} = \sum_{m=n}^{j-1}\alpha_m^n\Gamma(cj - cm + 1)/\Gamma(j - m + 1)$. Hence, the reliability function under the δ-shock model is expressed as [50]:

$$R(t) = \sum_{n=0}^{\infty} P(\min(B_1, B_2, \ldots, B_n) > \delta, X(t) < H | N(t) = n) P(N(t) = n) \tag{13}$$

where B_k is the inter-arrival time between the $(k-1)$th shock and kth shock, $X(t)$ is the total degradation value, and H is the threshold. All B_k are independent and follow Weibull distribution.

Eryilmaz and Bayramoglu [51] investigated the system reliability under the δ-shock model by assuming the interarrival time follows a uniform distribution. This assumption is useful when the first-order effects of random changes are important to the result; in other words, when the difference between deterministic models and stochastic models is critical. Eryilmaz [52] studied the reliability properties of a discrete-time shock model. The inter-arrival time is assumed to follow a geometric distribution with a mean $1/p$.

In addition, the inter-arrival time in the δ-shock model in most studies is assumed to be independently and identically distributed. Some studies considered the inter-arrival time are dependent. For example, Eryilmaz [53] proposed a reliability model under the δ-shock model when the occurrence of shocks follows Polya process. In this case, $P\{N(t) = n\} = \binom{\alpha+n-1}{n} \left(\frac{t}{t+\beta}\right)^n \left(\frac{\beta}{\beta+t}\right)^\alpha$, for $n = 0, 1, \ldots$, where α and β are parameters. In other words, $N(t)$ follows a negative Binomial distribution with parameters α and $\beta/(\beta + t)$. The reliability function is derived as [53]:

$$R(t) = P(T > t) = \left(\frac{\beta}{\beta + t}\right)^\alpha \sum_{n=0}^{[\frac{t}{\delta}]} \binom{\alpha + n - 1}{n} \left(\frac{t - n\delta}{t + \beta}\right)^n \tag{14}$$

where $[x]$ is the integer part of x for $t \geq 0$.

Some studies consider the inter-arrival times are nonidentical. For example, Tuncel and Eryilmaz [54] described the inter-arrival times as a proportional hazard rate process which can apply to the situation that the inter-arrival time is stochastically increasing or decreasing. The reliability function of the interarrival time X_i is thus expressed as:

$$R_i(t) = P(X_i > t) = \left(\overline{G}(t)\right)^{\alpha_i}, \alpha_i > 0 \tag{15}$$

where $\overline{G}$ is the reliability function of a baseline random variable.

2.2.5 System Reliability Models with Mixed Shock Model

Parvardeh and Balakrishnan [35] proposed a mixed shock model which is the combination of extreme shock model and δ-shock model (extreme-δ mixed shock model). The system fails when the magnitude of any shock is larger than a threshold γ or the inter-arrival time between two consecutive shocks is smaller than another threshold

δ. In the extreme shock model, the time lapse between the $(k-1)$th shock and the kth shock X_k has the marginal distribution F and the magnitude of the kth shock, Z_k, has the marginal distribution G. X_k and Z_k are assumed to be dependent and has a joint distribution $\mathcal{H}$. The reliability function is derived as [35]:

$$R(t) = (F(\delta) - F(t))I_{[0,\delta)}(t) \\ + \sum\nolimits_{n=2}^{\infty}\left[\overline{F}(\delta) - \overline{\mathcal{H}}(\delta,\gamma)\right]^{n-1}\int_0^{\infty} P\left(S_{n-1}^* > t - x\right)dF(x) \\ + \overline{\mathcal{H}}(\max\{\delta, t\},\gamma) - \sum\nolimits_{n=2}^{\infty}\left[\overline{F}(\delta) - \overline{\mathcal{H}}(\delta,\gamma)\right]^{n} P(S_n^* > t) \quad (16)$$

where $\{S_n^*, n \geq 1\}$ is a renewal process with the time between successive renewals whose CDF is $F_{\delta,\gamma}(x) = (\mathcal{H}(x,\gamma) - \mathcal{H}(\delta,\gamma))/(G(\gamma) - \mathcal{H}(\delta,\gamma)), x > \delta$.

Lorvand et al. [55] proposed another extreme-δ mixed shock model by setting a new threshold δ_2, which can switch the system to a lower partially working state. Thus, there are three situations that can cause system failure: (1) the classic δ-shock model; (2) the classic extreme shock model; (3) when k out of interarrival times between two successive shocks are in (δ_1, δ_2). The extreme-δ mixed shock model has also been investigated by studies [56, 57].

Some studies considered the mixed shock models in the combination of more than two shock models. For example, Rafiee et al. [36] discussed the system failures can be caused by the internal degradation, or fatal shocks, in which the shock falls into any of three shock models, run shock model, extreme shock model, and δ-shock model. The system reliability function without degradation-based failure is expressed as [36]:

$$R(t) = \sum\nolimits_{m=0}^{\infty} P(S > N(t), X(t) < H|N(t) = m)P(N(t) = m) \quad (17)$$

where S is the number of fatal shocks, and H is the threshold of degradation failure.

Moreover, some studies considered the degradation rate and failure threshold can be changed multiple times as the changes of three mixed shocks patterns [59]. Jiang et al. [58] assumed the failure threshold will decrease as the increase of shocks. Specifically, when the inter-arrival time is smaller than δ or there are m shocks whose magnitude is larger than γ, the threshold will decrease. Zhao et al. [60] incorporated the system self-healing mechanism into random shock modeling to predict the system reliability. The system is assumed to have two stages by incorporating a change point, which is defined as the time when the cumulative number of valid shocks exceeds a threshold. Before the change point, the system is capable of self-healing from shocks. However, the system cannot recover from the damage after the change point.

3 System Reliability Models with Uncertainty

Generally, three sources result in the uncertainty of complex engineering systems, temporal variability, item-to-item variability, and measurement error [19–28]. Temporal variability represents the inherent uncertainty changed with the degradation progression over time [17]. Item-to-item variability refers to the diversity of degradation paths induced by manufacturing processes and service conditions. Measurement error represents the difference between the observed degradation data and the true degradation data [19]. This error is mainly due to the imperfect instrument, random environment, or imperfect inspection which is inevitable in the measurement process. In this section, we review system reliability models with uncertainty based on stochastic process, Wiener process [19, 20], gamma process [21–23], and inverse Gaussian process [24]. Typically, Wiener process is utilized when the degradation is non-monotonic, while gamma process and inverse Gaussian process are used to analyze the monotonic degradation processes [21, 61, 62].

The following notations are defined in this section. $Y(t)$ is the measured degradation by time t. $X(t)$ is the true degradation by time t. $\varepsilon(t)$ is the measurement error by time t.

3.1 System Reliability Models Based on Wiener Process

In Sect. 3.1, we define the following notations. $X(0)$ is the initial degradation value. θ is the drift parameter of Wiener process. δ_B is the volatility parameter of Wiener process. $B(t)$ is the standard Brownian motion.

In general, a Wiener-based degradation model is expressed as [18]:

$$X(t) = X(0) + \theta t + \sigma_B B(t) \tag{18}$$

Si et al. [18] considered three sources of uncertainty in Wiener process to reliability estimation. Stochastic dynamics of the degradation process is represented by $\delta_B B(t) \sim N\big(0, \delta_B^2 t\big)$, $t > 0$. Item-to-item variation is illustrated by assuming parameter θ as a random variable that follows a specific distribution. In most cases, θ is assumed to follow a normal distribution, denoted as $\theta \sim N(\mu_\theta, \sigma_\theta^2)$, which is s-independent of $\{B(t), t \geq 0\}$. System reliability models developed in studies [19, 20, 26] are also based on Eq. (18).

According to the property of Wiener process, the first passage time exceeding the critical threshold follows an inverse Gaussian distribution [63]. Thus, the probability density function (PDF) of the lifetime T is:

$$f_T(t) = \frac{H}{\sqrt{2\pi t^3 \sigma_B^2}} exp\left(-\frac{(H-\theta t)^2}{2\sigma_B^2 t}\right), t > 0, \theta > 0 \tag{19}$$

where H is a threshold. Then, system reliability models can be obtained.

Equation (18) is the general function of the linear Wiener process. For complex systems, it is necessary to take the degradation nonlinearity into account; thus, Liu and Fan [64] used a nonlinear Wiener-based degradation model to model the degradation process, $X(t) = X(0) + \theta\Lambda(t; \gamma) + \sigma_B B(t)$, where $\Lambda(t; \gamma)$ is a nonlinear function with unknown parameter γ. In their study [64], $\Lambda(t; \gamma)$ is represented by a power function, t^{γ}. Zheng et al. [65] developed a generalized form of Wiener process: $X(t) = X(0) + f(t; \theta_1)^T\theta_2 + \sigma_B B(t)$, where $f(t; \theta_1)^T$ is a n-dimensional vector with a group of fundamental functions, θ_1 and θ_2 are parameter vectors, and $\theta_2 \in \mathcal{R}^n$. The temporal variability and the item-to-item variation are represented by $B(t)$ and θ_2, respectively. Moreover, Wiener-based degradation model can be used to model measurement error $\varepsilon(t)$ following a normal distribution, Eq. (18) will be [64]:

$$Y(t) = X(t) + \varepsilon(t) \tag{20}$$

The widely used assumptions [19, 20, 26–28] to model measurement error are: (1) $\varepsilon(t)$ follows a normal distribution; (2) all measurement error terms are mutually independent and independent with the true degradation. Similarly, by using the concept of the first passage time, the lifetime of a system is modeled as [65]:

$$T = \inf\{t : X(t) \geq w|X(0) < w\} \tag{21}$$

where w is the predetermined degradation-based failure threshold. Then, system reliability models and remaining useful life models can be obtained [64, 65].

3.2 *System Reliability Models Based on Gamma Process*

A continuous-time stochastic process $\{X(t), t \geq 0\}$ is defined as a gamma process with the shape function $\eta(t)$ and scale parameter θ if the following properties can be satisfied [66]:

(1) $P(X(0) = 0) = 1$;
(2) The increment $\Delta X(t_1, t_2) = X(t_2) - X(t_1)$, for all $t_2 > t_1 \geq 0$, follows a gamma distribution with shape parameter $\Delta\eta(t_1, t_2) = \eta(t_2) - \eta(t_1)$ and scale parameter θ;
(3) The increments are independent.

Gamma process can be used to represent the degradation path function with the uncertainty, for example, temporal variability [21–23]. Moreover, a group of studies further develop the gamma process-based degradation model to capture the item-to-item variation. For example, Lawless and Crower [21] incorporated a random variable z into $X(t)$; thus, $X(t)$ follows a gamma distribution with shaped parameter $\eta(t)$ and scale parameter $z\theta$. Liu et al. [67] introduced the parameter vector θ_{Ga} following a gamma distribution with hyper-parameters $\theta_{Ga}^{H} = \left(\delta_{\mu_{Ga}}, \gamma_{\mu_{Ga}}, \delta_{\lambda_{Ga}}, \gamma_{\lambda_{Ga}}\right)$ to model random effects.

Gamma process can also be used to model degradation path function with measurement error [68]:

$$Y(t) = X(t) + \varepsilon(t) \tag{22}$$

where $\varepsilon(t)$ is assumed to follow a Gaussian distribution with the mean value as 0. The true increment is expressed as $\Delta X(t) = X(t + \Delta t) - X(t)$. The measured value of increment $\Delta Y(t)$ expressed as [68]

$$\Delta Y(t) = \Delta X(t) + \varepsilon(t + \Delta t) - \varepsilon(t) \tag{23}$$

where $\Delta X(t)$ follows a gamma distribution, denoted as, $X(t) \sim Ga(\alpha, 1/\lambda)$, and $\varepsilon(t + \Delta t) - \varepsilon(t)$ follows a normal distribution, denoted as $\varepsilon(t + \Delta t) - \varepsilon(t) \sim N(0, 2\sigma^2)$. Similar models are also studied in references [23, 69–73]. System reliability models and remaining useful life predictions can be obtained based on the assumption of gamma process [21–23, 69–73].

Measurement error is commonly assumed to be independent with degradation; however, it may not be realistic in practice [74]. For example, Pulcini [66] proposed a perturbed gamma process in which the measurement error is statistically dependent on the degradation state. The error term in Eq. (22) is assumed to follow a normal distribution with the zero mean and the variance equal to $\sigma^2(x_t)$, where x_t is the current degradation level. Under the condition that the true degradation is $x_t = X(t)$, the conditional PDF of the measurement error $\varepsilon(t)$ given the measured degradation level y_t is obtained [66]:

$$\begin{aligned} f_{\varepsilon(t)}(\varepsilon_t|y_t) &= \int_0^{\infty} f_{\varepsilon(t)}(\varepsilon_t|x_t) f_{X(t)}(x_t|y_t) dx_t \\ &= \frac{1}{\sqrt{2\pi}} \frac{\int_0^{\infty} \frac{x_t^{\eta(t)-1}}{\sigma_\varepsilon^2(x_t)} exp\left[-\frac{1}{2(\varepsilon_t/\sigma(x_t))^2} - \frac{1}{2}\left(\frac{y_t - x_t}{\sigma(x_t)}\right)^2 - \frac{x_t}{\theta}\right] dx_t}{\int_0^{\infty} \frac{x_t^{\eta(t)-1}}{\sigma(x_t)} exp\left[-\frac{1}{2}\left(\frac{y_t - x_t}{\sigma(x_t)}\right)^2 - \frac{x_t}{\theta}\right] dx_t} \end{aligned} \tag{24}$$

The system reliability model is then proposed in consideration of the false alarm caused by the degradation measurement error [66].

3.3 System Reliability Models Based on Inverse Gaussian Process

An inverse Gaussian process $\{X(t); t \geq 0\}$ with function $\Lambda(t)$ and parameters β and λ has the following properties [24, 75]:

(1) $P(X(0) = 0) = 1$;
(2) Each increment follows an inverse Gaussian distribution, expressed as $X(t + \Delta t) - X(t) \sim IG(\beta\Delta\Lambda(t), \lambda\Delta\Lambda(t)^2)$;
(3) Increments are independent.

Inverse Gaussian process can be used to model monotonic degradation process with uncertainty. For example, Pan et al. [75] assumed β as a random parameter to denote the variability among products. They assume the prior distribution of $1/\beta$ follows the normal distribution, expressed as $1/\beta \sim N(\mu_\beta, 1/\sigma_\beta^2)$, which is statistically independent of λ. By using the concept of the first passage time, the lifetime T of a system can be obtained. The CDF of the lifetime T with the random effect of β is formulated by using the monotonicity property of inverse Gaussian process [75]:

$$F_T(t) = P(X(t) > \mathrm{H}) = \Phi\left(\sqrt{\frac{\lambda}{H}} \cdot \frac{(\sigma_\beta t - \mu_\beta \sigma_\beta H)}{\sqrt{\sigma_\beta^2 + \lambda H}}\right) - exp\left(2\mu_\beta \lambda t + \frac{2\lambda^2 t^2}{\sigma_\beta^2}\right) \times \Phi\left(-\sqrt{\frac{\lambda}{H}} \cdot \frac{(\sigma_\beta^2 + 2\lambda H)t + \mu_\beta \sigma_\beta^2 H}{\sqrt{\sigma_\beta^4 + \lambda H \sigma_\beta^2}}\right) \tag{25}$$

where H is a threshold.

Peng [76] established a normal-gamma mixture of inverse Gaussian degradation model to incorporate the heterogeneity among products. To be specific, λ is assumed to have a gamma density function, expressed as $f(\lambda) = [\lambda^{\alpha-1}/\Gamma(\alpha)\tau^\alpha]\exp(-\lambda/\tau), \lambda, \alpha, \tau > 0$. Let δ, $\delta = \beta^{-1}$, have a conditional normal PDF with mean ξ and variance σ_β^2/λ. The PDF of δ is $f(\delta|\lambda) = \sqrt{\lambda/2\pi\sigma_\beta^2}\exp\left(-\lambda(\delta - \xi)^2/2\sigma_\beta^2\right), \delta, \xi \in R, \sigma_\beta^2 > 0$. Later, Hao et al. [77] relaxed the normal assumption of δ and assumed δ follows a skew-normal distribution, in which $\delta \sim SN(\mu, \sigma^2, \alpha)$. The PDF of δ is presented as [77]:

$$f_\delta(x) = \frac{2}{\sigma}\Phi\left(\frac{x - \mu}{\sigma}\right)\Phi\left(\alpha\frac{x - \mu}{\sigma}\right) \tag{26}$$

where μ is the location parameter, σ is the scale parameter, and α is the shape parameter. They [77] further model the CDF of the lifetime based on the skew-normal distribution assumption of δ. Recently, Sun et al. [78] predicted the system remaining useful life with the inverse Gaussian degradation model with measurement errors and further applied to the hydraulic piston pump. The errors are assumed to follow a normal distribution conditioning on the degradation level.

Meanwhile, the degradation path can be modeled by other distributions. For example, Zhai and Ye [29] discussed that Gaussian distribution has low probabilities in large values, which may result in a misleading result when some fatal errors are introduced during the observation process. Thus, the measurement error is assumed to follow a student's *t*-distribution. Shen et al. [30] assumed the measurement error follows a logistic distribution. This distribution has relatively heavier tails compared with the normal distribution, which is more suitable to use when there are large errors in the degradation data. Li et al. [32] considered that measurement errors are time-series data, which has the auto-correlation due to modeling errors or environmental changes especially when the time interval is short. Thus, a Wiener process degradation model with one-order autoregressive (AR(1)) measurement errors is established. The AR(1) measurement error is also considered in studies [31, 33]. Giorgio et al. [34] modeled $\varepsilon(t)$ as a three-parameter inverse gamma distributed random variable that is conditionally distributed on the degradation level.

4 Conclusion

Reliability evaluation of complex engineering systems is a critical task in many safety–critical applications. System failure is generally caused by random shocks and internal degradation. Typically, five random shock models are commonly used in the field of Reliability Engineering, cumulative shock model, extreme shock model, run shock model, δ-shock model, and mixed shock model. In addition, the uncertainty in the degradation process can influence the accuracy of the reliability estimation. In general, there are three sources of variability that can result in uncertainty, temporal variability in the degradation process, unit-to-unit variability, and measurement error caused by imperfect instruments or imperfect inspection. Considering the importance and popularity of considering random shocks and uncertainty in modeling system reliability, in this chapter, we first review system reliability models with random shock models and then system reliability models with uncertainty in terms of three classic stochastic processes, Wiener process, gamma process, and inverse Gaussian process.

References

1. Pham H (2022) Statistical reliability engineering: methods, models and applications. Springer
2. Esary J, Marshall A (1973) Shock models and wear processes. Ann Probab 1(4):627–649
3. Gut A (1990) Cumulative shock models. Adv Appl Probab 22(2):504–507
4. Che H, Zeng S, Guo J, Wang Y (2018) Reliability modeling for dependent competing failure processes with mutually dependent degradation process and shock process. Reliab Eng Syst Saf 180:168–178

5. Dong W, Liu S, Bae SJ, Cao Y (2021) Reliability modelling for multi-component systems subject to stochastic deterioration and generalized cumulative shock damages. Reliab Eng Syst Saf 205:107260
6. Shanthikumar JG, Sumita U (1983) General shock models associated with correlated renewal sequences. J Appl Probab 20(3):600–614
7. Wang J, Han X, Zhang YA, Bai G (2021) Modeling the varying effects of shocks for a multi-stage degradation process. Reliab Eng Syst Saf 215:107925
8. Hao S, Yang J (2018) Reliability analysis for dependent competing failure processes with changing degradation rate and hard failure threshold levels. Comput Ind Eng 118:340–351
9. Gut A (2001) Mixed shock models. Bernoulli 7:541–555
10. Mallor F, Omey E (2001) Shocks, runs and random sums. J Appl Probab 38(2):438–448
11. Li Z, Kong X (2007) Life behavior of δ-shock model. Statist Probab Lett 77(6):577–587
12. Gong M, Eryilmaz S, Xie M (2020) Reliability assessment of system under a generalized cumulative shock model. Proc Inst Mech Eng Part O J Risk Reliab 234(1):129–137
13. Ozkut M, Eryilmaz S (2019) Reliability analysis under Marshall-Olkin run shock model. J Comput Appl Math 349:52–59
14. Wu B, Cui L, Qiu Q (2021) Two novel critical shock models based on Markov renewal processes. Nav Res Logist (NRL). 69(1):163–176
15. Wang GJ, Peng R (2017) A generalised δ-shock model with two types of shocks. Int J Syst Sci Oper Logistics 4(4):372–383
16. Lorvand H, Nematollahi AR, Poursaeed MH (2020) Assessment of a generalized discrete time mixed δ-shock model for the multi-state systems. J Comput Appl Math 366:112415
17. Giorgio M, Guida M, Pulcini G (2011) An age-and state-dependent Markov model for degradation processes. IIE Trans 43(9):621–632
18. Si XS, Wang W, Hu CH, Zhou DH (2014) Estimating remaining useful life with three-source variability in degradation modeling. IEEE Trans Reliab 63(1):167–190
19. Si XS, Wang W, Hu CH, Zhou DH (2011) Remaining useful life estimation–a review on the statistical data driven approaches. Eur J Oper Res 213(1):1–14
20. Peng CY, Tseng ST (2009) Mis-specification analysis of linear degradation models. IEEE Trans Reliab 58(3):444–455
21. Lawless J, Crowder M (2004) Covariates and random effects in a gamma process model with application to degradation and failure. Lifetime Data Anal 10(3):213–227
22. Park SH, Kim JH (2016) Lifetime estimation of LED lamp using gamma process model. Microelectron Reliab 57:71–78
23. Hazra I, Pandey MD, Manzana N (2020) Approximate Bayesian computation (ABC) method for estimating parameters of the gamma process using noisy data. Reliab Eng Syst Saf 198:106780
24. Wang X, Xu D (2010) An inverse Gaussian process model for degradation data. Technometrics 52(2):188–197
25. Yuan XX, Pandey MD (2009) A nonlinear mixed-effects model for degradation data obtained from in-service inspections. Reliab Eng Syst Saf 94(2):509–519
26. Si XS, Chen MY, Wang W, Hu CH, Zhou DH (2013) Specifying measurement errors for required lifetime estimation performance. Eur J Oper Res 231(3):631–644
27. Meeker WQ, Escobar LA, Pascual FG (2021) Statistical methods for reliability data. Wiley, p 639
28. Ye ZS, Wang Y, Tsui KL, Pecht M (2013) Degradation data analysis using Wiener processes with measurement errors. IEEE Trans Reliab 62(4):772–780
29. Zhai Q, Ye ZS (2017) Robust degradation analysis with non-Gaussian measurement errors. IEEE Trans Instrum Meas 66(11):2803–2812
30. Shen Y, Shen L, Xu W (2018) A Wiener-based degradation model with logistic distributed measurement errors and remaining useful life estimation. Qual Reliab Eng Int 34(6):1289–1303
31. Li J, Wang Z, Liu C, Qiu M (2019) Accelerated degradation analysis based on a random-effect Wiener process with one-order autoregressive errors. Eksploatacja i Niezawodność 21(2)
32. Li J, Wang Z, Zhang Y, Liu C, Fu H (2018) A nonlinear Wiener process degradation model with autoregressive errors. Reliab Eng Syst Saf 173:48–57

33. Lin JG, Wei BC (2007) Testing for heteroscedasticity and/or autocorrelation in longitudinal mixed effect nonlinear models with AR (1) errors. Commun Stat Theory Methods 36(3):567–586
34. Giorgio M, Mele A, Pulcini G (2019) A perturbed gamma degradation process with degradation dependent non-Gaussian measurement errors. Appl Stoch Model Bus Ind 35(2):198–210
35. Parvardeh A, Balakrishnan N (2015) On mixed δ-shock models. Statist Probab Lett 102:51–60
36. Rafiee K, Feng Q, Coit DW (2015) Condition-based maintenance for repairable deteriorating systems subject to a generalized mixed shock model. IEEE Trans Reliab 64(4):1164–1174
37. Hao S, Yang J, Ma X, Zhao Y (2017) Reliability modeling for mutually dependent competing failure processes due to degradation and random shocks. Appl Math Model 51:232–249
38. Wang Y, Pham H (2011) Modeling the dependent competing risks with multiple degradation processes and random shock using time-varying copulas. IEEE Trans Reliab 61(1):13–22
39. Rafiee K, Feng Q, Coit DW (2014) Reliability modeling for dependent competing failure processes with changing degradation rate. IIE Trans 46(5):483–496
40. Song S, Coit DW, Feng Q (2014) Reliability for systems of degrading components with distinct component shock sets. Reliab Eng Syst Saf 132:115–124
41. Fan M, Zeng Z, Zio E, Kang R (2017) Modeling dependent competing failure processes with degradation-shock dependence. Reliab Eng Syst Saf 165:422–430
42. Song S, Coit DW, Feng Q (2016) Reliability analysis of multiple-component series systems subject to hard and soft failures with dependent shock effects. IIE Trans 48(8):720–735
43. Wang R, Zhu M (2022) Shock-loading based method for modeling dependent competing risks with degradation processes and random shocks. Int J Reliab Qual Saf Eng. In press
44. Ranjkesh SH, Hamadani AZ, Mahmoodi S (2019) A new cumulative shock model with damage and inter-arrival time dependency. Reliab Eng Syst Saf 192:106047
45. Cha JH, Finkelstein M (2011) On new classes of extreme shock models and some generalizations. J Appl Probab 48(1):258–270
46. Eryilmaz S, Kan C (2019) Reliability and optimal replacement policy for an extreme shock model with a change point. Reliab Eng Syst Saf 190:106513
47. Eryilmaz S, Kan C (2021) A new shock model with a change in shock size distribution. Probab Eng Inf Sci 35(3):381–395
48. Gong M, Xie M, Yang Y (2018) Reliability assessment of system under a generalized run shock model. J Appl Probab 55(4):1249–1260
49. Poursaeed MH (2021) Reliability analysis of an extended shock model. Proc Inst Mech Eng Part O J Risk Reliab 1748006X20987794
50. Liu H (2019) Reliability and maintenance modeling for competing risk processes with Weibull inter-arrival shocks. Appl Math Model 71:194–207
51. Eryilmaz S, Bayramoglu K (2014) Life behavior of delta-shock models for uniformly distributed interarrival times. Stat Pap 55(3):841–852
52. Eryilmaz S (2013) On the lifetime behavior of a discrete time shock model. J Comput Appl Math 237(1):384–388
53. Eryilmaz S (2017) δ-shock model based on Polya process and its optimal replacement policy. Eur J Oper Res 263(2):690–697
54. Tuncel A, Eryilmaz S (2018) System reliability under δ-shock model. Commun Stat Theory Methods 47(19):4872–4880
55. Lorvand H, Nematollahi A, Poursaeed MH (2020) Life distribution properties of a new δ-shock model. Commun Stat Theory Methods 49(12):3010–3025
56. Wang GJ, Zhang YL (2005) A shock model with two-type failures and optimal replacement policy. Int J Syst Sci 36(4):209–214
57. Doostmoradi A, Akhoond MR, Zadkarami MR (2021) Reliability of a system under a new mixed shock model. Commun Stat Theory Methods 1–15
58. Jiang L, Feng Q, Coit DW (2012) Reliability and maintenance modeling for dependent competing failure processes with shifting failure thresholds. IEEE Trans Reliab 61(4):932–948
59. Rafiee K, Feng Q, Coit DW (2017) Reliability assessment of competing risks with generalized mixed shock models. Reliab Eng Syst Saf 159:1–11

60. Zhao X, Guo X, Wang X (2018) Reliability and maintenance policies for a two-stage shock model with self-healing mechanism. Reliab Eng Syst Saf 172:185–194
61. Rodríguez-Picón LA, Flores-Ochoa VH, Méndez-González LC, Rodríguez-Medina MA (2017) Bivariate degradation modelling with marginal heterogeneous stochastic processes. J Stat Comput Simul 87(11):2207–2226
62. Ye ZS, Chen N (2014) The inverse Gaussian process as a degradation model. Technometrics 56(3):302–311
63. Folks JL, Chhikara RS (1978) The inverse Gaussian distribution and its statistical application—a review. J Roy Stat Soc Ser B (Methodol) 40(3):263–275
64. Liu S, Fan L (2022) An adaptive prediction approach for rolling bearing remaining useful life based on multistage model with three-source variability. Reliab Eng Syst Saf 218:108182
65. Zheng JF, Si XS, Hu CH, Zhang ZX, Jiang W (2016) A nonlinear prognostic model for degrading systems with three-source variability. IEEE Trans Reliab 65(2):736–750
66. Pulcini G (2016) A perturbed gamma process with statistically dependent measurement errors. Reliab Eng Syst Saf 152:296–306
67. Liu D, Wang S, Zhang C, Tomovic M (2018) Bayesian model averaging based reliability analysis method for monotonic degradation dataset based on inverse Gaussian process and Gamma process. Reliab Eng Syst Saf 180:25–38
68. Wei Q, Xu D (2014) Remaining useful life estimation based on gamma process considered with measurement error. In: 2014 10th international conference on reliability, maintainability and safety (ICRMS). IEEE, pp 645–649
69. Liu X, Matias J, Jäschke J, Vatn J (2022) Gibbs sampler for noisy transformed Gamma process: inference and remaining useful life estimation. Reliab Eng Syst Saf 217:108084
70. Lu D, Pandey MD, Xie WC (2013) An efficient method for the estimation of parameters of stochastic gamma process from noisy degradation measurements. Proc Inst Mech Eng Part O J Risk Reliability 227(4):425–433
71. Kallen MJ, Van Noortwijk JM (2005) Optimal maintenance decisions under imperfect inspection. Reliab Eng Syst Saf 90(2–3):177–185
72. Le Son K, Fouladirad M, Barros A (2012) Remaining useful life estimation on the non-homogenous gamma with noise deterioration based on Gibbs filtering: a case study. In: 2012 IEEE conference on prognostics and health management. IEEE, pp 1–6
73. Bordes L, Paroissin C, Salami A (2016) Parametric inference in a perturbed gamma degradation process. Commun Stat Theory Methods. 45(9):2730–2747
74. Rabinovich SG (2013) The international vocabulary of metrology and the guide to the expression of uncertainty in measurement: analysis, criticism, and recommendations. Evaluating measurement accuracy. Springer, New York, NY, pp 269–285
75. Pan D, Liu JB, Cao J (2016) Remaining useful life estimation using an inverse Gaussian degradation model. Neurocomputing 185:64–72
76. Peng CY (2015) Inverse Gaussian processes with random effects and explanatory variables for degradation data. Technometrics 57(1):100–111
77. Hao S, Yang J, Berenguer C (2019) Degradation analysis based on an extended inverse Gaussian process model with skew-normal random effects and measurement errors. Reliab Eng Syst Saf 189:261–270
78. Sun B, Li Y, Wang Z, Ren Y, Feng Q, Yang D (2021) An improved inverse Gaussian process with random effects and measurement errors for RUL prediction of hydraulic piston pump. Measurement 173:108604

A Hybrid Approach for Evaluation and Prioritization of Software Vulnerabilities

Vivek Kumar, Misbah Anjum, Vernika Agarwal, and P. K. Kapur

Abstract A software vulnerability is a technical flaw or glitch in the software which might be exploited to contravene the security policy of the system. The intensity of software vulnerabilities amplifies at an exponential rate, which is a tedious task for software testers. The removal of these vulnerabilities is an important task for software developers. With the constraint on the cost and time limitations, it becomes important to prioritize the software vulnerabilities and identify those vulnerabilities which are most severe. In this study, we have sub-grouped software vulnerability types into two categories: code execution vulnerabilities and improper authentication vulnerabilities. In this view, the present study focuses on assessing the vulnerabilities which are most prone to attacks. The study utilizes a hybrid methodology comprising of the fuzzy Best Worst Method to prioritize the identified software vulnerabilities, followed by a two-way analysis to integrate the opinion of decision-makers. The research findings show that the vulnerabilities that are caused because of improper code execution are more severe than those of authentication error vulnerabilities. The present framework is validated by using the case of an Indian software testing company situated in the National capital region of India.

Keywords Software vulnerabilities · Multi-criteria decision-making (MCDM) · Fuzzy best–worst method (BWM) · Two-way analysis

V. Kumar
Department of Operational Research, University of Delhi, Delhi, India

M. Anjum
Amity Institute of Information Technology, Amity University, Noida, Uttar Pradesh, India

V. Agarwal (✉)
Amity International Business School, Amity University, Noida, Uttar Pradesh, India
e-mail: vagarwal1@amity.edu

P. K. Kapur
Amity Center for Inter-Disciplinary Research, Amity University, Noida, Uttar Pradesh, India

V. Kumar and H. Pham (eds.), *Predictive Analytics in System Reliability*,
Springer Series in Reliability Engineering,
https://doi.org/10.1007/978-3-031-05347-4_3

1 Introduction

The digital management system has massive and severe security requirements. The main requirements are the continuous disclosure of software vulnerabilities that can jeopardize their systems in certain aspects [1]. In the field of security and risk management, vulnerability assessment plays a vital role. Vulnerability refers to defects or weaknesses in the design, operation, or implementation of a system that might be exploited to contravene the security policy of the system [2]. To access, harm, or breach the information system illegally, any defect or weakness in an information system might be exploited [3]. It is of paramount significance to classify and understand common software vulnerabilities that lead to safety risks. It is believed that 90% of the incidents reported are caused by the exploitation of design or software code [4]. A single exploited software vulnerability can cause severe damage to an organization. Yearly damages of up to $226 billion from cyber-attacks have been recorded worldwide [5]. It is important to incorporate measures for the integrated safety of the enabling software in order to assure system stability, integrity and safety [6]. Before the deployment of software, it is essential to discover and mitigate software vulnerabilities.

New vulnerabilities in software security are found nearly every day and have caused major financial damage [7]. It is therefore essential that they can be detected and solved as soon as possible. The security team is responsible for recognizing and addressing these vulnerabilities using different software and hardware platforms [8]. These vulnerabilities must be addressed and assessed to meet business deadlines and operate within limited financial resources [9]. Security managers generally operate under a restricted budget so that the identified vulnerabilities have to be prioritized and the mitigating measures are taken into account [10]. A response procedure proportional to its seriousness should address a vulnerability and priority should be given to more serious vulnerabilities than less serious [11]. The reaction processes for vulnerability are not standardized and may differ with regards to response speed, roles involved, the effect of production and operations, and in a particular overall cost of response [12].

Due to security flaws left during program developments software becomes susceptible. The rate of vulnerability exploitation will increase. The process of assessing and prioritizing vulnerability is thus a genuine difficulty and sensitive task [13]. In literature, the priority of vulnerability has been explored, and the necessity for priority vulnerability is generally acknowledged in organizations [14]. The evaluation should be done in such a way that the vulnerabilities that offer the largest threat are fixed first [15]. Vulnerability priority contains different features that developers and testing professionals need to take into account in selecting the order to remedy the vulnerability [16]. It is thus a crucial responsibility for developers and testers to identify these vulnerabilities according to the degree of their severity so that they can be managed correctly and a fix can be delivered in good time [11].

A critical vulnerability can allow malicious code to be run without user interactions, which may lead to a security breakdown if exploited by attackers [17].

The literature has investigated the number of qualitative and quantitative assessment approaches to attribute sensitivity ratings [18]. Several vulnerability scores have been identified, endorsed, and implemented, qualitatively [19] or quantitatively [20], by a diverse range of technological and non-profit providers. In the past vulnerability assessment studies, prioritization has been done using several "Multi-Criteria Decision-Making" (MCDM) approaches such as [16] prioritized software vulnerabilities using "Analytic Hierarchy Process", "Normalized Criteria Distance" and "DEMATEL". [21] presented software vulnerability prioritization with "AHP" based on "Verbal Rating Scales". [22] analyzed prioritization using "Fuzzy Analytical Hierarchy Process" and "fuzzy synthetic decision-making approach". [19] prioritized vulnerabilities using the "Analytic Network Process" method. The authors are concerned with assessing the code smells, based on their effect on large-scale open-source (OSS) projects [23]. The study by Anjum et al. [11] has considered very few software vulnerabilities for the prioritization process by using the best–worst method. However, the crisp methodology involves ambiguity and uncertainty caused by decision-makers contextual judgment, crisp constraints cannot be used to construct multi-criteria decision-making (MCDM) issues in real-world settings. To defy the problem of uncertainty we have used fuzzy best worst method to rank the software vulnerabilities. Further, in this study, we have categorized 16 different software vulnerabilities into two groups: improper code execution vulnerabilities and improper authentication vulnerabilities. The categorization is being done based on highly severe vulnerability types which is the novelty of our work.

From the above discussion the following research objectives can be identified:

- To identify the software vulnerability types into two groups of code execution vulnerabilities and improper authentication vulnerabilities.
- To prioritize the evaluate the code execution vulnerabilities and improper authentication vulnerabilities.
- To incorporate the ambiguity in decision-makers opinions by assessing the vulnerabilities groups.

To address the above-mentioned research questions, the present study focuses on assessing the vulnerabilities which are most prone to attacks. The study utilizes a hybrid methodology comprising of the fuzzy Best Worst Method to prioritize the identified software vulnerabilities, followed by a two-way analysis to integrate the opinion of decision-makers. We have included the vulnerabilities caused by code execution and incorrect authentication in this study. The vulnerabilities involving authentication bypass could allow attackers to execute different harmful activities by circumventing the device authentication system. The consequences of flaws in authentication can be quite serious. Once an attacker is either bypassed or brutalized into the account of a user, he may utilize all the data and features of the afflicted account. If they can hack a highly privileged account, such as a system administrator, the entire program may be monitored and the core infrastructure may be accessed. In this article, we suggest the hybrid approach involving the best–worst technique of classifying and prioritizing these vulnerabilities, and the total measure of severity of vulnerabilities is then analyzed using the two-way technique.

The paper is structured accordingly: The research technique is covered in Sect. 2. Section 3 displays the analysis of the data followed by the conclusion of the paper and the possible future work in Sect. 4.

2 Research Methodology

This section focuses on research methodologies. In this article, we have devised a dual approach to measure different software vulnerabilities. FBWM prioritizes vulnerabilities in the initial stage. During the second step, the overall severity of the vulnerabilities is calculated employing a two-way evaluation approach relying on FBWM weights.

2.1 *Dataset Description*

Before the investigation begins, the literature survey initially finds various software vulnerabilities that are either caused due to an authentication problem or improper code execution. Remote code execution is caused by attackers creating malicious code and injecting it into the server via input points. The server unknowingly executes the commands, and this allows an attacker to gain access to the system. On the other hand, authentication vulnerabilities are among the most apparent conceptual concerns. However, they can be one of the most essential because of the link between authentication and security and because attackers can access vital data and functions directly. Additional attack surfaces are also displayed for future exploitation. That is why it is essential to comprehend how code execution and authentication vulnerabilities are identified and exploited, including how typical protective measures may be avoided. The authentic database for data collecting is the "National Vulnerability Database" [24]. This study is aimed at industry professionals and academicians (managers/developers, testers/stakeholders). The panel is made up of professionals with minimum expertise in the relevant disciplines for 7–11 years (Table 1).

2.2 *Fuzzy Best Worst Method*

Considering the paucity of complete information and the uncertainty caused by decision-makers contextual judgment, crisp constraints cannot be used to construct multi-criteria decision-making (MCDM) issues in real-world settings. In this section, the fuzzy best–worst approach is used to tackle such situations. The elaborate steps of fuzzy BWM are as follows [25]:

Table 1 Description of software vulnerabilities

Vulnerability category	Notations	Vulnerability type
Code execution	CDR1	SQL injection (SQLI)
	CDR2	Cross-site scripting (XSS)
	CDR3	Buffer overflow (B0)
	CDR4	File inclusion (FI)
	CDR5	Code execution (CE)
	CDR6	Race condition (RC)
	CDR7	Memory corruption (MC)
	CDR8	Http response splitting (HTTPRS)
Improper authentication	AUE1	Cross site request forgery (CSRF)
	AUE2	Information gain (IG)
	AUE3	Gain of privileges (GP)
	AUE4	Bypass something (BS)
	AUE5	Denial of service (DoS)
	AUE6	Directory traversal (DT)
	AUE7	Improper authentication (IA)
	AUE8	Insufficient entropy (IE)

Step 1. Establishment of decision criterion layout. In the present scenario, improper code execution vulnerability collection (group 1) reflects the area of decision set {D1, D2 …, Dn}.

Step 2. Identification of the most and least severe vulnerability with the help of decision-makers opinions and are represented as D_B, and D_z respectively.

Step 3. Determination of the severity of the most crucial vulnerability over others using fuzzy comparisons. The fuzzy comparisons are done using the linguistic terms as "Highly Severe [HS]" having a fuzzy value of (1, 1, 1), "Severe [Se]" of value (0.6, 1, 1.5), "Medium Severe [MS]" as the value of (1.5, 2, 2.5), "Slightly Severe [SS]" with value (2.5, 3, 3.5) and "Least Severe [LS]" of fuzzy value as (3.5, 4, 4.5). The vector of fuzzy Best-to-Others is:

$$\tilde{B}_B = \left(\tilde{b}_{B1}, \tilde{b}_{B2}, \ldots, \tilde{b}_{Bn}\right)$$

where $\tilde{B}_B$ represents the fuzzy Best-to-Others vector; $\tilde{b}_{Bj}$ represents the fuzzy preference of D_B over criterion j, j = 1, 2…, n, also $\tilde{b}_{BB} = (1,\ 1,\ 1)$.

Step 4. Determination of the severity of other vulnerabilities over the least crucial vulnerability. The resulting vector of fuzzy worst-to-others is:

$$\tilde{B}_z = \left(\tilde{b}_{1z},\ \tilde{b}_{2z}, \ldots, \tilde{b}_{nz}\right)$$

where $\tilde{B}_z$ represent the fuzzy Others-to-Worst vector; $\tilde{b}_{iz}$ over D_z, i = 1, 2…, n, also $\tilde{b}_{zz} = (1,\ 1,\ 1)$.

Step 5. Calculation of the optimal fuzzy weights

$$(\tilde{z}*_1, \tilde{z}*_2, \ldots, \tilde{z}*_n)$$

The optimal fuzzy weight for each criterion is the one where for each fuzzy pair $\tilde{Z}_B/\tilde{Z}_j$ and $\tilde{Z}_j/\tilde{Z}_z$ it should have $\frac{\tilde{Z}_B}{\tilde{Z}_j} = \tilde{b}_{Bj}$ and $\frac{\tilde{Z}_j}{\tilde{Z}_z} = \tilde{b}_{jz}$. To satisfy these conditions for all j, it should determine a solution where the maximum absolute gaps $\left|\frac{\tilde{Z}_B}{\tilde{Z}_j} = \tilde{b}_{Bj}\right|$ and $\left|\frac{\tilde{Z}_j}{\tilde{Z}_z} = \tilde{b}_{jz}\right|$ for all j are minimized. We use the following constrained optimization problem for determining the optimal fuzzy weights $(\tilde{z}*_1, \tilde{z}*_2, \ldots, \tilde{z}*_n)$ as follows:

$$\min \max_j \left\{ \left| \frac{\widetilde{Z}_B}{\widetilde{Z}_i} - \widetilde{b}_{Bj} \right|, \left| \frac{\widetilde{Z}_j}{\widetilde{Z}_z} - \widetilde{b}_{jZ} \right| \right\}$$

$$\text{s.t.} \left\{ \begin{array}{c} \sum_{j=1}^{n} \mathrm{R}_e(\widetilde{Z}_j) = 1 \\ e_j^z \le f_j^z \le g_j^z \\ e_j^z \ge 0 \\ j = 1, 2, \ldots, n \end{array} \right\} \tag{1}$$

where $\widetilde{Z}_B = (e_B^z,\ \mathrm{f}_B^z,\ g_B^z)$, $\widetilde{Z}_j = (e_j^z,\ \mathrm{f}_j^z,\ g_j^z)$, $\widetilde{Z}_z = (e_Z^z,\ \mathrm{f}_Z^z,\ g_Z^z)$, $\tilde{b}_{Bj} = (e_{Bj}, f_{Bj}, g_{Bj})$, $\tilde{b}_{jz} = (e_{jz}, f_{jz}, g_{jz})$ and e, f and g represent the lower, middle and upper values respectively. Equation (1) can then be transferred to the following nonlinearly constrained optimization problem.

$$\min \tilde{\psi}$$

$$\text{s.t.} \left\{ \begin{array}{l} \left| \frac{\tilde{Z}_B}{\tilde{Z}_j} - \widetilde{b}_{Bj} \right| \le \tilde{\psi} \\ \left| \frac{\tilde{Z}_j}{\tilde{Z}_Z} - \widetilde{b}_{jz} \right| \le \tilde{\psi} \\ \sum_{j=1}^{n} R_e(\tilde{z}_j) = 1 \\ e_j^z \le f_j^z \le g_j^z \\ e_j^z \ge 0 \\ j = 1, 2, \ldots, n \end{array} \right\} \tag{2}$$

Table 2 Consistency index table for FBWM

$\tilde{a}_{Bw}$	(1, 1, 1)	(2/3, 1, 3/2)	(3/2, 2, 5/2)	(5/2, 3, 7/2)	(7/2, 4, 9/2)
Consistency index	3.00	3.80	5.29	6.69	8.04

where $\tilde{\psi} = (e^{\psi}, f^{\psi}, g^{\psi})$.

Considering

$$\min \tilde{\psi}*$$

$$s.t. \begin{Bmatrix} \left| \frac{(e_B^z, f_B^z, g_B^z)}{(e_j^z, f_j^z, g_j^z)} - (e_{Bj}, f_{Bj}, g_{Bj}) \right| \le (t*, t*, t*) \\ \left| \frac{(e_j^z, f_j^z, g_j^z)}{(e_j^z, f_j^z, g_j^z)} - (e_{jz}, f_{jz}, g_{jz}) \right| \le (t*, t*, t*) \\ \sum_{j=1}^{n} R_e(\widetilde{Z}_j) = 1 \\ e_j^z \le f_j^z \le g_j^z \\ e_j^z \ge 0 \\ j = 1, 2, \ldots, n \end{Bmatrix} \tag{3}$$

Step 6. Check the consistency of the solution.

The closer the consistency ratio is to the zero value, the more consistent is the comparison system provided by the DM. We check the consistency of the solution by calculating the consistency ratio:

Consistency Ratio = $\frac{\psi^*}{\text{Consistency Index}}$. Table 2 is used to get the value of the consistency index.

Step 7. The above steps from 1 to 6 are being repeated for another group of vulnerabilities.

2.3 Two Way Assessment

We take account of the views of the stakeholders and developers concerning the crucial nature of every vulnerability attribute and compute their criticality concerning utility values [26]. Participants are requested to give priority to vulnerabilities based on their severity. In this study, we take five severity values for each vulnerability as determined $K = (k_1, k_2, \ldots, k_5)$. The acceptability value of this study is 2, 4, 6, 8, and 10, where k_1 is the high severity of value 10, k_2 has the severity of 8, whereas k_3, k_4, and k_5 have moderate, slight, and low degree of severities. The stakeholders' opinions are gathered in the form of pairwise comparisons indicated by F_{ij}, which are expressed by a percentage depending on the intervener's reaction to

Table 3 Overall utility measure for FBWM

Vulnerabilities	FBWM weights	Levels					Expected weight level	Contribution to total expected utility (Ui)
		Highly severe	Severe	Medium severe	Slightly severe	Least severe		
		k_1	k_2	k_3	k_4	k_5		
AUE1	Ow_1	F_{11}	F_{12}	F_{13}	F_{14}	F_{15}	$Ex_1 = \sum_j F_{1j} k_j$	$Ex_1 ow_1$
AUE2	Ow_2	F_{21}	F_{22}	F_{23}	F_{24}	F_{25}	$Ex_2 = \sum_j F_{2j} k_j$	$Ex_2 ow_2$
⋮	⋮	⋮	⋮	⋮	⋮	⋮	⋮	⋮
AUEn	Ow_n	F_{n1}	F_{n2}	F_{n3}	F_{n4}	F_{n5}	$Ex_n = \sum_j F_{nj} k_j$	$Ex_n ow_n$
Total utility								$\sum_i Ex_i ow_i$

the criticality. The anticipated level weight (Ex) is then calculated by multiplying the F_{ij} values of each characteristic with their corresponding acceptance level (k) and by adding each vulnerability attribute to the total value. We multiply the weight of all attributes by their corresponding predicted level weight and the sum of all individual utilities for measuring the individual perception, providing an overall utility measure as set out in Table 3.

3 Data Analysis

As stated earlier, we address the vulnerabilities associated with an authentication error as well as a code execution error. These assaults are slightly tougher to evaluate automatically because programs find it almost impossible to identify whether a small authorization error has been made by applications. The numerical description in this part prioritizes the severity of the vulnerabilities to take the appropriate action as quickly as possible.

3.1 Prioritizing Vulnerabilities Using FBWM

Following the stages of FBWM, we classified vulnerabilities of each group according to their severity, as described in Sect. 2.2. Because FBWM needs fewer comparisons, decision-makers (DMs) are thus utilizing step 2 to pick the best/most severe and worst/least severe criteria. In this study, we collected data from five different decision-makers (1 software developer, 2 testers, and 2 academicians). From group 1 (i.e., improper code executed vulnerabilities) three out of four decision-makers selected CDR1 as most severe and CDR8 as least severe. From group 2 (i.e., improper

Table 4 Best-to-others (BO) and others-to-worst preference matrix

Vulnerability category	Notations	Preference of decision-makers									
		DM1		DM2		DM3		DM4		DM5	
		BO	OW	BO	OW	BO	OW	BO	OW	BO	OW
Improper code execution	CDR1	Se	SS	HS	LS	HS	SS	HS	LS	Se	MS
	CDR2	MS	Se	SS	MS	MS	Se	MS	SS	MS	SS
	CDR3	Se	SS	Se	SS	MS	SS	SS	Se	Se	SS
	CDR4	SS	MS	MS	Se	SS	MS	MS	MS	SS	Se
	CDR5	HS	LS	Se	MS	Se	MS	Se	Se	HS	LS
	CDR6	MS	SS	LS	HS	SS	LS	MS	SS	SS	Se
	CDR7	Se	MS	MS	SS	MS	MS	LS	HS	MS	MS
	CDR8	LS	HS	SS	MS	LS	HS	SS	Se	LS	HS
Improper authentication	AUE1	SS	Se	MS	MS	SS	Se	SS	Se	MS	MS
	AUE2	HS	LS	LS	LS	Se	LS	Se	LS	LS	LS
	AUE3	Se	SS	MS	Se	MS	Se	Se	SS	MS	Se
	AUE4	LS	HS	SS	HS	SS	SS	LS	HS	LS	HS
	AUE5	MS	MS	SS	Se	LS	HS	MS	MS	SS	MS
	AUE6	Se	MS	MS	SS	Se	MS	MS	SS	Se	SS
	AUE7	Se	SS	Se	SS	LS	LS	HS	LS	Se	LS
	AUE8	Se	SS	MS	MS	Se	MS	Se	SS	MS	MS

authentication) AUE2 is identified as most severe and AUE4 as least severe. Steps 2 and 3 allow DM to provide its fuzzy preference of most crucial vulnerability to other vulnerabilities (BO) and fuzzy preference of other vulnerabilities to the least crucial vulnerability (OW) as outlined in Table 4.

We use the non-linear constrained programming problem given in Sect. 2.2 of step 5 to discover the appropriate weights. On solving Eq. 1, the ideal weight value is calculated.

As can be seen from Table 5, from group 1 the vulnerability CDR1 has the highest weight of 0.223 followed by CDR5 with 0.207 and are therefore holding the rank Ist and 2nd. From group 2 AUE2 has the highest weight 0.232 and is in rank Ist, AUE7 is ranked second with weights 0.178. Vulnerabilities CDR8 and AUE4 from both the groups are having the least weight and are ranked at number 8th. From Table 4 it is shown that CDR1, CDR5 from group 1, and AUE2 and AUE7 vulnerabilities from group 2 are extremely severe. Also, on checking the consistency our resulted value comes out to be closer to zero which makes our data collection table matrix consistent. Consequently, we utilize a two-way assessment to focus on the overall impact of vulnerabilities in the decision-making process and evaluate the stakeholder perspective.

Table 5 Weights of the vulnerabilities calculated

Vulnerability category	Notations	Crisp weights	Ranking
Improper code execution	CDR1	0.223	I
	CDR2	0.103	IV
	CDR3	0.132	III
	CDR4	0.092	VI
	CDR5	0.207	II
	CDR6	0.086	VII
	CDR7	0.098	V
	CDR8	0.065	VIII
Improper authentication	AUE1	0.086	VI
	AUE2	0.232	I
	AUE3	0.111	V
	AUE4	0.065	VIII
	AUE5	0.078	VII
	AUE6	0.118	IV
	AUE7	0.178	II
	AUE8	0.131	III

3.2 *Two-Way Assessment Technique*

The severity is computed using a two-way evaluation technique for each group of vulnerabilities. The DM-rated vulnerabilities are based on specified weights. For example, the weight of vulnerability CDR1 computed for two-way analysis is 22.375, with 100% of respondents ranking it as extremely severe. To obtain the expected level weight, we multiply $(10 * 1) + (8 * 0) + (6 * 0) + (4 * 0) + (2 * 0)$ which is equal to 10. Also, multiplying weights calculated from FBWM the expected level weights i.e., $22.375 * 10 = 223.750$ gives us the individual criticality of SQL injection. Similarly, we calculate the individual criticality and overall severity score of all the remaining vulnerabilities belonging to both groups. The severity measure of vulnerabilities related to code execution error listed in group 1 and authentication issues from group 2 are represented in Tables 6 and 7 respectively.

The total severity value of improper code error vulnerabilities (group 1) comes out to be 796.889 while as the overall severity value of improper authentication vulnerabilities (group 2) comes out as 765.211. The individual severity of each vulnerability in the descending order from group 1 are as CDR1 (223.750) > CDR5 (207.430) > CDR3 (84.499) > CDR2 (70.667) > CDR6 (69.142) > CDR7 (66.769) > CDR4 (59.022) > CDR8 (15.610). Likewise, in group 2 the descending order of vulnerabilities are as AUE2 (222.956) > AUE7 (192.370) > AUE8 (99.797) > AUE3(84.470) > AUE1 (65.178) > AUE6 (56.715) > AUE5 (28.027) > AUE4 (15.698). The findings derived from Tables 5 and 6 suggest that our overall severity value for code

Table 6 Overall severity measure of group 1 vulnerabilities

Vulnerabilities	FBWM weights	Levels					Expected level weight	Contribution to total expected criticality (U_i)
		Highly severe	Severe	Medium severe	Slightly severe	Least severe		
		10	8	6	4	2		
CDR1	22.375	1.0	0.0	0.0	0.0	0.0	10	223.750
CDR2	10.392	0.0	0.4	0.6	0.0	0.0	6.8	70.667
CDR3	13.203	0.0	0.4	0.4	0.2	0.0	6.4	84.499
CDR4	9.222	0.0	0.4	0.4	0.2	0.0	6.4	59.022
CDR5	20.743	1.0	0.0	0.0	0.0	0.0	10	207.430
CDR6	8.643	0.2	0.4	0.4	0.0	0.2	8	69.142
CDR7	9.819	0.0	0.4	0.6	0.0	0.0	6.8	66.769
CDR8	6.504	0.0	0.0	0.0	0.2	0.8	2.4	15.610
Total severity								796.889

Table 7 Overall severity measure of group 2 vulnerabilities

Vulnerabilities	FBWM weights	Levels					Expected level weight	Contribution to total expected criticality (U_i)
		Highly severe	Severe	Medium severe	Slightly severe	Least severe		
		10	8	6	4	2		
AUE1	8.576	0.2	0.4	0.4	0.0	0.0	7.6	65.178
AUE2	23.225	0.8	0.2	0.0	0.0	0.0	9.6	222.956
AUE3	11.114	0.2	0.4	0.4	0.0	0.0	7.6	84.470
AUE4	6.541	0.0	0.0	0.0	0.2	0.8	2.4	15.698
AUE5	7.785	0.0	0.0	0.2	0.4	0.4	3.6	28.027
AUE6	11.816	0.0	0.0	0.4	0.6	0.0	4.8	56.715
AUE7	17.812	0.8	0.2	0.0	0.0	0.6	10.8	192.370
AUE8	13.131	0.2	0.4	0.4	0.0	0.0	7.6	99.797
Total severity								765.211

executed vulnerabilities from group 1 (Ui) = 796.889 is higher than the recommended threshold (i.e., 600) and closer to the ideal best (i.e., 1000). It may thus be argued that utmost efforts should be made to address these vulnerabilities first.

4 Conclusion

Vulnerabilities associated with coding errors and authentication errors pose a major challenge to software security systems. Software testing must identify effective measures to reduce the risks of software vulnerability to such attacks. In this context, our study identifies the two primary vulnerability groups of code execution and authentication errors. The novelty of the study lies in understanding the two groups individually and further prioritizing these vulnerabilities based on their severity levels. The fuzzy best-worst method is applied for both groups individually to prioritize software vulnerabilities. The fuzzy methodology is incorporated to include the complexity and the subjectivity of decision-makers and to calculate the individual, as well as overall utility of each vulnerability Two-way assessment technique, is used. Our results show that the code executed vulnerabilities are having overall severity more than authentication error vulnerabilities therefore, they need to be tackled first to reduce the loss. Among all the 16 vulnerabilities that were selected for this study, SQL Injection (CDR1), Code execution (CDR5), Information gain (AUE 2) and Improper Authentication (AUE 7) vulnerabilities are highly severe vulnerabilities. While the vulnerabilities like File inclusion (CDR4), HTTP Response splitting (CDR8), Bypass Something (AUE 4), and Denial of service (AUE 5) are least severe so they can be tackled later after finished with the highly severe ones. For validating the same a case of an Indian software testing company situated in the National capital was used. The future scope of the study will be to include the mathematical modeling in discovering and patching of these software vulnerabilities as well as other MCDMs can also be explored.

References

1. Bozorgi M, Saul LK, Savage S, Voelker GM (2010) Beyond heuristics: learning to classify vulnerabilities and predict exploits. In: Proceedings of the 16th ACM SIGKDD international conference on Knowledge discovery and data mining, pp 105–114
2. Vallentin M (2008) Software vulnerabilities lecture notes
3. Kapur PK, Pham H, Gupta A, Jha PC (2011) Software reliability assessment with OR applications. Springer, London, p 364
4. Wang JA, Wang H, Guo M, Xia M (2009) Security metrics for software systems. In: Proceedings of the 47th annual southeast regional conference, pp 1–6
5. Cashell B, Jackson WD, Jickling M, Webel B (2004) The economic impact of cyber-attacks. Congressional research service documents, vol 2, CRS RL32331, Washington DC
6. Kapur PK, Yadavali VS, Shrivastava AK (2015) A comparative study of vulnerability discovery modeling and software reliability growth modeling. In: 2015 international conference on futuristic trends on computational analysis and knowledge management (ABLAZE). IEEE, pp 246–251
7. Pham NH, Nguyen TT, Nguyen HA, Nguyen TN (2010) Detection of recurring software vulnerabilities. In: Proceedings of the IEEE/ACM international conference on automated software engineering, pp 447–456
8. Shi L et al (2012) Developing an evaluation approach for software trustworthiness using combination weights and TOPSIS. JSW 7(3):532–543

9. Viega J, McGraw G (2005) Building secure software. Addition Wesley
10. Agrawal A, Khan RA (2009) A framework to detect and analyze software vulnerabilities: development phase perspective. Int J Recent Trends Eng 2(2):82
11. Anjum M, Kapur PK, Agarwal V, Khatri SK (2020) Assessment of software vulnerabilities using best-worst method and two-way analysis. Int J Math Eng Manag Sci 5(2):328–342
12. Fruhwirth C, Mannisto T (2009) Improving CVSS-based vulnerability prioritization and response with context information. In: 2009 3rd international symposium on empirical software engineering and measurement. IEEE, pp 535–544
13. Lin Z, Jiang X, Xu D, Mao B, Xie L (2007) AutoPaG: towards automated software patch generation with source code root cause identification and repair. In: Proceedings of the 2nd ACM symposium on information, computer and communications security, pp 329–340
14. Bhatt N, Anand A, Yadavalli VSS, Kumar V (2017) Modeling and characterizing software vulnerabilities
15. Manzuik S, Pfeil K, Gold A (2006) Network security assessment: from vulnerability to patch. Elsevier
16. Sibal R, Sharma R, Sabharwal S (2017) Prioritizing software vulnerability types using multi-criteria decision-making techniques. Life Cycle Reliab Saf Eng 6(1):57–67
17. Jimenez W, Mammar A, Cavalli A (2009) Software vulnerabilities, prevention and detection methods: a review1. In: Security in model-driven architecture, vol 215995, p 215995
18. Schiffman M, Cisco C (2005) A complete guide to the common vulnerability scoring system (cvss). White paper. Identification of basic measurable security components in software intensive systems
19. Kansal Y, Kapur PK, Kumar U, Kumar D (2017) User-dependent vulnerability discovery model and its interdisciplinary nature. Life Cycle Reliab Saf Eng 6(1):23–29
20. Symantec Corporation [US] (2018) Internet security threat report. https://www.symantec.com/content/dam/symantec/docs/reports/istr-23-executive-summary-en.pdf
21. Liu Q, Zhang Y (2011) VRSS: a new system for rating and scoring vulnerabilities. Comput Commun 34(3):264–273
22. Sarfaraz A, Mukerjee P, Jenab K (2012) Using fuzzy analytical hierarchy process (AHP) to evaluate web development platform. Manag Sci Lett 2(1):253–262
23. Tandon S, Kumar V, Singh VB (2021) Empirical evaluation of code smells in open-source software (OSS) using Best Worst Method (BWM) and TOPSIS approach. Int J Qual Reliab Manag
24. National Vulnerability Database, N. http://nvd.nist.gov/
25. Guo S, Zhao H (2017) Fuzzy best-worst multi-criteria decision-making method and its applications. Knowl Based Syst 121:23–31
26. Kapur PK, Singh G, Sachdeva N, Tickoo A (2014) Measuring software testing efficiency using two-way assessment technique. Paper presented at the Proceedings of 3rd international conference on reliability, Infocom technologies and optimization. IEEE, pp 1–6

Investigating Bad Smells with Feature Selection and Machine Learning Approaches

Aakanshi Gupta, Rashmi Gandhi, and Vijay Kumar

Abstract Code Smell is a piece of code that is designed and implemented poorly and it gives adverse effect on the software quality and maintenance. Now, a day's machine learning based techniques have been extensively used towards code smell research. The main objective of this research is to optimise the features of Android code smells in terms of software metrics using feature selection technique based on Correlation on 2896 instances of open-source projects which are extracted from GitHub. Further, we have examined the performance measures like accuracy, precision, F-measure and execution time etc. with the reduced features data set of Android code smells. This paper also discussed about implementation of correlation-based feature selection algorithm to reduce the features of code smells. Then, the data has been analyzed with 4 machine learning algorithms that are Logistic Regression, Stochastic Gradient Descent (SGD), Simple Logistic and Sequential minimal optimization (SMO). The performance metrics for the above-mentioned machine learning algorithms with and without performing the feature selection have been compared. The computed outcome shows that the best accuracy and lesser execution time for all 3 considered Android code smells have been achieved using Logistic Regression algorithm. After feature selection the accuracy has increased up to 16%, 25% and 4.7% for NLMR, MIM and DTWC code smells respectively. Meanwhile, the other performance measures have also been increased.

Keywords Feature selection · Android code smell · Logistic regression · GITHUB

A. Gupta · R. Gandhi (✉)
Computer Science & Engineering, Amity School of Engineering and Technology, Amity University Uttar Pradesh, Noida, India
e-mail: rashmibehal@gmail.com

V. Kumar
Department of Mathematics, Amity Institute of Applied Sciences, Amity University Uttar Pradesh, Noida, India

V. Kumar and H. Pham (eds.), *Predictive Analytics in System Reliability*,
Springer Series in Reliability Engineering,
https://doi.org/10.1007/978-3-031-05347-4_4

1 Introduction

Recently, most explored market is the mobile application market and will remain in future as well [1]. In comparison with other software products; the mobile application software are the most enriched software products [1]. The Android operating system usage is 76.23% and the spare world constitutes another operating system [2]. In contrast to other operating systems, cost's and GUI's of Android operating systems is not so rib off so easily accessible to the world of billions. Thus, Android application market is spreading and is persistent. In the Android software systems; there are code smells that have adverse effect on the software in the future perspective [2–6]. Fowler [7] states that deeper problems in software indicated as code smells. The poor implementation methods are being used by software developer during the implementation phase [1, 8–11]. Specifically, the software developer will not avoid source code from its main output in spite they affect the performance and maintenance of the software product [1, 7, 8]. Even so, they can work if design rules are not in technical debts and maintainability problems [3, 5, 7, 8].

Ward Cunningham [12, 13] states technical debt as disparity among the selected software product and its design, which is successful and flawless in any modus operation. This disparity set in time delivery emphasis instead of optimal code consignment. Distinction in desktop application development and mobile application is conveyed with different programming and development approaches [14, 15]. Due to refrained resources of mobile phones such as battery, memory, CPU mobile applications are confining. Still, in comparison to desktop application; it is ruled with proper deadlines [6, 16].

Other factors are also impacting on mobile application such as screen size, interactive GUI's, and device fragmentation. Developers and researchers in practice observed code smells acting as an immoral characteristic of maintainability affair which diminishes the performance and resources of the Android mobile applications [15, 17, 18]. Literature focused that amount of work done on manual detection is more than automatic detection of code smells [19]. Specific 30 categories of code smells impacting Android application performance are proposed in Reimann et al. [6]. Still, there were many techniques that emphasized for examining the code smells in Android applications [20, 21]. Though in literature code smells are examined and explored footprint of code smells on energy consumption [21], resource usage (battery, memory, CPU) [1]. Software quality is always inspected while exploring the Android applications and multi-objective approach is tried to identify code smells.

The feature selection is one of the machine learning tool which is used to analyze the performance. The identification of most relevant features with maintaining the characteristic of data for prediction and analysis is feature selection. Hall [22] stated that with concept of correlation in data, it is useful to eliminate enquired data. Feature Selection on the basis of correlation bound the evaluation with a meaningful measure and searching strategy. It is a process to present the advantages of learning methods without considering the size of data. There are two different feature selection techniques popular known as wrapper and filter.

Code smells can work with supervised filter technique as it is two-class data. In this work, it is applied due to its fast execution time [7] as well as the benefits of wrapper technique. To determine the responsible features in code smell identification correlation based feature selection is employed. Identification results of 4 classifiers are examined in the context of 3 code smells. There description is given below:

- Data Transmission Without Compressing (DTWC): Evaluates the worth of individual predictive features follow when original file is not compressed over network transmission.
- No Low Memory Resolver (NLMR): Evaluates the worth of individual predictive features follow when outer class is referred by non-static inner class.
- Member Ignoring Method (MIM): Evaluates the worth of individual predictive features follow when only static methods accessing internal properties of the class.

Definitely, this work presents some specific Android code smells completely. Besides considering complete code smells with all its features, some lights up the performance. It includes three prime steps, all are individually marked with tools named as Understand, aDoctor and Weka. Initially, Data collection is primarily done of 10 android applications. With the support of aDoctor, java-based source codes are analyzed for exposure of specific code smells in android applications. To bring about static metrics Understand tool is used. To select features in considered code smells correlation feature selection is applied. Weka is imposed by using distinct classification performance metrics on complete or selected features of code smells. Hence, the subsequent sections constitute work to explore the android code smells with all and selected features.

Applying feature selection algorithm and selecting the appropriate features of code smells.

- Measuring performance on complete and selected features using classification algorithms.
- Specifying the best classification algorithm for exploring the Android smells.
- The results are validated with 10-folds cross validation technique.
- The following Performances: Accuracy, Kappa Statistics, Precision, Recall etc. are computed and analyzed.
- Logistic Regression algorithm computed the best accuracy for the considered Android code smells.

The objective of this paper is the Android specific smell identification and detection in a hi-tech era are to maintain mobile phones with efficacy and efficiency. So, to achieve this, elimination of irrelevant code with maintaining the structure of the data is done. Among existing Feature selection methods, correlation is taken into account for code smell detection. In this paper, the work explores correlation based feature selection on three distinct types of purely dedicated Android detailed code smells. Therefore, very less number of Android code smell type detection in comparison to original ones.

The paper is organized as follows: Sect. 2 illustrates Motivation and description of all the concerned terminology referred in current topic. Section 3 presents the empirical study with data creation and data processing in detailed way. Section 4 describes the various experiments analysis for used in result discussion. Section 5 conclude the work for current and future reference.

2 Motivation and Related Work

The immense usage of mobile applications is raising the size of code smells and degrading the performance measures. As every feature of the code smells has different weightage in computation, so, rather than computing complete data of code smells, some features can be selected to enhance the performance metrics. The motivation is to select the efficient and correlated features to make a good prediction model. Feature selection from the code smells is being performed using correlation statistics. This will select features which are linearly dependent and impacting with same centre to decide the code smell.

2.1 Feature Selection

The presence of irrelevant features in the data set may hamper the performance of classification algorithms [23], while the predicting features may enhance the effectiveness of the prediction with more effect and reliability [24]. Di Nucci et al. [25] addressed that the existence of irrelevant features in original data set is a prime issue. Their analysis investigates that most of the independent features are irrelevant, which may cause to over fitting in classification performance. Literature stated that many feature selection algorithm available for different applications [26]. The data distribution metrics represent smelly and non-smelly codes using the Goal Question Metrics [27]. Their experiments described to examine stacking heterogeneous ensemble model for detection of code smells in context performance metrices like f-measure, precision, recall, kappa statistics and classification accuracy.

2.2 Involvement of Android Smells

Conventionally, code smells symbolised that it is not perfect in software design and require a lot of attention to maintain the software quality and maintenance cost. Systems software may also impact on presence and effectiveness of code smells. In Android system software, code smells ruled out when the software design was not optimal. It happens when developers focus on deadlines rather than optimal software design [1, 14]. Thus, code smells do not impact exactly on the functioning of an application instead impact on poor performance. This may lead to major constraint in the maintenance phase of application [20]. Earlier, Fowler introduced 22 code smells but these were not considered for Android applications. Now a days, research is conducting on the studies of improving the poor performance measures in Android applications. Till now, Reimann et al. [6] presented an archive of 30 Android related code smells. Software clone detection survey researched for redundant code in software system [14]. Gupta et al. [28] presented a systematic literature review for code smells. Tufano et al. [29] reviewed the occurrence of bad smells in the existing software's. They also conformed durability of the bad smells over the revision of 200 projects (open source) from distinct software systems and explore the time of injection of code smells by programmers, besides the reasons and conditions responsible for their occurrence. Gupta et al. [1] deployed a prediction model with the concept of different entropies: Renyi, Shannon and Tsallis and Habchi et al. [30] study show that during the advancement who should be pointed for the emergence of Android code smells.

2.3 Empiricism Tools and Techniques

A famous tool aDoctor has been designed to observe 15 android related code smells with 98% recall rate as well as precision rate for master code of 18 distinct android applications [4]. Literature study explored the effect of 3 code smells on latest smartphones with the resource consumption like memory and CPU by involving refactoring techniques [16]. Besides, earlier study observed with different quality metrics which may also referred for approximation of the quality of master code with different existing tools like Paprika [31] and Infusion [6]. Dustin Lim [6] tested conventional smell identification tool for android application's code smell. Hecht et al. [31] proposed automatic identification approach to detect 4 Android-related antipatterns and 3 object-oriented from digital smartphone applications to trace the software quality of smartphone applications during their advancements. Kessentini et al. [20] introduced the identification rules of code smells in android applications applying a multi-objective programming.

3 Empirical Study

With the provision of feature selection algorithms, the identification of android related code smells datasets are formalized intentionally. The data sets must be relatively legitimate in order to obtain the necessary identification features. However, once the features are identified they will remotely enhance the performance of software product and diminishes the over hanged software maintenance cost. Figure 1 depicts an overview of the proposed work in the research work.

Initially, the range of Android applications may be considered for the appropriate extraction of steady data sets. For experimental analysis, the freely available java source code on GitHub is taken. With the help of aDoctor tool android related

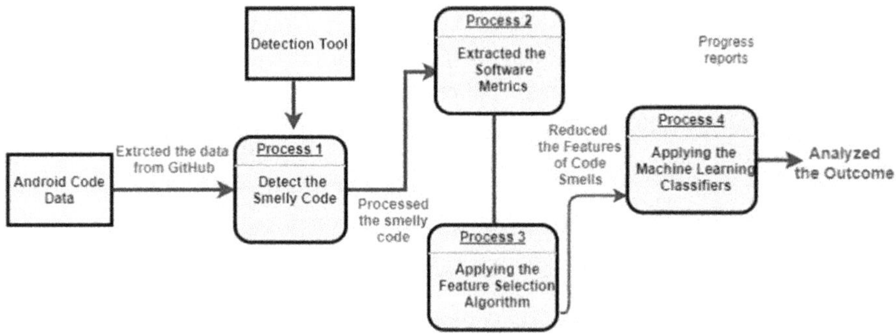

Fig. 1 Overview of the proposed methodology

code smells are extracted and a metrics and performance met rices [4]. Further, Co relation feature selection method is applied to find the co relation between features for better prediction. Later it is figured out that these both data sets can be evaluated using different classification algorithms. The sub sections are followed as Sect. 3.1 describe data creation and Sect. 3.2 elaborate android related smell detection approach. Section 4.1 presents the correlation feature selection in association with code smells. Section 3.3 evaluates appropriate classification algorithm.

The proposed method worked on extracted data from GitHub [29]. Initially, smelly code is segregated from dataset with the help of detection tool. Further the processed the smelly code are précised with the software metrics. On the basis of co relation feature selection algorithm data is reduced. Finally, the original dataset and reduced data are validated with machine learning classifiers to achieve the desired outcome.

3.1 Data Sets

In this section, Authors describe how data collection and selection can be done from open-source platform, GitHub. These freely available open source codes were also used in earlier work for code smells investigation. With this data set distinct Java based android related application data sets are analyzed to select features for performance metrices. The criteria for selection task for android related code smells and pointing best features precisely. GitHub contains large number of varieties of software codes and is based on application. Feature selection is performed with the help of aDoctor tool to select android related code smells. The Scitools Understand and aDoctor provides selection of code smells with best performance measures. For each code smell there are around 1000 code listed. Each code smell is given a binary judgement whether the code smell is android or not for a particular code. A static metric in the available data set is represented using 20 metrics. The feature includes correlation function popular in feature selection to select some of the highly performed metrics.

3.2 Data Processing

Data pre-processing is the most valuable data mining techniques for the preparation of data for the machine learning algorithms. It focuses to reduce the amount of data size with different existing techniques like relationship between the data, normalization, handling missing values etc. As this cloud era, makes as available plenty of multiple data for the particular task. So, either programmer/data scientist can go with same but performance is not achieved. So, to grab the performance they must need to focus on the size of data.

Data reduction or dimension reduction acting as data reconstruction approach to reduce the data set size. To consider, only the relevant and non-redundant data the many techniques like linearity, input type, neighbourhood, inverse transform etc. are available [32]. Data pre-processing helps in removing unwanted effects from the data set so that meaningful information is being used for efficient modelling. It is mainly dependent on dealing artefacts.

3.3 Performance Metrics

The strength of the model can be evaluated using the performance metrics such as accuracy, precision, recall, and F1 score. The accuracy may be defined as the ratio of the number of correctly predicted images to the total number of predictions.

$$accuracy = (Number\ of\ correct\,predictions)/(Total\ number\ of\ predictions) \quad (1)$$

Another performance metrics are precision metrics which may be expressed as the proportion of correctly predicted positive results (TP) to the total number of positive results (TP + FP) predicted.

$$Precision = TP/(TP + FP) \quad (2)$$

The Recall metrics may be expressed as the proportion of a number of true positive results (TP) to the number of all samples (TP + FN).

$$Recall = TP/(TP + FN) \quad (3)$$

where TP-True Positive, FP-False Positive, TN-True Negative, FN-False Negative. The F1 score is used to calculate the model performance and it is found out taking the weighted harmonic mean between the precision and recall.

$$Flscore = 2 \times (Recall \times Precision)/(Recall + Precision). \quad (4)$$

4 Result and Discussions

4.1 Correlation Feature Selection in Code Smells

To confirm the strong correlation among Android related code smells and feature selection analysis on entire data set is performed. Correlation is a statistical evaluation method to analyse the relationship in two features. There are two correlation coefficients Spear man Rho [33] and Kendall tau [22] are prominent to find the stability of relationships between the features of non-normalize data. The correlation value near to 0 indicates no relationship among the features. For this reason, authors do not report co relation result table. Despite the literature results are not entirely consistent, they indicate some android code smells appear more annoying than others. In result analysis the tables summarize the approximate confirmation of these finding which may fruitful to developer, who was designing when implementing the code. Among the distinct available techniques feature subset selection is used in this paper to represent data set to reduced volume w.r.t. maintaining the integrity of original data set (Tables 1, 2 and 3).

$$\mathrm{r} = \frac{n(\sum ab) - (\sum a)(\sum b)}{\left[n\sum \boldsymbol{a}^2 - (\sum a)^2\right]\left[n\sum b^2 - (\sum b)^2\right]} \tag{5}$$

Here, n = Number of values
a = sum of corresponding values in a column
b = sum of corresponding values in b column
ab = sum of product of a and b values

Table 1 NLMR code smell with statistical measures on logistic, SGD, simple logistic, and SMO classifiers

NLMR	Time (s)	Accuracy	Precision	Recall	F measure	Kappa
Logistic						
Original	3.73	68.0939	0.695	0.681	0.688	0.0884
Reduced	1.96	79.005	0.745	0.790	0.718	0.0751
SGD						
Original	1.38	78.5912	0.729	0.786	0.720	0.0794
Reduced	2.13	78.8674	0.738	0.789	0.717	0.0723
Simple logistics						
Original	3.09	78.5912	0.728	0.786	0.718	0.0731
Reduced	3.21	77.6243	0.665	0.776	0.0695	0.0003
SMO						
Original	1.25	78.8614	0.739	0.789	0.723	0.0913
Reduced	2.02	78.9293	0.733	0.787	0.717	0.0694

Table 2 DTWC code smell with statistical measures on logistic, SGD, simple logistic, and SMO classifiers

DTWC	Time (s)	Accuracy	Precision	Recall	F measure	Kappa
Logistic						
Original	0.397	84.9448	0.827	0.849	0.836	0.2162
Reduced	0.6	88.9503	0.875	0.890	0.879	0.4154
SGD						
Original	1.61	87.8453	0.861	0.878	0.834	0.1302
Reduced	1.28	87.7072	0.856	0.877	0.831	0.1138
Simple logistic						
Original	2.34	87.7072	0.848	0.877	0.842	0.1861
Reduced	2.68	87.8453	0.851	0.878	0.847	0.2113
SMO						
Original	1.31	87.5691	0.845	0.876	0.834	0.1354
Reduced	1.31	88.5359	0.873	0.885	0.877	0.4157

Table 3 MIM code smell with statistical measures on logistic, SGD, simple logistic, and SMO classifiers

MIM	Time (s)	Accuracy	Precision	Recall	F measure	Kappa
Logistic						
Original	5.608	61.326	0.615	0.613	0.612	0.2275
Reduced	0.198	76.7956	0.776	0.768	0.766	0.5349
SGD						
Original	0.329	75.9669	0.768	0.760	0.757	0.5183
Reduced	0.07	77.6243	0.778	0.776	0.776	0.552
Simple logistic						
Original	3.8	76.3852	0.781	0.775	0.773	0.5489
Reduced	0.29	77.4862	0.771	0.764	0.762	0.5267
SMO						
Original	0.82	74.7238	0.755	0.747	0.745	0.4934
Reduced	0.3	76.9337	0.775	0.769	0.768	0.5379

a^2 is the sum of squares of a column values
b^2 is the sum of squares of b column values.

The data set experimented with 4 machine learning algorithms: Logistic Regression, Stochastic Gradient Descent (SGD), Simple Logistic and Sequential minimal optimization (SMO) and compared the performance metrics with and without performing the feature selection. The computed outcome shows that the best accuracy for the Android smell DTWC, NLMR and MIM have been achieved using the Logistic Regression classifier. After feature selection; the accuracy of Logistic

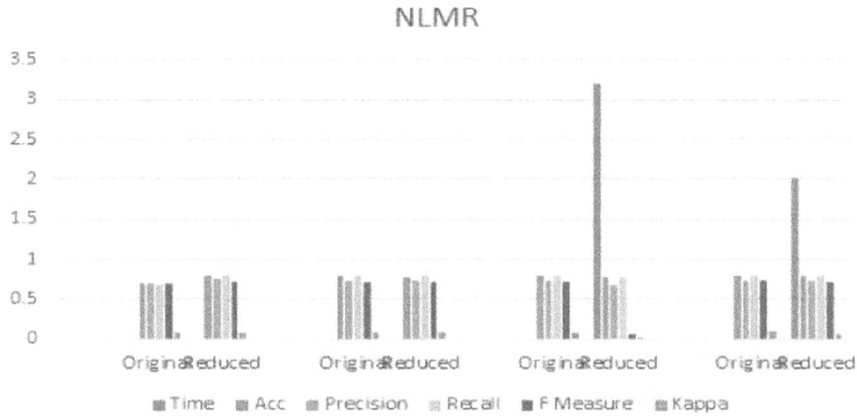

Fig. 2 Analysis of original and reduced NLMR code smells in time, accuracy, precision, recall, F measure and kappa statistics

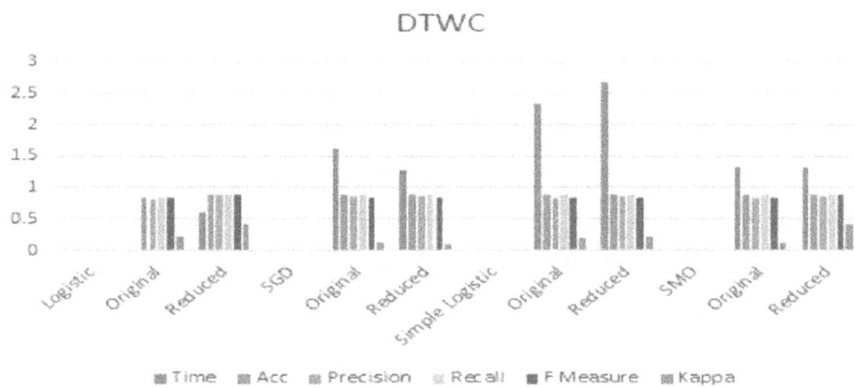

Fig. 3 Analysis of original and reduced DTWC code smells in time, accuracy, precision, recall, F measure and kappa statistics

Regression model has increased upto 16% for NLMR code smell, 25.23% for MIM code smell and 4.7% for DTWC code smell (Figs. 2, 3 and 4).

5 Conclusion

Based on analytical information of code smells with feature selection techniques; we have investigated and evaluated the extent of code smell detection process with the reduced feature data set. Secondly, the evaluation metrics like accuracy, precision,

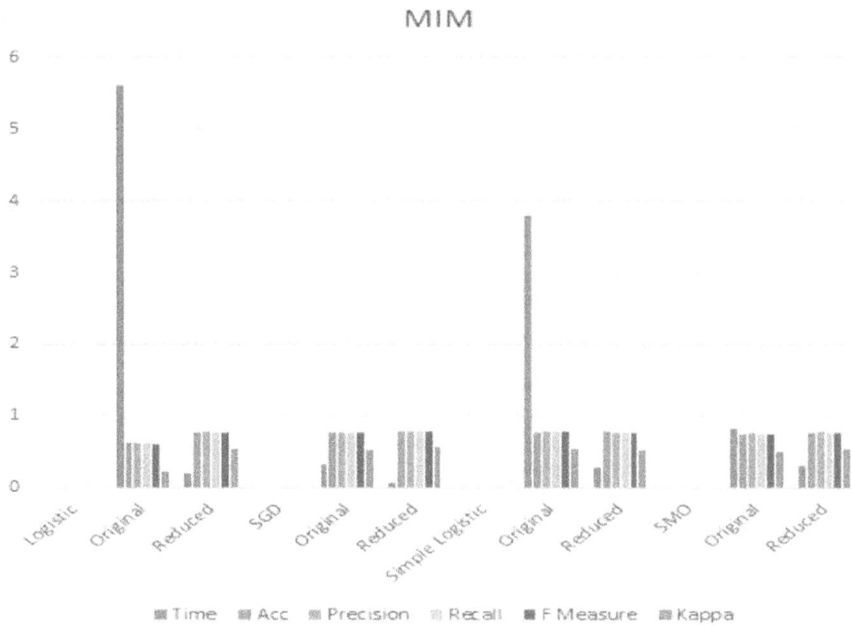

Fig. 4 Analysis of original and reduced MIM code smells in time, accuracy, precision, recall, F measure and kappa statistics

f-measure and kappa statistics are compared by using 4 classifiers between without reducing and with reducing the features of Android code smells.

The results obtained in this research reflects the potential work to improve the detection procedure of code smells with reduced features in terms of software metrics. These outcomes can provide a guidance in improving the overall software quality and software maintenance of the Android software systems. Furthermore, the various code smells of different languages can be considered with other techniques of feature selection can be analysed and compared in the future.

References

1. Palomba F, Di Nucci D, Panichella A, Zaidman A, De Lucia A (2019) On the impact of code smells on the energy consumption of mobile applications. Inf Softw Technol 105:43–55
2. Ozkaya I, Kruchten P, Nord RL, Brown N (2011) Managing technical debt in software development: report on the 2nd international workshop on managing technical debt, held at ICSE 2011. ACM SIGSOFT Softw Eng Notes 36(5):33–35
3. Gupta A, Suri B, Bhat V (2019) Android smells detection using ML algorithms with static code metrics. In: International conference on recent developments in science, engineering and technology. Springer, pp 64–79

4. Gupta A, Suri B, Kumar V, Misra S, Blazauskas T, Damasevicius R (2018) Software code smell prediction model using Shannon, Rényi and Tsallis entropy. Entropy 20(5):372
5. Husien HK, Harun MF, Lichter H (2017) Towards a severity and activity based assessment of code smells. Procedia Comput Sci 116:460–467
6. Reimann J, Brylski M, Aßmann U (2014) A tool-supported quality smell catalogue for android developers. In: Proceedings of the conference Modellierung 2014 in the Workshop Modellbasierte und modellgetriebene Softwaremodernisierung–MMSM, vol 2014
7. Habchi S, Moha N, Rouvoy R (2019) The rise of android code smells: who is to blame? In: 2019 IEEE/ACM 16th international conference on mining software repositories (MSR). IEEE, pp 445–456
8. Lim D (2018) Detecting code smells in android applications
9. Tandon S, Kumar V, Singh VB (2021) Empirical evaluation of code smells in open-source software (OSS) using best worst method (BWM) and TOPSIS approach. Int J Qual Reliab Manag. https://doi.org/10.1108/IJQRM-02-2021-0045
10. Gupta A, Suri B, Kumar V, Jain P (2021) Extracting rules for vulnerabilities detection with static metrics using machine learning. Int J Syst Assur Eng Manag 12(1):65–76
11. Kumar V, Ram M (eds) (2021) Predictive analytics: modeling and optimization, 1st edn. CRC Press. https://doi.org/10.1201/9781003083177
12. Hecht G, Rouvoy R, Moha N, Duchien L (2015) Detecting antipatterns in android apps. In: 2015 2nd ACM international conference on mobile software engineering and systems. IEEE, pp 148–149
13. Kumar NA, Krishna K, Manjula R (2016) Challenges and best practices in mobile application development. Imp J Interdiscip Res 2(12):1607–1611
14. Parikh G (1982) The guide to software maintenance. Winthrop, Cambridge, MA
15. Zhang M, Hall, T, Baddoo N. Code bad smells: a review of current knowledge. J Softw Maint Evolut: Res Pract 23(3):179–011
16. Hecht G, Benomar O, Rouvoy R, Moha N, Duchien L (2015) Tracking the software quality of android applications along their evolution (t). In: 2015 30th IEEE/ACM international conference on automated software engineering (ASE). IEEE, pp 236–247
17. Roy CK, Cordy JR (2007) A survey on software clone detection research. Queen's School Comput TR 541(115):64–68
18. Saifan AA, Al-Rabadi A (2017) Evaluating maintainability of android applications. In: 2017 8th international conference on information technology (ICIT). IEEE, pp 518–523
19. Fowler M, Beck K, Brant J, Opdyke W, Roberts D (1999) Refactoring: improving the design of existing code. Addison-Wesley Professional, Berkeley, CA
20. Kessentini M, Ouni A (2017) Detecting android smells using multi-objective genetic programming. In: 2017 IEEE/ACM 4th international conference on mobile software engineering and systems (MOBILESoft). IEEE, pp 122–132
21. Tufano M, Palomba F, Bavota G, Oliveto R, Di Penta M, De Lucia A, Poshyvanyk D (2015) When and why your code starts to smell bad. In: 2015 IEEE/ACM 37th IEEE international conference on software engineering, vol 1. IEEE, pp 403–414
22. Alazba A, Aljamaan H (2021) Code smell detection using feature selection and stacking ensemble: an empirical investigation. Inf Softw Technol 106648
23. John GH, Kohavi R, Pfleger K (1994) Irrelevant features and the subset selection problem. In: Machine learning proceedings. Elsevier, pp 121–129
24. Aldehim G, Wang W (2017) Determining appropriate approaches for using data in feature selection. Int J Mach Learn Cybern 8(3):915–928
25. Di Nucci D, Palomba F, Tamburri DA, Serebrenik A, De Lucia A (2018) Detecting code smells using machine learning techniques: are we there yet? In: 2018 IEEE 25th international conference on software analysis, evolution and reengineering (SANER). IEEE, pp 612–621
26. Gandhi R, Ghose U, Thakur HK (2021) Revisiting feature ranking methods using information-centric and evolutionary approaches: survey. Int J Sens Wirel Commun Control 11. https://doi.org/10.2174/2210327911666210204142857

27. Basili VR, Rombach HD (1988) The tame project: towards improvement-oriented software environments. IEEE Trans Softw Eng 14(6):758–773
28. Gupta A, Suri B, Misra S (2017) A systematic literature review: code bad smells in Java source code. In: International conference on computational science and its applications. Springer, Cham, pp 665–682
29. Ganesh S, Sharma T, Suryanarayana G (2013) Towards a principle-based classification of structural design smells. J Object Technol 12(2):1–1
30. Palomba F, Di Nucci D, Panichella A, Zaidman A, De Lucia A (2017) Lightweight detection of android-specific code smells: the adoctor project. In: 2017 IEEE 24th international conference on software analysis, evolution and reengineering (SANER). IEEE, pp 487–491
31. Oliveira J, Viggiato M, Santos MF, Figueiredo E, Marques-Neto H (2018) An empirical study on the impact of android code smells on resource usage. In: SEKE, pp 314–313
32. Nonato LG, Aupetit M (2018) Multidimensional projection for visual analytics: linking techniques with distortions, tasks, and layout enrichment. IEEE Trans Vis Comput Gr 25(8):2650–2673
33. Gupta H, Kumar L, Neti LBM (2019) An empirical framework for code smell prediction using extreme learning machine. In: 2019 9th Annual information technology, electromechanical engineering and microelectronics conference (IEMECON). IEEE, pp 189–195

SDE Based SRGM Considering Irregular Fluctuation in Fault Introduction Rate

Deepika, Adarsh Anand, Shinji Inoue, and Prashant Johri

Abstract Software debugging is complicated and can be considered as stochastic in nature. During fault removal, debuggers at-times introduce new faults. Thereafter, the fault introduction process can be said to be non-linear in nature. In this study, we have proposed a software reliability growth model considering the irregular fluctuation of fault introduction rate over time with non-constant fault detection rate. We assume that fault introduction changes non-linearly over time and the fault introduction rate fluctuates irregularly. Ito's process is used for solving the differential equation to find the analytical solution. The model is fitted on two real world data sets from two open-source project: Mozilla and Gnome. The experimental findings show that present model exhibit estimation result and having strong prediction skill.

Keywords Brownian motion · Irregular fluctuation · Ito's integral · Fault removal · Statistical analytical software (SAS) · Stochastic differential equation (SDE) · Wiener process

1 Introduction

Software failure is considered as not only financial loss but also it is a major reputational loss for the company since customers are always in need and want to take a failure free product for their use. The percentage of software reliability is actually identified by the project managers with the remnant faults during software testing

Deepika · A. Anand (✉)
Department of Operational Research, Faculty of Mathematical Sciences, University of Delhi, Delhi 110007, India
e-mail: adarsh.anand86@gmail.com

S. Inoue
Kansai University, Osaka, Japan
e-mail: ino@kansai-u.ac.jp

P. Johri
School of Computing Science & Engineering, Galgotias University, Gautam Budh Nagar, Greater Noida, UP, India

V. Kumar and H. Pham (eds.), *Predictive Analytics in System Reliability*, Springer Series in Reliability Engineering,
https://doi.org/10.1007/978-3-031-05347-4_5

process. This is actually a complex process as it involves some inadequate resource information apart from previously recorded fault data during the testing of the software. Hence, we can say the fault detection of software is an uncertain process. Development of model to check the software reliability in an actual situation can help to study its practical use and efficiency during the testing time.

Non-Homogenous Poisson Process (NHPP) based Software Reliability Growth Models (SRGMs) have been the major thrust area of research in the field of Software Reliability Engineering. Over years, various aspects of reliability growth modelling have been studied in great detail. Although, most models consider a deterministic behaviour for the fault detection process but latest research has also considered randomness in this rate. Randomness in the detection rate arises due to changes in debugging process caused by changes in testing effort expenditure, testing efficiency, etc. In large sized software prolonged testing takes place where numerous faults are detected before software release. This extensive testing leaves very few latent faults in the software in comparison to the original fault content. Such a situation can be modelled well using a stochastic process [13, 14]. The uncertainty or randomness in the testing efficiency, efforts, team skills, strategies, etc. in such a stochastic process is described using a *noise factor* in the debugging process.

2 Related Work

In the past years, a plenty of NHPP based software reliability models have been developed by the researchers. They generally assume that the software testing is a perfect debugging process in which no new faults are introduced during fault removal. For example, the models recognized by Jelinski and Moranda [8] and Goel and Okomuto (G–O) [6] assumed that the fault detection rate is constant and the fault intensity is proportional to the number of remnant faults. By contrast, the delayed S-shaped [29] and the inflection S-shaped models [18] assume that the fault detection rate is an increasing variable. However, in realistic testing, fault detection and removal are complicated and can be affected by many factors, such as testing resources, testing tools, and tester's skill. Therefore, the assumption of imperfect debugging is reasonable in the development of a software reliability model. Yamada et al. [30] and Pham and Zhang [19] proposed different imperfect software debugging models that consider the function of the fault content following exponential distribution. Kapur et al. [13, 14] assumed that the function of fault content is a linear function of the mean function and the fault detection or removal distributes differently. Moreover, they presented numerous imperfect software debugging models. Given that the function of the fault content is a linear function of time, some researchers established related imperfect software debugging models [7, 30]. Although these models can be applied effectively to specific testing environments, but they cannot be adapted to the other testing situations due to the unrealistic assumption that the rate of fault introduction is constant. In practice, software debugging is complicated and stochastic [3, 25, 32]. During fault removal, debuggers can introduce new faults and Deepika et al. [4] have developed software reliability models with testing domain concept. It was Yamada et al. [28] who first introduced the exponential based SDE based SRGM. According to them fault detection rate is linear along with noise factor. On the contrary, Yamada et al. confirmed from their study using the probability distributions that fault detection rate varied depending on applied software reliability measures during the process. Similar studywas done by Shyur et al. [21] related to stochasticity with imperfect debugging and change point. Lee et al. [17] described the stochastic differential equation to illustrate per-fault detection rate that advocates random fluctuation instead of a non-homogeneous poisson process. In 2006, Tamura and Yamada [26] have extended their work and developed a flexible irregular fluctuation model considering distribution development environment. They also discussed an optimum release time considering the reusable rate of software components. Tamura and Yamada [27] implemented the stochastic based reliability model to assess the active state of the open-source project. They considered the failure intensity as a function of time, and the bug tracking system that report the software faults account an irregular state. In addition, Kapur et al. [11] have worked on determination of logistic error detection rate in the modelling. The same team also studied [10] on a unified approach for SRGMs related to SDE. In this similar time period Singh et al. [24] inculcated the impact of randomness in the mathematical construction in this domain. After that some researchers proposed the allocate the resources in an optimal manner to minimize the cost during testing phase using FDP and FCP under dynamic environment [16]. Recently, Singh et al. [22] modelled a multi up-gradation framework that deploys the concept of randomness with learning effect and impact of faults severity. Later, Anand et al. [1] proposed fault severity based modeling. Then they proposed multi release Stochastic models with the concept of convolution ([23], Anand et al. 2018). In 2016, Kumar et al. assumed that there is a time lag between fault detection and fault correction. Thus, removal of a fault is performed after a fault is detected. In addition, detection process and correction process are taken to be independent simultaneous activities with different budgetary constraints. A structured optimal policy based on optimal control theory is proposed for software managers to optimize the allocation of the limited resources with the reliability criteria. Some authors develop a vulnerability discovery model that accumulate the vulnerabilities due to the influence of previously discovered vulnerabilities. Further, they evaluate the proportion of previously discovered vulnerabilities along with the fraction additional vulnerabilities detected [2]. In 2019, Shakshi et al., developed SDE based innovation diffusion model. Recently, Deepika et al., came up with the concept of entropy prediction related to stochasticity [5]. Moreover, it can be state that fault introduction is a non-linear changing process where rate of change of fault introduction varies irregularly over time. More recently work done by Kumar et al. [15] in which they talked that hybrid approach identifies the need of the relative importance of criteria for a given application without which inter-criterion comparison cannot be accomplished. It requires a set of model selection criteria along with a set of SRGMs and their level of criteria for optimal selection. It successfully displays the result in terms of a merit value which is used to rank the SRGMs.

In the current chapter, a new model with imperfect software debugging considering fault introduction to be a non-linear procedure and the rate of fault introduction to be a factor that fluctuates irregularly over time during software debugging has been proposed. In this sense, the proposed model is in line with the actual software debugging situation. The experimental results also confirm that our model has good fitting capability and significantly better predictive performance than other imperfect software debugging models.

2.1 Basic Assumptions

A reasonable assumption is necessary in building a good software reliability model. Thus, the models proposed in this chapter has basic assumptions as follows:

1. The fault detection is based on NHPP.
2. Software is subjected to failure at random times as result of remaining faults in the software.
3. Faults are introduced in to software at random times affected by remaining faults in software debugging.
4. Fault introduction rate fluctuate irregularly over time. Change in total fault content in small span is non-deterministic because of stochastic nature.
5. Each time a fault is detected, it is removed immediately and new faults can then be introduced.
6. The function of the fault content is non-linear and time dependent.

2.2 Notations

$m(t)$	Expected number of faults removed by time 't'.
β	Learning parameter.
α	Error generation rate.
a	Initial fault content.
b	Hazard rate function for fault removal.
σ	Fluctuation rate in SDE.
a_1	Total number of faults eventually introduced due to fluctuation.
$\gamma(t)$	Standard Gaussian white noise.
$a(t)$	Time dependent stochastic fault content function.

3 Model Development

Kapur and Garg Model [9] is created on the postulation that some added errors are also removed by debugging team while eliminating some error without any failure.

The faults which are recognized on a failure are called as independent faults while the faults removed in addition are named as dependent faults [13, 14].

$$\frac{f(t)}{1-F(t)} = p + qF(t) \tag{1}$$

where $f(t)$ is the probability density function of detection; $F(t)$ is a cumulative distribution function i.e., $F(t) = m(t)/a$; p symbolizes the fault detection rate for independent faults and q denotes the fault detection rate for dependent faults which is considered constants for this formulation. Additionally, let a denote the fault content. Then, Kapur and Garg Model [9] (fault removal phenomenon) that represents the cumulative number of faults which is attained by the following differential equation (DE):

$$\frac{dm(t)}{dt} = \left(p + q\frac{m(t)}{a}\right)(a - m(t)) \tag{2}$$

where $m(t)$ defines the cumulative number of faults by time t, and $(a - m(t))$ denotes the residual number of faults.

With the initial condition $m(0) = 0$, The above DE (Eq. (2)) is further solved, to accomplish the mean value function i.e.,

$$m(t) = a\left(\frac{1 - e^{-(p+q)t}}{1 + (q/p)e^{-(p+q)t}}\right) \tag{3}$$

Equation (3) shows an S-Shaped form over the entire software lifespan.

According to Kapur et al. [13, 14] the alternative DE can be expressed as follow:

$$\frac{dm(t)}{dt} = b(t)(a(t) - m(t)) \tag{4}$$

here $b(t)$ signifies the time-dependent detection rate and can be expressed as:

$$b(t) = \frac{b}{1 + \beta e^{-bt}} \tag{5}$$

In Eq. (5), b means the detection parameter and β denotes the learning parameter. In accumulation, Kapur et al. [12] presumed that fault content of a software develops exponentially.

$$a(t) = a(t) = ae^{\alpha t} \quad \alpha > 0 \tag{6}$$

where a is the original fault content and α denotes the error generation rate. Thus, with the seed value $m(0) = 0$, Eq. (4) becomes:

$$m(t) = a\left(\frac{b}{\alpha + b}\right)\left[\frac{e^{\alpha t} - e^{-bt}}{1 + \beta e^{-bt}}\right] \tag{7}$$

where $F(t) = \left(\frac{e^{\alpha t} - e^{-bt}}{1+\beta e^{-bt}}\right)$. Considering, (i) $b = p+q$, (ii) $\beta = \frac{q}{p}$, and (iii) $\alpha = 0$, i.e. fault content is constant. We can conclude that the K-G model alternatively comes from the Kapur et al. [12] model.

4 Proposed Methodology

The fault removal rate equations discussed in Eqs. (2) and (4) interprets in terms of certainty. Thus, their behaviour can be forecast in deterministic form. Due to the non-deterministic behaviour of various factors such as testing effort expenditure, testing efficiency and skill, testing method and strategy, testing goes in uncertain way.

Hence, the fault content of software is described in probabilistic terms and based on above mentioned assumption (Eq. 2), the differential equation can be built as:

$$\frac{dm(t)}{dt} = b(t)[Ea(t) - m(t)] \tag{8}$$

$m(t)$ denotes the cumulative number of faults in the software by the time t, $b(t)$ denotes the hazard rate function for fault removal; $a(t)$ represents the time dependent stochastic fault content. So, hazard rate function is described as follow:

$$b(t) = \frac{f(t)}{1 - F(t)} \tag{9}$$

Again, the detection rate $b(t)$ following logistic rate can be advocated

$$b(t) = \frac{b}{1 + \beta e^{-bt}} \tag{10}$$

as prearranged by Kapur et al. [12]. Here b indicates the detection or measure parameter and β signifies the learning parameter. Increase value of b signifies the decrease the value of β and hence make the possibility of rapid fault removal phenomenon. Thus, (Eq. (10) in Eq. (8)), the instantaneous fault removal rate equation converts to:

$$\frac{dm(t)}{dt} = \frac{b}{1 + \beta e^{-bt}}(E(a(t) - m(t)) \tag{11}$$

Equation (11) portrays the proposed fault removal rate equation with variable fault content. As per sixth postulation, the DE representing total fault content can be stated using Eq. (12):

$$\frac{da(t)}{dt} = \alpha(t)[(a + a1) - a(t)] \tag{12}$$

where $\alpha(t)$ denotes time dependent function in respect of fault content. $a1$ is the upper bound of increase in the fault content. In this proposed model, the behaviour of fault removal is framed using Brownian motion process during the software development life cycle. And fault content $a(t)$ is a random variable following the stochastic process.

If, $h(t)$, probability density function and $H(t)$, cumulative distribution function of fault content by time t, then the mathematical expression for fault content using the postulation (4) is:

$$\alpha(t) = \frac{h(t)}{1 - H(t)} + \sigma\gamma(t) \tag{13}$$

$h(t)/(1 - H(t))$ time dependent rate of fault content,
σ constant representing the scale of irregular fluctuation, and
$\gamma(t)$ standard gaussian white noise that is stochastic in nature.
On substituting Eq. (13) in Eq. (12):

$$\frac{da(t)}{dt} = \left(\frac{h(t)}{1 - H(t)} + \sigma\gamma(t)\right)(a + a1 - a(t)) \tag{14}$$

Equation (14) symbolizes the SDE, a modified of ordinary differential equations that are parameterized by Weiner processes. Itô stochastic calculus is applied to formulate the equation.

$$\left.\begin{array}{l} da(t) = \left(\dfrac{h(t)}{1 - H(t)} - \dfrac{\sigma^2}{2}\right)(a + a1 - a(t))dt \\ +\sigma(a + a1 - a(t))dW(t) \end{array}\right\} \tag{15}$$

In this Eq. (15), $W(t)$ symbolizes a Brownian motion or wiener process and it hold the following axioms:

(i) Continuous process with $W(0) = 0$
(ii) Have independent increments $\forall t > 0$
(iii) Have Gaussian increments, i.e., $W(t + dt) - W(t) \sim N(0, dt)$; where $N(0, dt)$ is a normal distribution centred at zero.

Now, if x is a random variable then, $f_{w_t}(x) = \frac{1}{\sqrt{2\pi t}} e^{-\frac{x^2}{2t}}$.

Using the seed value, $a(0) = 0$ Eq. (15) can be combined using the Itô formula to obtain the cumulative fault content function:

$$a(t) = a + a1(1 - (1 - H(t))e^{-\sigma W(t)} \tag{16}$$

From Eq. (16), it can be derived that $a(0) = a$ when $t = 0$ and $a(\infty) = a + a1$, when $t \to \infty$. That is, the fault content at entry time is a and eventual fault content over its life cycle will be $a + a1$. Considering expectation on both sides, Eq. (16) is assumed as:

$$E[a(t)] = a + a1\Big((1 - (1 - H(t))e^{\frac{\sigma^2 t}{2}}\Big) \tag{17}$$

Equation (17) belong the uncertainty and random fluctuations in fault introduction rate. It was designed using two types of distribution functions to represent two different behaviour of fault introduction function in the system.

5 General Framework

Fault introduction function following Erlang-k distribution function can be expressed as:

$$H(t) \sim Erlang(k, \alpha) \;\; i.e. \;\; H(t) = 1 - \sum_{n=0}^{k-1} \frac{(\alpha t)^n}{n!} e^{-\alpha t} \tag{18}$$

It is a continuous distribution function of the summation of k and identically distributed random variables. Two parameters associated with Erlang distribution function are shape (k) and scale (α) parameter.

For $k = 1$ and $k = 2$ in Eq. (18):

Case I: In this case, Erlang distribution becomes exponential distribution function. i.e. $H(t) \sim \exp(\alpha)$

It defines a continuous process where increments occur independently at a constant rate.

$$H(t) = (1 - e^{-\alpha t}) \tag{19}$$

Here, α is a scalar parameter signifying the introduction rate of fault content. Thus, using Eq. (19), the expected value of fault content is:

$$E[a(t)] = a + a1\Big((1 - e^{-\alpha t + \frac{\sigma^2 t}{2}}\Big) \tag{20}$$

Case II: If shape parameter is taken as $k = 2$, the distribution function for the fault content can be written as:

$$H(t) \sim Erlang(2, \alpha)$$

$$H(t) = (1 - (1 + \alpha t)e^{-\alpha t}) \tag{21}$$

Thus, the expected value becomes:

$$E[a(t)] = a + a1(1 - (1 + \alpha t)e^{-\alpha t + \frac{\sigma^2 t}{2}}) \tag{22}$$

Equation (22) symbolizes the S-shaped pattern of fault content function.

Using Eqs. (20) and (22), the DE (11) representing fault removal rate function under $t = 0$, $N(t) = 0$ to attain the mean value function.

In Table 1, two diverse stochastic software reliability growth models are reported using the proposed modeling framework.

In Table 1, model-1 depicts the cumulative fault removal with stochastic exponential nature. Besides, model-2 signifies the fault removal phenomenon with S-shaped pattern. The most important characteristic of a stochastic model is converging to its deterministic nature. So, when α and σ equals to zero, that is, the fault content held deterministic and constant.

Then the proposed SDE based models will reduce to the following logistic fault removal model. As stated in Eq. (15):

$$m(t) = a\left(\frac{1 - e^{-bt}}{1 + \beta e^{-bt}}\right) \tag{23}$$

In Eq. (23), if $b = p + q$ and $\beta = q/p$ then model convert to widely-known Kapur and Garg [9] fault removal phenomenon model.

6 Data Analysis and Comparison Criteria

We have carried out the data analysis of real software data sets. The parameters of the models have been estimated using SAS(SAS)/ETS user's guide 9.1 [20]. The model is fitted on two real world data sets from two open-source project: Mozilla and Gnome. In the total epoch, first data comprises of 53 in which 1497 faults have been detected and second data has 17 weeks in which 354 faults has been detected [31]. The experimental findings show that present model exhibit estimation result and having strong prediction skill. The parameter estimation and comparison criteria result for data set of the models (Model-I (M-I) and Model-II (M-II)) under consideration can be viewed through Tables 2 and 3 respectively.

Table 1 Proposed stochastic models

SDE based models	Fault content function	Mean failure function
Model 1	$E[a(t)] = a + a1\left((1 - e^{-\alpha t + \frac{\sigma^2 t}{2}}\right)$	$m(t) = \frac{1}{1+\beta e^{-bt}}\left(\begin{array}{l}(a + a1)\left(1 - e^{-bt}\right)\\ + \frac{2ba1}{-\sigma^2 + 2\alpha - 2b}\left(e^{-\alpha t + \frac{\sigma^2 t}{2}} - e^{-bt}\right)\end{array}\right)$
Model 2	$E[a(t)] = a + a1(1 - (1 + \alpha t)$ $e^{-\alpha t + \frac{\sigma^2 t}{2}})$	$m(t) = \frac{1}{1+\beta e^{-bt}}\left(\begin{array}{l}(a + a1)\left(1 - e^{-bt}\right)\\ + \frac{2ba1}{-\sigma^2 + 2\alpha - 2b}\left(\begin{array}{l}\left(1 + \alpha t - \frac{2\alpha}{\sigma^2 - 2\alpha + 2b}\right)e^{-\alpha t + \frac{\sigma^2 t}{2}}\\ -\left(1 - \frac{2\alpha}{\sigma^2 - 2\alpha + 2b}\right)e^{-bt}\end{array}\right)\end{array}\right)$

Table 2 Parameters estimation

Parameter	DS-I		DS-II	
Models	M-I	M-II	M-I	M-II
a	1904.112	1769.354	600	468
β	2.393689	1.968125	4.034	2.685231
b	0.008146	0.006859	3.803E-6	0.000032
α	0.041684	0.036854	0.056976	0.001253
σ	0.258942	0.317386	0.337555	0.1675832
al	3.950815	4.321563	500	437

Table 3 Comparison criterion of proposed models

Parameter	DS-I			DS-II		
Models	M-I	M-II	Kapur and Garg [9]	M-I	M-II	Kapur and Garg [9]
SSE	2471.2	2089.6	4167.7	458.7	401.3	474.5
MSE	51.4842	45.2316	83.354	32.7618	46.7219	63.15
Root MSE	7.1752	8.1374	9.1299	5.7238	4.3653	8.777
R-square	0.9998	0.9887	0.99	0.9979	0.9956	0.99
Adj. R-square	0.9998	0.9789	0.99	0.9976	0.9896	0.99

7 Model Validation

In this chapter, widely known model Kapur and Garg [9] is considered as benchmark. This model is considered due to its comprehensive applicability and flexibility. Five statistical criteria sum of squared error (SSE), Mean squared error (MSE), Root mean squared error (RMSE), the coefficient of determination (R^2) and adjusted R-square are evaluated for model validation. The fitness measures MSE measures the average of squares of error between the observed and predictive number of faults, SSE is the deviation of predicted faults from the observed faults, RMSE measures how much error there in the predicted faults and the observed faults, R-square measures the percentage of entire discrepancy and Adjusted R^2 is a modified version of R^2 that is adjusted based on the independent terms that affects the dependent variables. If $m(t)$ is the observed faults value, $\widehat{m}(t)$ is the estimated value, T is the number of data points and n is the number of model parameters, then.

Mean Squared error, $MSE = \left(\sum_{t=1}^{T}\left(m(t) - \widehat{m}(t)\right)^2\right) \Big/ T.$

Sum of squared error, $SSE = \sum_{t=1}^{T}\left(m(t) - \widehat{m}(t)\right)^2.$

Root mean squared error, $RMSE = \sqrt{\left(\sum_{t=1}^{T}\left(m(t) - \widehat{m}(t)\right)^2\right) \Big/ T}.$

Coefficient of determination, $R^2 = 1 - \frac{\sum_{t=1}^{T} (m(t)-\widehat{m}(t))^2}{\sum_{t=1}^{T} (m(t)-\overline{m}(t))^2}$.

Adjusted R-square, $\overline{R}^2 = 1 - (1 - R^2)(T - 1)/(T - n - 1)$.

where $\widehat{m}(t)$ signifies the mean value of predicted faults.

As per the results stated in Table 3, model 1and model 2 exhibited better results with respect to Kapur and Garg [9] model for both the data sets. All five performance measures MSE, SSE, RMSE, R-square and Adjusted R-Square displayed that the Model 1 and Model 2 have a better fitting capability. The pictographic (Fig. 1 for Mozilla (DS1) and Fig. 2 for Gnome (DS2)) representation indicates that the proposed models fit the real fault removal curves excellently. Figures 1 and 2 show the fitting and forecasting cases between the actual and estimated fault data. These data indicate that our model can effectively fit historical fault data and precisely forecast the number of software faults in the test.

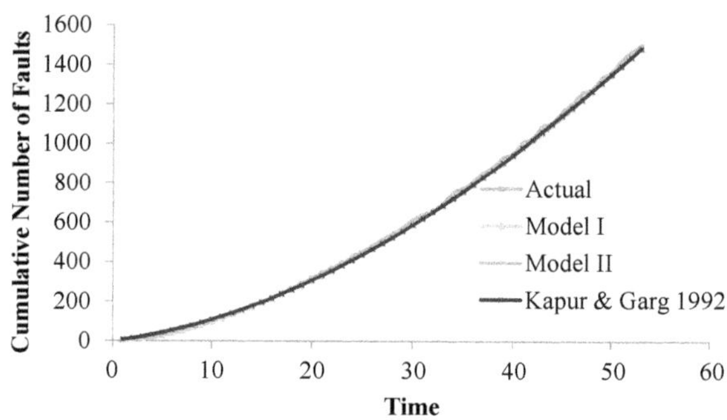

Fig. 1 Goodness of fit curve for DS-1

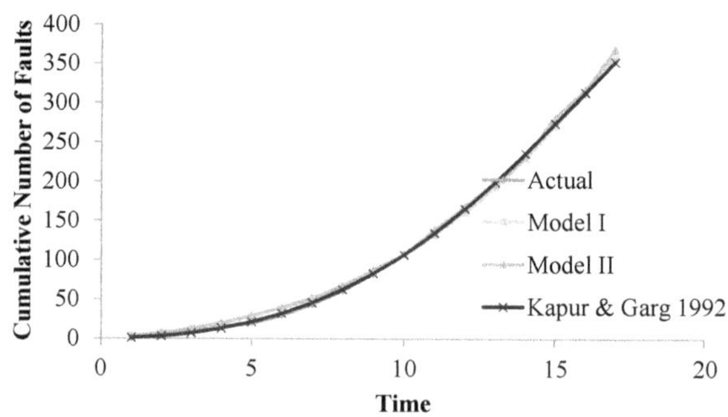

Fig. 2 Goodness of fit curve for DS-2

8 Conclusion

This chapter advocates new SDE based models that can characterize the fault removal phenomenon. Further, different type of distributions has been inculcated in the fault removal process. We have reviewed standard distributions such as Exponential and Erlang 2-stage for fault removal behaviour. The proposed mathematical modeling is implemented on the open-source software data of Mozilla and Gnome. The proposed models have produced reliable parameter estimates and goodness of fit curve. The findings show that the proposed models have stronger prediction skills than the benchmark model. In future, we can broaden the methodology to capture the more realistic scenario such as imperfect debugging, testing effort etc.to yield a generalized framework. Further, the more use of Itô's process in fault introduction rate apart from the fault detection rate can also be inculcated with stochastic differential equation. The mean value function with both rate (introduction and detection) scenario will formulated and hence it will be interesting to note the behavior. Also, we wish to study two-dimensional aspect of proposed methodology i.e. the role of testing efforts and time, in estimating the final count of bugs present in the software system.

References

1. Anand A, Singh O, Das S (2015) Fault severity based multi up-gradation modeling considering testing and operational profile. Int J Comput Appl 124(4):9–15
2. Bhatt N, Anand A, Yadavalli VSS, Kumar V (2017) Modeling and characterizing software vulnerabilities. Int J Math Eng Manag Sci 2(4):288–299
3. Bregon A, Alonso-González CJ, Pulido B (2014) Integration of simulation and state observers for online fault detection of nonlinear continuous systems. IEEE Trans Syst Man Cybern Syst 44(12):1553–1568
4. Deepika OS, Anand A, Singh JN (2017) Testing domain dependent software reliability growth models. Int J Math Eng Manag Sci 2(3):140–149
5. Deepika SO, Anand A, Singh J (2021) SDE based unified scheme for developing entropy prediction models for OSS. Int J Math Eng Manag Sci 6(1):207–222
6. Goel AL, Okumoto K (1979) Time-dependent error-detection rate model for software reliability and other performance measures. IEEE Trans Reliab 28(3):206–211
7. Pham H, Nordmann L, Zhang X (1999) A general imperfect software debugging model with S-shaped fault detection rate. IEEE Trans Reliab R-48(2):169–175
8. Jelinski Z, Moranda P (1972) Software reliability research. Statistical computer performance evaluation. Academic Press, pp 465–484
9. Kapur PK, Garg RB (1992) A software reliability growth model for an error-removal phenomenon. Softw Eng J 7(4):291–294
10. Kapur PK, Anand S, Yadav K, Singh J (2012) A unified scheme for developing software reliability growth models using stochastic differential equations. Int J Oper Res 15(1):48–63
11. Kapur PK, Anand S, Yamada S, Yadavalli VS (2009) Stochastic differential equation-based flexible software reliability growth model. Math Probl Eng
12. Kapur PK, Bardhan AK, Jha PC (2004) An alternative formulation of innovation diffusion model and its extension. Math Inf Theory 17–23
13. Kapur PK, Pham H, Anand S, Yadav K (2011a) A unified approach for developing software reliability growth models in the presence of imperfect debugging and error generation. IEEE Trans Reliab 60(1):331–340

14. Kapur PK, Pham H, Gupta A, Jha PC (2011b) Software reliability assessment with OR applications. Springer, London, p 364
15. Kumar V, Saxena P, Garg H (2021) Selection of optimal software reliability growth models using an integrated entropy–technique for order preference by similarity to an ideal solution (TOPSIS) approach. Math Methods Appl Sci. https://doi.org/10.1002/mma.7445
16. Kumar V, Khatri SK, Dua H, Sharma M, Mathur P (2014) An assessment of testing cost with effort-dependent FDP and FCP under learning effect: a genetic algorithm approach. Int J Reliab Qual Saf Eng 21(06):1450027
17. Lee CH, Kim YT, Park DH (2004) S-shaped software reliability growth models derived from stochastic differential equations. IIE Trans 36(12):1193–1199
18. Ohba M (1984) Inflection S-shaped software reliability growth model. Stochastic models in reliability theory. Springer, Berlin, Heidelberg, pp 144–162
19. Pham H, Zhang X (1997) An NHPP software reliability model and its comparison. Int J Reliab Qual Saf Eng 4(03):269–282
20. SAS Institute Inc. (2004) SAS/ETS user's guide version 9.1. SAS Institute, Cary, NC
21. Shyur HJ (2003) A stochastic software reliability model with imperfect-debugging and change-point. J Syst Softw 66(2):135–141
22. Singh J, Singh O, Kapur PK (2015) Multi up-gradation software reliability growth model with learning effect and severity of faults using SDE. Int J Syst Assur Eng Manag 6(1):18–25
23. Singh O, Kapur PK, Anand A (2011) A stochastic formulation of successive software releases with faults severity. In: 2011 IEEE international conference on industrial engineering and engineering management. IEEE, pp 136–140
24. Singh O, Kapur PK, Anand A, Singh J (2009) Stochastic differential equation based modeling for multiple generations of software. In: Proceedings of fourth international conference on quality, reliability and infocom technology (ICQRIT), trends and future directions. Narosa Publications, pp 122–131
25. Singh P, Pal NR, Verma S, Vyas OP (2016) Fuzzy rule-based approach for software fault prediction. IEEE Trans Syst Man Cybern Syst 47(5):826–837
26. Tamura Y, Yamada S (2006) A flexible stochastic differential equation model in distributed development environment. Eur J Oper Res 168(1):143–152
27. Tamura Y, Yamada S (2009) Optimisation analysis for reliability assessment based on stochastic differential equation modelling for open source software. Int J Syst Sci 40(4):429–438
28. Yamada S, Kimura M, Tanaka H, Osaki S (1994) Software reliability measurement and assessment with stochastic differential equations. IEICE Trans Fundam Electron Commun Comput Sci 77(1):109–116
29. Yamada S, Ohba M, Osaki S (1983) S-shaped reliability growth modeling for software error detection. IEEE Trans Reliab 32(5):475–484
30. Yamada S, Tokuno K, Osaki S (1992) Imperfect debugging models with fault introduction rate for software reliability assessment. Int J Syst Sci 23(12):2241–2252
31. Yang J, Liu Y, Xie M, Zhao M (2016) Modeling and analysis of reliability of multi-release open source software incorporating both fault detection and correction processes. J Syst Softw 115:102–110
32. Yin X, Li Z (2015) Reliable decentralized fault prognosis of discrete-event systems. IEEE Trans Syst Man Cybern Syst 46(11):1598–1603

Ant Colony Optimization Algorithm with Three Types of Pheromones for the Component Assignment Problem in Linear Consecutive-k-out-of-n:F Systems

Taishin Nakamura, Isshin Homma, and Hisashi Yamamoto

Abstract The ant colony optimization (ACO) algorithm is a meta-heuristic optimization method used to solve challenging optimization problems. Notably, the pheromone model of ACO impacts algorithmic performance. Hence, this paper presents an ACO algorithm with three types of pheromones for solving the component assignment problem of the linear consecutive-k-out-of-n:F system. This configuration can be used to represent a real system in which consecutive failed components cause system failures. Moreover, the component assignment problem seeks a component arrangement in which system reliability is maximized. The proposed algorithm is incorporated with either adjacence-, position-, or k-interval-wise pheromones that are compared using a numerical experiment. The results indicate that the ACO algorithm with the position-wise pheromone performs well within the scope of the experiment.

1 Introduction

The ant colony optimization (ACO) algorithm is a meta-heuristic optimization method that is often employed to solve challenging optimization problems [4, 6]. The ACO algorithm was inspired by observing the social behavior of ants as they search for food. Specifically, ants utilize pheromones as an indirect communication channel to indicate the shortest paths between the nest and food sources. This behavioral characteristic of real ants enables us to algorithmically solve several combinatorial optimization problems. The first ACO algorithm was implemented to solve the traveling salesperson problem (i.e., "Ant System").

T. Nakamura (✉)
Tokai University, 1-4-1 Kitakaname, Hiratsuka, Japan
e-mail: nakamura@tsc.u-tokai.ac.jp

I. Homma · H. Yamamoto
Tokyo Metropolitan University, 6-6 Asahigaoka, Hino, Japan
e-mail: honma-issin@ed.tmu.ac.jp

H. Yamamoto
e-mail: yamamoto@tmu.ac.jp

V. Kumar and H. Pham (eds.), *Predictive Analytics in System Reliability*,
Springer Series in Reliability Engineering,
https://doi.org/10.1007/978-3-031-05347-4_6

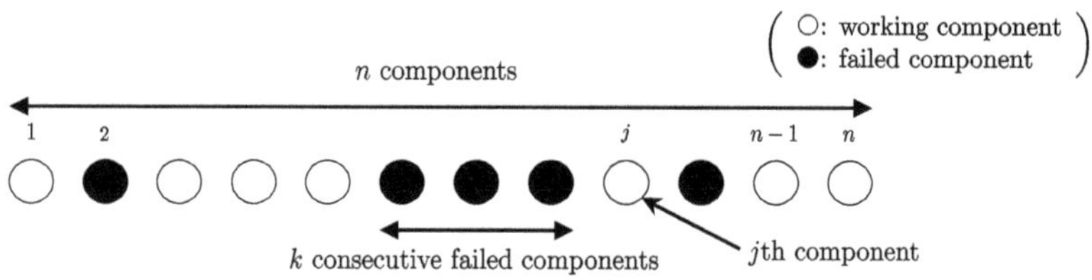

Fig. 1 Lin/Con/k/n:F system

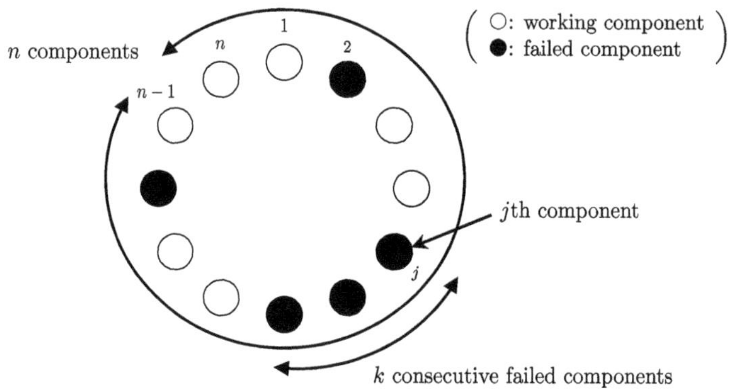

Fig. 2 Cir/Con/k/n:F system

The ACO algorithm has also been applied to reliability optimization problems, e.g., the component assignment problem of the consecutive-k-out-of-n:F system, which can represent a real system in which consecutive failed components cause system failures. This system can be classified into two types according to the arrangements of components: a linear consecutive-k-out-of-n:F system (i.e., Lin/Con/k/n:F system) and a circular consecutive-k-out-of-n:F system (i.e., Cir/Con/k/n:F system). The Lin/Con/k/n:F system comprises of n components arranged in a line and fails if there exist k consecutive failed components, as shown in Fig. 1. The Cir/Con/k/n:F system comprises of n components arranged in a circle, as shown in Fig. 2. The applications of these systems are displayed as follows:

Example 1 Production Monitoring System (*Zhao et al.* [22]) Consider a production monitoring system equipped with n monitors. The monitors, which are equally spaced along the production line, can observe k units. When a single monitor fails, the entire production line can be monitored by adjacent monitors that are still in operation. However, if k monitors break down consecutively, a dead zone will appear in the monitoring system. As a result, failure occurs in the production monitoring system. Therefore, such a system can be expressed as a Lin/Con/k/n:F system.

Example 2 Road Lights along a Highway (*Dafnis et al.* [5], *Peng et al.* [13]) For safety reasons, a highway is illuminated at night by lights installed at regular intervals. If k consecutive road lights are not lit, there will be a particular area where there is not enough light in a particular area. As a result, traffic may be affected. This scenario can be represented as a Lin/Con/k/n:F system.

The component assignment problem (CAP) is a classic reliability engineering optimization problem. Solving it requires a component arrangement in which system reliability is maximized (i.e., the optimal arrangement). Here, reliability is defined as the probability that the system and its components will perform the required function. Obtaining an optimal arrangement is expected to reduce the occurrence of system failures and ensure that the system operates with stability and longevity. The CAP has been proven to belong to the NP-hard class [10], and, in recent decades, several heuristic and meta-heuristic algorithms have been proposed to solve the CAP of the Lin and Cir/Con/k/n:F systems.

Shingyochi et al. [16] proposed a genetic algorithm for this purpose. This algorithm preferentially assigns reliable components to every k-th position in the system, because system failure cannot occur if there exist working components at every k-th position. Furthermore, Shingyochi et al. [17] proposed standard and improved simulated annealing algorithms. The improved algorithm allows us to reduce the solution space by effectively removing the equivalent arrangements.

In recent years, many researchers have utilized importance measures to efficiently find a pseudo-optimal arrangement for CAP. Birnbaum importance (B-importance), first proposed by Birnbaum [1], is a key index for measuring the influence of component reliability on system reliability. Thus, the B-importance is able to identify the most or least important component in a system; hence, it can provide insight into a system and its components. For this reason, it has been applied to various algorithms to enhance their performance. Si et al. [18] presented a summary of methods for reliability optimization that utilize importance measures.

Using the B-importance, Zuo and Kuo [24] proposed two types of ZK-type heuristics (i.e., ZKA and ZKB), and Lin and Kuo [11] presented LKA heuristics. Yao et al. [20] outlined the B-importance-based two-stage approach (BITA) for seeking the optimal arrangement. The BITA includes LKA, LKB, ZKB, and ZKD.

To design efficient metaheuristics, two contradictory criteria should be considered: diversification and intensification. Several studies have used genetic algorithms (GAs) for diversification and B-importance for intensification. Yao et al. [21] built a B-importance-based genetic local search algorithm (BIGLS). A gradual reduction of the solution space can be achieved using a local search with ZK-type heuristics; consequently, the BIGLS can find the pseudo-optimal arrangement efficiently. Subsequently, Cai et al. [2] developed a B-importance-based GA (BIGA), which integrates the advantages of BITA and GAs to obtain the pseudo-optimal arrangement of the Lin/Con/k/n:F system. Numerical experiments showed that BIGA outperformed BIGLS for large Lin/Con/k/n:F systems. Accordingly, BIGA appears to currently be one of the most efficient metaheuristics for solving CAP.

Cai et al. [3] first proposed an ACO algorithm for solving the CAP of the Cir/Con/k/n:F system. Subsequently, Wang et al. [19] developed an ACO algorithm for the Lin/Con/k/n:F system, which incorporates B-importance. Wang et al. [19] showed that the algorithm outperformed the BIGLS algorithm in seeking optimal arrangements of large Lin/Con/k/n:F systems.

In ACO, the pheromone trail creates a probability distribution over the search space, which determines which part of the search space is effectively selected [7]. In the CAP example, the pheromone determines the probability of a component being assigned. Cai et al. [3] and Wang et al. [19] determined the probability of component assignment based on the information of the adjacent component. A pheromone that determines probabilities based on the adjacent component is called an *adjacence-wise pheromone* in this study. In Cir/Con/k/n:F systems, only the relative order of the components is important because they are arranged in a circle. For example, arrangements $(1, 2, 3, 4, 5)$ and $(3, 4, 5, 1, 2)$ have the same system reliability, which is the objective function of CAP. On the other hand, in Lin/Con/k/n:F systems, $(1, 2, 3, 4, 5)$ and $(3, 4, 5, 1, 2)$ represent two different solutions, each with different reliabilities. Therefore, the positions of components may be more informative in seeking the optimal arrangement of the Lin/Con/k/n:F system. A pheromone that determines probabilities based on the positions of components is called a *position-wise pheromone* in this study. Shingyochi et al. [16] also suggested that assigning reliable components at k intervals may increase system reliability. Thus, we also consider a pheromone that determines the probabilities based on the components at k intervals, which is called a *k-interval-wise pheromone* in this study.

If we select the optimal pheromone model, it is expected that we will obtain a high-performance ACO algorithm. However, identifying the best pheromone model is challenging. Thus, the study develops an ACO algorithm with three types of pheromones for the purpose of solving the CAP of the Lin/Con/k/n:F system. A numerical experiment is then performed to investigate the efficacy of the three different types of pheromones. Additionally, the proposed ACO algorithm is compared to BIGA, which is currently the most popular and efficient metaheuristic for solving CAP.

The remainder of this paper is organized as follows: A mathematical description of the CAP of the Lin/Con/k/n:F system is provided in Sect. 2. In Sect. 3, detailed descriptions of the proposed ACO algorithm and its constituent parts are presented. Section 4 presents the results of a numerical experiment and evaluates the efficacy of the proposed ACO algorithms. Finally, the conclusion is presented in Sect. 5.

2 CAP of the Lin/Con/k/n:F System

2.1 Definition of the CAP

Herein, the CAP of the Lin/Con/k/n:F system is defined. This paper assumes that

- each component of the system can have only two states: working or failed,
- the component reliabilities are given,

- all components are mutually statistically independent, and
- the components are functionally interchangeable.

We next define notation. Recall that the Lin/Con/k/n:F system comprises of n components arranged in a line and fails if the system has k consecutive failed components. For $j = 1, 2, \ldots, n$, let j be the index of the components. The reliability of component j is denoted by p_j. The vector, $\boldsymbol{p}$ $(= (p_1, p_2, \ldots, p_n))$, represents an n-vector with component reliabilities as its entries. Note that, without loss of generality, we take $p_1 \leq p_2 \leq \cdots \leq p_n$ to be true, meaning that p_j represents the reliability of the jth least reliable component. Let $a = 1, 2, \ldots, n$ be the index of component positions and $\pi(a)$ be the index of the component assigned to position a. Here, position a represents the a-th position from the left side in an arrangement. Additionally, an arrangement of n components in which component $\pi(a)$ to position a for $a = 1, 2, \ldots, n$ is expressed as the n vector $\boldsymbol{\pi} = (\pi(1), \pi(2), \ldots, \pi(n))$. Let $R_L(k, n, \boldsymbol{p}; \boldsymbol{\pi})$ denote the reliability of the Lin/Con/k/n:F system which has $\boldsymbol{p}$ under the arrangement of $\boldsymbol{\pi}$.

The CAP aims to determine the arrangements of components so that the reliability of the Lin/Con/k/n:F system can be maximized. Thus, we formulate the CAP of the Lin/Con/k/n:F system as follows: Please generate equation alttext for display equation MathID103

$$\boldsymbol{\pi}^* = \arg\max_{\boldsymbol{\pi} \in S} \{R_L(k, n, \boldsymbol{p}; \boldsymbol{\pi})\}, \tag{1}$$

where S represents the set of all arrangements. In this study, we call the arrangement $\boldsymbol{\pi}^*$ that satisfies Eq. (1) an "optimal arrangement."

2.2 B-Importance

In this subsection, we present the B-importance, which was first introduced by Birnbaum [1] to measure the relative importance of a component to the overall system reliability. Herein, the definition of the B-importance of a Lin/Con/k/n:F system is presented, although it is worth noting that the B-importance can be applied to any coherent system.

Definition 1 (*Birnbaum* [1]) The B-importance of a component assigned to position a in a Lin/Con/k/n:F system having component reliabilities $\boldsymbol{p}$ and component arrangement $\boldsymbol{\pi}$, is denoted by $I_B(a; \boldsymbol{p}, \boldsymbol{\pi})$ and defined as

$$I_B(a; \boldsymbol{p}, \boldsymbol{\pi}) = R_L(k, n, (1_a, \boldsymbol{p}); \boldsymbol{\pi}) - R_L(k, n, (0_a, \boldsymbol{p}); \boldsymbol{\pi}), \tag{2}$$

where $(\cdot_i, \boldsymbol{p}) = (p_1, p_2, \ldots, p_{i-1}, \cdot, p_{i+1}, \ldots, p_n)$.

The B-importance can increase the exploration ability of algorithms for solving the CAP.

2.3 Necessary Condition

This subsection describes a necessary condition that needs to be met by the optimal arrangement. Kuo et al. [9] provided this necessary condition for the CAP of the Lin/Con/k/n:F system.

Theorem 1 (Kuo et al. [9]) *Given the reliability of n components with $p_1 \leq p_2 \leq \cdots \leq p_n$, the optimal arrangement of the Lin/Con/k/n:F systems with $n \geq 2k$ satisfies the following condition:*

$$\pi^*(1) < \pi^*(2) < \cdots < \pi^*(k), \tag{3}$$

$$\pi^*(n-k+1) > \pi^*(n-k+2) > \cdots > \pi^*(n). \tag{4}$$

As aforementioned, the components are numbered in ascending order of their reliabilities. Thus, Eq. (3) states that components from positions 1 to k are in nondecreasing order of the reliabilities of components.

Let us consider the arrangement $(2, 1, 4, 6, 5, 3)$ of the Lin/Con/3/6:F system. This arrangement does not satisfy the necessary condition in Eq. (3), meaning that it will not be optimal. This is because there exists an arrangement with higher reliability (e.g., the arrangement $(1, 2, 4, 6, 5, 3)$), which can be obtained by swapping components $\pi(1)$ and $\pi(2)$. Therefore, the necessary condition can be used to reduce the solution space. As a result, the computational load to find the optimal arrangement can be reduced.

2.3.1 Remarks

This study considers the CAP for Lin/Con/k/n:F systems with $n \geq 2k$. According to Kuo et al. [9], if $n < 2k$, then the $(2k - n)$ most reliable components should be assigned between positions $(n - k + 1)$ and k in any order. Subsequently, the remaining components are optimally assigned to the Lin/Con/$(n-k)$/$2(n-k)$:F systems; as a result, the optimal arrangement is obtained. In summary, the CAP in the case of $n < 2k$ can be reduced to that of the $n \geq 2k$ case.

3 Proposal of the ACO Algorithm for Solving the CAP

This section presents the ACO algorithm for solving the CAP of the Lin/Con/k/n:F system. The existing ACO algorithm used the *adjacence-wise pheromone* to determine the probability of a component to be assigned. However, this pheromone does not account for the positions of components and those positioned at k intervals. Therefore, we build the ACO algorithm with not only *adjacence-wise pheromones* but also *position-wise* and *k-interval-wise pheromones*.

The notation used in the algorithm is as follows:

j	:	the index of components ($j = 1, 2, \ldots, n$)
a	:	the index of positions ($a = 1, 2, \ldots, n$), where position a indicates the a-th position from the left side in each arrangement
m	:	the number of ants
h	:	the index of ants ($h = 1, 2, \ldots, m$)
I_{max}	:	the maximum number of iterations ($0 \leq I_{max}$)
t	:	the index of iterations ($t = 1, 2, \ldots, I_{max}$)
$\tau^{adj}_{\pi(a-1),j}$	:	the value of the adjacence-wise pheromone, which indicates whether component j should be assigned based on component $\pi(a-1)$ (i.e., the component in the right adjacent position a)
$\tau^{pos}_{a,j}$	:	the value of the position-wise pheromone, which indicates whether component j should be assigned based on position a
$\tau^{k}_{\pi(a-k),j}$	:	the value of the k-interval-wise pheromone, which indicates whether component j should be assigned based on component $\pi(a-k)$ (i.e., the component at the position $(k-1)$ components away from position a)
τ_0	:	the initial value of the pheromone
$p_h(x, y)$	:	the probability that ant h at node x visits node y
J_h	:	the set of nodes that have not been visited by ant h
LS_{start}	:	the number of iterations to start the local search
ρ	:	a parameter of pheromone evaporation ($0 < \rho < 1$)
R_h	:	the system reliability corresponding to an arrangement that ant h constructs
R_{max}	:	the system reliability corresponding to an arrangement that an elite ant constructs

The pseudocode of the proposed ACO algorithm is presented as follows:

Algorithm 1 Pseudocode of the proposed ACO algorithm

Initialize the pheromone values
Generate the initial solutions using BITA [20].
$t \leftarrow 1$
while termination condition is not met ($t < I_{max}$) **do**
    Construct the solutions using pheromones as per Eq. (5), (6), or (7)
    Apply the local search based on ZKD [20]
    Evaluate the solutions
    Update the best solution
    Update the pheromone values according to Eq. (8)
    $t \leftarrow t + 1$
end while
return Best solution

In the following, each step is explained along with an overview of the underlying theory.

3.1 Initializing the Pheromone Values and Generating the Initial Solutions

The algorithm begins by initializing pheromone values to a constant value. The initial values of pheromone trails are set to be τ_0 per Cai et al. [3].

Subsequently, the initial solutions are generated using the BITA proposed by Yao et al. [20]. The BITA procedure is as follows [10]:

STEP 1 : Generate two initial arrangements by using both the LKA and LKB heuristics

STEP 2 : If the instance being solved contains only low-reliability components, ($p_j \leq 0.2\ \forall j$), choose ZKB; otherwise, choose ZKD. ZKB or ZKD is applied separately to the arrangements generated at **STEP 1**; then, the better of the two arrangements is regarded as the solution

3.2 Constructing the Solutions

Herein, we describe the solution construction, which is the main part of the proposed algorithm. In this algorithm, m ants construct m arrangements of the Lin/Con/k/n:F systems. Specifically, the components corresponding to the nodes visited by an ant are assigned one by one, starting from position 1. The n components are assigned in the order of positions $1, 2, \ldots, n$. It should be noted that, in this algorithm, the method for determining which component is assigned depends on positions 1 to k, positions $k+1$ to $n-k$, and positions $n-k+1$ to n.

3.2.1 Assigning Components into Positions 1 to k

Herein, we explain the assignment of the components to positions 1 to k. It is assumed that, at iteration $(t-1)$, an ant will have constructed an arrangement denoted by

$$\boldsymbol{\pi}^{(t-1)} = (\pi^{(t-1)}(1), \pi^{(t-1)}(2), \ldots, \pi^{(t-1)}(n)).$$

Let

$$\boldsymbol{\pi}^{(t)} = (\pi^{(t)}(1), \pi^{(t)}(2), \ldots, \pi^{(t)}(n))$$

be an arrangement constructed by the ant at iteration t. At iteration t, k components assigned to positions $n-k+1$ to n of $\boldsymbol{\pi}^{(t-1)}$ are assigned to positions 1 to k of $\boldsymbol{\pi}^{(t)}$ by reversing their positions. That is, for $a = 1, 2, \ldots, k$, $\pi^{(t)}(a) = \pi^{(t-1)}(n-a+1)$. It is worth noting that, if components $\pi(n-k+1), \ldots, \pi(n)$ assigned at iteration $(t-1)$ satisfy Eq. (4), then components $\pi(1), \ldots, \pi(k)$ assigned at iteration t satisfy Eq. (3). Consequently, the arrangements that satisfy Eq. (3) can be easily generated.

3.2.2 Assigning Components into Positions $k+1$ to $n-k$

Herein, we explain the assignment of the components to positions $k+1$ to $n-k$. The m ants construct arrangements probabilistically by iteratively adding components to partial arrangements based on the pheromone type. There are three types of pheromones available in the proposed ACO algorithm.

(1) Adjacence-wise Pheromone [3]

Let $\tau^{adj}_{\pi(a-1),j}$ be the value of the adjacence-wise pheromone, which indicates whether component j should be assigned based on component $\pi(a-1)$. This pheromone was defined by Cai et al. [3]. Using $\tau^{adj}_{\pi(a-1),j}$, the probability that component j is assigned to position a based on ant h, denoted by $p_h(\pi(a-1), j))$, is computed as

$$p_h(\pi(a-1), j) = \begin{cases} \frac{\tau^{adj}_{\pi(a-1),j}}{\sum_{s\in J_h} \tau^{adj}_{\pi(a-1),s}} & \text{if } j \in J_h, \\ 0 & \text{otherwise,} \end{cases} \tag{5}$$

where J_h is the set of nodes that have not been visited by ant h and is implicitly defined by the solution building process performed by ant h.

(2) Position-wise Pheromone

Let $\tau^{pos}_{a,j}$ be the value of the position-wise pheromone, which indicates whether component j should be assigned based on position a. In this case, the positions of components are considered to be informative when seeking the optimal arrangement of the Lin/Con/k/n:F system. Thus, using $\tau^{pos}_{a,j}$, the probability that component j is assigned to position a based on ant h, denoted by $p_h(a, j)$, is given as

$$p_h(a, j) = \begin{cases} \frac{\tau^{pos}_{a,j}}{\sum_{s\in J_h} \tau^{pos}_{a,s}} & \text{if } j \in J_h, \\ 0 & \text{otherwise.} \end{cases} \tag{6}$$

(3) k-interval-wise Pheromone

Let $\tau^{k}_{\pi(a-k),j}$ be the value of the k-interval-wise pheromone, which indicates whether component j should be assigned based on component $\pi(a-k)$. A Lin/Con/k/n:F system fails if it has k consecutive failed components; thus, assigning reliable components at k intervals can avoid system failure. Using $\tau^{k}_{\pi(a-k),j}$, the probability that component j is assigned to position a based on ant h, denoted by $p_h(\pi(a-k), j)$, is given as

$$p_h(\pi(a-k), j) = \begin{cases} \frac{\tau^k_{\pi(a-k),j}}{\sum_{s \in J_h} \tau^k_{\pi(a-k),s}} & \text{if } j \in J_h, \\ 0 & \text{otherwise.} \end{cases} \quad (7)$$

Note that in standard ACO algorithms, the probability that an ant visits a node is determined by not only the pheromone but also by the heuristic information of the problem being solved. However, this study is devoted to comparing multiple pheromones; hence, heuristic information is not considered.

3.2.3 Assigning Components into Positions $n - k + 1$ to n

Herein, we consider a procedure for the assignment of components to positions $n - k + 1$ to n, which is dictated by the necessary condition [9]. In a situation in which $(n - k)$ components have been assigned to positions $1, 2, \ldots, n - k$, there exists only one component arrangement satisfying Eq. (4). Recall that Eq. (4) states that components in positions $n - k + 1$ to n are arranged in a nonincreasing order of component reliability. Therefore, assigning the remaining components can be uniquely determined based on the order of the reliabilities of the components. For example, let us consider the arrangement $(1, 2, 4, \cdot, \cdot, \cdot)$ of the Lin/Con/3/6:F system, where "$\cdot$" implies that a component has not been assigned; in this case, components 3, 5, 6 have not been assigned. The only arrangement that satisfies Eq. (4) is $(1, 2, 4, 6, 5, 3)$.

3.3 *Applying the Local Search*

Herein, the application of a local search is explained; local searches aim to seek a better solution in a neighborhood of the current solution. The local search in the proposed algorithm is based on ZKD, which was introduced by Yao et al. [20]. After a complete set of candidate solutions is obtained, if the number of iterations is greater than or equal to LS_{start} $(t \geq LS_{start})$, then the local search, based on ZKD, is performed on m arrangements. The arrangements constructed by m ants can be replaced by those obtained by the local search.

3.4 *Evaluating the Solutions and Updating the Best Solution*

The system's reliability is used as a measure for evaluating the arrangements. These arrangements are further improved by applying the local search based on ZKD [20]. We use the recursive equation derived by Hwang [8] to compute the reliability of a Lin/Con/k/n:F system. If the system reliability is higher than the maximum system reliability that is currently stored, the optimal solution is updated.

3.5 Updating the Pheromone Values

Herein, we explain the pheromone update (evaporation and deposit) for the purpose of generating quality solutions for the following iterations. First, at each iteration, the value of the pheromone is decreased, which mimics natural evaporation. The goal of this evaporation is to avoid overly-fast convergence, which could result in becoming trapped in sub-optimal solutions. Subsequently, the pheromone deposit increases the value of the pheromone, which is based on the quality of the obtained solution. The goal of this deposit is to make the elements of the solution more attractive to the ants in the following iterations. Additionally, the proposed ACO algorithm utilizes an elitist strategy in which the best solution that is found during the search contributes significantly to the pheromone deposit. In this study, for each iteration, an ant that constructs an arrangement with the maximum system reliability is called an "elite ant." By employing an elitist strategy, ants can find better solution quality in a small number of iterations.

For instance, we now outline the update of the adjacence-wise pheromone. Let $\tau^{adj}_{\pi(a-1),j}(t)$ be the adjacence-wise pheromone at iteration t. By evaporation and deposit, the adjacence-wise pheromone at iteration $(t+1)$ $(\tau^{adj}_{\pi(a-1),j}(t+1))$ can be computed as follows:

$$\tau^{adj}_{\pi(a-1),j}(t+1) = (1-\rho)\cdot\tau^{adj}_{\pi(a-1),j}(t) + \sum_{h=1}^{m}\Delta\tau^{h}_{\pi(a-1),j}(t) + \Delta\tau^{*}_{\pi(a-1),j}(t), \quad (8)$$

where ρ is a parameter called the "evaporation rate," $(0 < \rho < 1)$,

$$\Delta\tau^{h}_{\pi(a-1),j}(t) = \begin{cases} Q\cdot R_h & \text{if the comp. } j \text{ is right adjacent to comp. } \pi(a-1) \\ & \text{in an arrangement constructed by ant } h \text{ at iteration } t, \\ 0 & \text{otherwise,} \end{cases} \quad (9)$$

and

$$\Delta\tau^{*}_{\pi(a-1),j}(t) = \begin{cases} Q\cdot R_{max} & \text{if comp. } j \text{ is right adjacent to comp. } \pi(a-1) \text{ in an} \\ & \text{arrangement constructed by an elite ant at iteration } t, \\ 0 & \text{otherwise.} \end{cases} \quad (10)$$

Here, Q is a parameter called the "deposit rate," ($0 < Q$). The position-wise and k-interval-wise pheromones are updated in a similar manner to the adjacence-wise pheromone.

3.6 Checking the Termination Condition.

In this subsection, we describe the termination condition. If the maximum number of iterations has not been achieved, ($t < I_{max}$), return to the solution construction in Sect. 3.2; otherwise, the algorithm is terminated.

4 Numerical Experiment

This section presents the results of numerical experiments performed to evaluate the efficacy of the proposed ACO algorithms. First, we investigate the performance of the ACO algorithms with either adjacence-wise [3], position-wise, or k-interval-wise pheromones to identify the best model. Second, we compare the proposed ACO algorithm to the BIGA algorithm, proposed by Cai et al. [2]. BIGA was chosen for comparison with the proposed algorithm because numerical experiments [2] have confirmed that BIGA outperforms BIGLS for large Lin/Con/k/n:F systems. We used C++ for the implementation language. The algorithms were compared in terms of the quality of solutions (system reliability) and the computation time required to obtain the pseudo-optimal arrangement.

We prepared six Lin/Con/k/n:F systems for $n \in \{30, 80\}$ and $k \in \{5, 9, 12\}$. Additionally, to investigate the performance under different ranges of component reliability, three cases were considered for each system. The component reliabilities were randomly generated from a uniform distribution on (a) $[0.01, 0.99]$ (Arbitrary case), (b) $[0.01, 0.20]$ (Low-reliable case), or (c) $[0.80, 0.99]$ (High-reliable case). Each run of the algorithms was repeated 200 times, owing to the component reliabilities and the stochastic nature of ACO.

The ACO parameters are set as follows: the number of ants (m) is 20; the initial value of the pheromone (τ_0) is 10; the maximum number of iterations (I_{max}) is 100; the number of iterations required to start the local search (LS_{start}) is 99; the parameter of pheromone evaporation (ρ) is 0.05; and the parameter for deposit (Q) is 20.

Tables 1 and 2 show the comparisons of adjacence-wise [3], position-wise, and k-interval-wise pheromones for Lin/Con/k/n:F systems with $n = 30$ and 80, respectively. The best results are presented in bold for each case in these tables. As presented in Table 1, the ACO algorithm with the k-interval-wise pheromone achieved a fast convergence. However, the ACO algorithm with the position-wise pheromone outperformed those with the other pheromones in most cases. Notably, the computation times of the ACO algorithm with the position-wise pheromone were sufficiently

Table 1 Results of comparing adjacence-wise [3], position-wise, and k-interval-wise pheromones for Lin/Con/k/30:F systems

		Adjacence-wise [3]		Position-wise		k-interval-wise	
k	Comp. rel.	Reliability	Time (s)	Reliability	Time (s)	Reliability	Time (s)
	Arbitrary	0.900503	0.265	**0.900587**	0.231	0.900521	**0.185**
5	Low	**0.153258**	0.265	**0.153258**	0.215	**0.153258**	**0.197**
	High	0.997096	0.222	**0.997098**	0.190	0.997096	**0.167**
	Arbitrary	**0.998523**	0.475	0.998522	0.385	0.998522	**0.359**
9	Low	0.777554	0.444	**0.777559**	0.365	0.777557	**0.323**
	High	**0.999999**	0.451	**0.999999**	0.357	**0.999999**	**0.330**
	Arbitrary	**0.999937**	0.543	**0.999937**	0.404	**0.999937**	**0.395**
12	Low	0.942238	0.450	**0.942239**	0.351	**0.942239**	**0.342**
	High	**1.000000**	0.451	**1.000000**	**0.337**	**1.000000**	0.365

Table 2 Results of comparing adjacence-wise [3], position-wise, and k-interval-wise pheromones for Lin/Con/k/80:F systems

		Adjacence-wise [3]		Position-wise		k-interval-wise	
k	Comp. rel.	Reliability	Time (s)	Reliability	Time (s)	Reliability	Time (s)
	Arbitrary	0.776391	6.299	**0.776869**	5.690	0.776431	**5.155**
5	Low	0.001770	6.773	**0.001771**	6.069	0.001770	**5.755**
	High	0.990896	5.396	**0.990899**	5.004	0.990894	**4.506**
	Arbitrary	0.994364	14.854	**0.994365**	13.848	0.994361	**12.119**
9	Low	0.318706	13.858	**0.318707**	12.700	0.318706	**11.285**
	High	**0.999993**	12.570	**0.999993**	11.724	**0.999993**	**10.171**
	Arbitrary	**0.999662**	18.632	0.999661	19.368	0.999661	**17.091**
12	Low	**0.698446**	17.475	**0.698446**	**15.891**	0.698445	16.480
	High	**1.000000**	16.836	**1.000000**	**15.013**	**1.000000**	15.498

short. Table 2 shows that similar behavior was observed for the Lin/Con/k/80:F system. Therefore, it can be concluded that the ACO algorithm with the position-wise pheromone performed the best within the scope of the experiment.

Then, the ACO algorithm with the position-wise pheromone is compared to BIGA, which is known as one of the best metaheuristics for solving the CAP of Lin/Con/k/n:F systems. Here, the BIGA parameters are set as follows: the maximum generation is 200; the population size is 20; the mutation probability is 0.05; and the crossover probability is 0.8. Tables 3 and 4 show the comparisons for Lin/Con/k/n:F systems with $n = 30$ and 80, respectively, where the best results are presented in bold for each case in these tables. From the results, we observe that the proposed algorithm compares favorably to BIGA in the cases of $n = 30$ and $k = 9, 12$. Overall, it seems that BIGA outperformed the proposed algorithm. However, the proposed algorithm can determine the pseudo-optimal arrangement in

Table 3 Results of comparing the proposed algorithm and BIGA in the $n = 30$ case

		ACO with position-wise		BIGA [2]	
k	Comp. rel.	Reliability	Time (s)	Reliability	Time (s)
5	Arbitrary	0.900587	**0.231**	**0.900688**	1.194
	Low	0.153258	**0.215**	**0.153259**	0.416
	High	0.997098	**0.190**	**0.997108**	0.840
9	Arbitrary	**0.998522**	**0.385**	0.998521	1.108
	Low	0.777559	**0.365**	**0.777561**	0.720
	High	**0.999999**	0.357	**0.999999**	**0.327**
12	Arbitrary	**0.999937**	**0.404**	**0.999937**	0.801
	Low	**0.942239**	**0.351**	**0.942239**	0.684
	High	**1.000000**	0.337	**1.000000**	**0.034**

Table 4 Results of comparing the proposed algorithm and BIGA in the $n = 80$ case

		ACO with position-wise		BIGA [2]	
k	Comp. rel.	Reliability	Time (s)	Reliability	Time (s)
5	Arbitrary	0.776869	**5.690**	**0.777778**	41.909
	Low	**0.001771**	**6.069**	0.001770	9.784
	High	0.990899	**5.004**	**0.990949**	33.579
9	Arbitrary	0.994365	**13.848**	**0.994404**	72.642
	Low	0.318707	**12.700**	**0.318720**	35.902
	High	**0.999993**	**11.724**	**0.999993**	29.057
12	Arbitrary	0.999661	**19.368**	**0.999662**	63.969
	Low	0.698446	**15.891**	**0.698458**	52.451
	High	**1.000000**	15.013	**1.000000**	**4.585**

a shorter time. In the proposed ACO algorithm, an improvement in the quality of the solutions could be achieved by adding heuristic information into the CAP of the Lin/Con/k/n:F systems and appropriately adjusting the parameter of the ACO.

5 Conclusion

In this paper, we presented the ACO algorithm for solving the CAP of the Lin/ConPlease generate equation alttext for inline equation MathID351/ k/n:F system. The key feature of the proposed algorithm is to consider not only adjacence-wise pheromones but also position-wise and k-interval-wise pheromones. The performance of these pheromones was compared through a numerical experiment. Results concluded that the ACO algorithm with the position-wise pheromone performed well within the scope of the experiment. However, numerical experiments showed that the proposed algorithm was not as good as the existing one in terms of solution quality. The ACO algorithm with the position-wise pheromone could be enhanced by adding heuristic information into the CAP of Lin/Con/k/n:F systems and appropriately adjusting the parameter of the ACO; this is a challenge for further research.

Zhu et al. [23] defined the multi-type component assignment problem, which is an extension of CAP. The problem aims to determine the component arrangement with maximized system reliability under the assumption that each component needs to be assigned only to positions that belong to the same type of component. Several studies [12, 14, 15] have proposed exact and heuristic algorithms for solving this problem so far. One direction for our future work is to apply the ACO algorithm proposed in this study to solve multi-type component assignment problems.

Acknowledgements This work was supported by the Nakajima Foundation, Grant-in-Aid for JSPS KAKENHI Grant Numbers 21K14370 and 20K04964.

References

1. Birnbaum ZW (1969) On the importance of different components in a multicomponent system. In: Krishnaiah PR (ed) Multivariate analysis, vol II. Academic Press, pp 581–592
2. Cai Z, Si S, Sun S, Li C (2016) Optimization of linear consecutive-k-out-of-n system with a Birnbaum importance-based genetic algorithm. Reliab Eng Syst Saf 152:248–258. https://doi.org/10.1016/j.ress.2016.03.016
3. Cai ZQ, Wang W, Zhang S, Jiang ZY (2017) Ant colony optimization for component assignment problems in circular consecutive-k-out-of-n systems. In: Proceedings of the 2017 IEEE international conference on industrial engineering and engineering management, pp 954–958
4. Colorni A, Dorigo M, Maniezzo V (1992) A genetic algorithm to solve the timetable problem. Politecnico di Milano, Milan, Italy TR, pp 90-060
5. Dafnis SD, Makri FS, Philippou AN (2019) The reliability of a generalized consecutive system. Appl Math Comput 359:186–193
6. Dorigo M, Maniezzo V, Colorni A (1996) Ant system: optimization by a colony of cooperating agents. IEEE Trans Syst Man Cybern Part B (Cybern) 26(1):29–41
7. Dorigo M, Stützle T (2019) Ant colony optimization: overview and recent advances. Handbook of metaheuristics, pp 311–351
8. Hwang FK (1982) Fast solutions for consecutive-k-out-of-n: F system. IEEE Trans Reliab 31(5):447–448. https://doi.org/10.1109/TR.1982.5221426
9. Kuo W, Zhang W, Zuo MJ (1990) A consecutive-k-out-of-n: G system: the mirror image of a consecutive-k-out-of-n: F system. IEEE Trans Reliab 39(2):244–253
10. Kuo W, Zhu X (2012) Importance measures in reliability, risk, and optimization: principles and applications. Wiley
11. Lin FH, Kuo W (2002) Reliability importance and invariant optimal allocation. J Heuristics 8(2):155–171. https://doi.org/10.1023/A:1017908523107
12. Nakamura T, Yamamoto H (2021) Branch-and-bound-based algorithm for solving the multi-type component assignment problem in a consecutive-k-out-of-n:F system. Qual Reliab Eng Int
13. Peng R, Wu D, Gao K (2019) Reliability of a dual linear consecutive system with three failure modes. In: Li QL, Wang J, Yu HB (eds) Stochastic models in reliability, network security and system safety. Springer, pp 259–269

14. Qiu S, Ming X, Sallak M, Lu J (2022) A birnbaum importance-based two-stage approach for two-type component assignment problems. Reliab Eng Syst Saf 218:108051
15. Shi Y, Guan Z, Qiu S (2021) A heuristic for a special kind of multi-type component assignment problem. In: IOP conference series: materials science and engineering, vol 1043. IOP Publishing, p 022054
16. Shingyochi K, Yamamoto H, Tsujimura Y, Kambayashi Y (2006) Genetic algorithm for solving optimal component arrangement problem of circular consecutive-k-out-of-n:F system. In: Proceedings of the 2nd Asian international workshop, pp 176–184
17. Shingyochi K, Yamamoto H, Yamachi H (2012) Comparative study of several simulated annealing algorithms for optimal arrangement problems in a circular consecutive-k-out-of-n: F system. Qual Technol Quant Manag 9(3):295–303. https://doi.org/10.1080/16843703.2012.11673293
18. Si S, Zhao J, Cai Z, Dui H (2020) Recent advances in system reliability optimization driven by importance measures. Front Eng Manag 7:335–358
19. Wang W, Cai Z, Zhao J, Si S (2019) Optimization of linear consecutive-k-out-of-n systems with Birnbaum importance based ant colony optimization algorithm. J Shanghai Jiaotong Univ (Sci) 1–8
20. Yao Q, Zhu X, Kuo W (2011) Heuristics for component assignment problems based on the Birnbaum importance. IIE Trans 43(9):633–646. https://doi.org/10.1080/0740817X.2010.532856
21. Yao Q, Zhu X, Kuo W (2014) A Birnbaum-importance based genetic local search algorithm for component assignment problems. Ann Oper Res 212(1):185–200. https://doi.org/10.1007/s10479-012-1223-1
22. Zhao J, Si S, Cai Z (2019) A multi-objective reliability optimization for reconfigurable systems considering components degradation. Reliab Eng Syst Saf 183:104–115
23. Zhu X, Fu Y, Yuan T, Wu X (2017) Birnbaum importance based heuristics for multi-type component assignment problems. Reliab Eng Syst Saf 165:209–221
24. Zuo MJ, Kuo W (1990) Design and performance analysis of consecutive-k-out-of-n structure. Naval Res Logist 37(2):203–230

Reliability Assessment and Profit Analysis of Automated Teller Machine System Under Copular Repair Policy

Ibrahim Yusuf and Abdullahi Sanusi

Abstract The evaluation of reliability is a critical factor that ensures reliable system operation, improving quality of product, and cutting production losses. The structure for analyzing Automated Teller Machines (ATMs) failures is presented in this chapter, which also allows for the identification of the most appropriate methods for removing them. It is also essential to have sufficient information on failure and repair in order to assess system availability and reliability and calculate exact performance rates. The model under consideration is made up of three distinct subsystems, namely subsystems A, B, and C. All the components designed for this system have access to exponential failure and repair. Failure times are thought to have an exponential distribution, whereas repair times are thought to have a General distribution and Gumbel–Hougaard family Copula. The transition diagram's set of differential equations was solved using regenerative point techniques and Laplace transforms. For various assumed parameter values, different reliability characteristics such as availability, reliability, MTTF, sensitivity, and cost benefit are derived and validated to determine how the model under consideration affects them. A number of cases are used to demonstrate the analysis in depth. Tables are used to present computed results and while figures depicts the computed results. According to the computed results, Copula is a superior repair technique for improving the efficiency of repairable systems. Furthermore, the findings of the study are crucial for the banking sector's development.

Keywords Automated teller machine (ATM) · Availability · Reliability · Profit · Copula repair · General repair

I. Yusuf (✉)
Department of Mathematical Sciences, Bayero University, Kano, Nigeria
e-mail: iyusuf.mth@buk.edu.ng

A. Sanusi
School of Continuing Education (SCE), Bayero University, Kano, Nigeria
e-mail: asanusi.sce@buk.edu.ng

V. Kumar and H. Pham (eds.), *Predictive Analytics in System Reliability*, Springer Series in Reliability Engineering,
https://doi.org/10.1007/978-3-031-05347-4_7

1 Introduction

Over the years, banking sector has seen significant growth. Information and Communication Technology (ICT) changes and develops. Among them are the implementation of Automated Teller Machines (ATMs) which is a breakthrough that aims to decongest the banking system. Customers no longer need to go to bank to complete their transactions; instead, they can go to any nearby ATM. Cash withdrawals, cash transfers, payment of bills and deposit are examples of financial transactions that can be carried out using ATM. All researchers agree that ATMs are important for the banking sector's future development, however some of them have discovered a lack of proportionality between increasing the extent of technology use and increasing bank profitability.

Many banks recognize that service quality contributes to strategic competitiveness in a volatile business environment. To get into this situation, system failures must be addressed in a cost-effective and efficient manner. Thus, the necessity for Automated Teller Machine system (ATMs) evaluation and profit analysis is unavoidable.

Several reliability research studies have been reported in survey articles under various operating conditions and with assumptions, including as, Niwas and Garg [23] proposed a cost-warranty repair policy for analyzing the reliability, dependability and profitability of an industrial system, while Niwas and Kadyan [24] presented a method for analyzing the dependability and profitability of a single-unit system with multiple vacations. Singh et al. [36] investigated the performance of a CBT network system composed of four subsystems connected in series via the Copula approach. Chopra and Ram [4] applied Gumbel-Hougaard Copula to study reliability measures for two dissimilar units connected in parallel. Garg [10] presented a performance metric for an industrial system using a hybridized soft computing methodology. Kumar et al. [16] discussed the availability and profitability of engineering systems connected in series. Sanusi et al. [33] have recently studied the industrial systm performance where the system has serial arrangement. The performance of the sugar industry's A-pan crystallization technology was investigated using the RAMD technique by Dahiya et al. [6]. Ram and Goyal [30] investigated the reliability of a manufacturing system using a coverage and copula approach, discovering that the mixed coverage-copula technique improves system reliability. Garg [11] used credibility theory and several types of intuitionistic fuzzy numbers to present a fresh technique for analyzing the dependability of series–parallel systems. Ibrahim et al. [13] applied family of copula to assess the reliability of a complex system consisting two subsystems coupled in series connection. Kumar and Lather [17] applied a hybridized technique to determine the reliability of a robotic system. Garg [12] discussed on how to use Kolmogorov fuzzy differential equations to analyze the reliability of industrial systems. Yusuf et al. [40] have recently demonstrated the effectiveness of a multi client computer system consisting of three subsystems in serial connection utilizing the Copula repair strategy. Lado and Singh [19] have analyzed the cost of a sophisticated repairable system with two subsystems connected in series via copula linguistic approach. Berk et al. [2] explored how to analyze the reliability of safety–critical sensor data. Pourhassan et al. [28] proposed a simulation approach/technique for assessing the reliability of systems subjected to degradation and random shock. Abubakar and Singh [1] investigated the system clothing industry's performance. Yang et al. [41] investigated system reliability in the presence of inconsistencies in priors and multi-level data. Saini and Kumar used RAMD analysis to investigate the performance of an evaporation system in the sugar industry [38]. Gahlot et al. [7] investigated the performance of a repairable serial system with multiple failure types and two types of repairs. Zhao et al. [42] investigated the compressor rotor system of an aero-engine. Gulati et al. [8] used the copula approach to examine the performance of a repairable complex system with a serial connection under multiple failures and repair disciplines. Singh and Ayagi [34] used Copula to investigate the performance of a system under a proactive resume repair strategy. Jain et al. [14] analyzed the performance of a machining fault-tolerant system (FTMS) with standbys and a skilled or trained repairman. Kumar [18] develoed stochastic computer system models with maintenance and maximum operation time. Mehta et al. [21] used a copula family approach to assess the availability of an industrial system. Tyagi et al. [39] have presented copula analysis of a parallel system with fault coverage. Choudhary et al. [3] evaluated a cement plant based on its maintenance, availability, and reliability. Malik and Tewari [22] investigated the water flow system maintenance priorities of a coal-fired thermal power plant. Raissi and Ebadi [32] investigated a computer simulation model for assessing the reliability of a complex system. Potapov et al. [26] concentrated on simulating the dependability of a client–server information system. Kadyan and Kumar [15] conducted research on the availability and profitability of feeding systems in the sugar industry. Pandey et al. [27] assessed the dragline's critical subsystem's reliability and failure rate.

Besides the above works, numerous academics have previously examined various types of ATM systems using various techniques. Gupta et al. [9] used reliability analysis to determine the operational behavior of the ATM using three subsystems: the bank computer, the ATM machine, and the central computer, all of which were subjected to standby configuration. Cheong et al. [5] presented daily unattended ATM failures and then used forecasted results to optimize the number of field services to develop in each geographical zone in order to reduce the number of daily unattended ATM failures. Ram and Goyal [31] used stochastic modeling for reliability investigation and sensitivity analysis of ATM repairable system. In (2015), Menna provided a thorough understanding of ATM and its benefits. According to [25], fingerprint biometric authentication schemes for ATMs are more redeemable than personal identification numbers (PINs) for identification and security clearance. Meena [20] highlighted the importance of ATM in daily appears requesting monetary transactions. Pandy et al. [27] analysed and determine the criticality and reliability of the subsystems of three draglines and overall reliability of each dragline. Pourhassan et al. [29] dealt with reliability modelling and analysis of power station exposed to fatal and nonfatal. Ram and Goyal [31] dealt with modelling and evaluation of reliability metrics ATM.

Given the preceding empirical investigation, it's understandable that some researchers have conducted research in the area of reliability engineering. However, it is important to note that electronic payment has gained popularity in the banking sector especially with the usage of automated teller machine (ATM) system. This creates a significant gap, which the present ATM research attempts to fill. In particular, it evaluates the ATM's availability and reliability, as well as estimating exact performance rates.

2 Notations, Assumptions, and Materials and Methods

2.1 Notations

t	To denote the time
s	To denote variable of Laplace transform
λ_1	To denote rate of failure of subsystem A
λ_2	To denote rate of failure of subsystem B
λ_3	To denote rate of failure of subsystem C
λ_4	To denote rate of failure of subsystem D
$h(x)/y(y)$	To denote service (repair) rate of subsystem A/subsystem B
$\alpha_0(x)/\alpha_0(y)$	To denote service (repair) rate for completely failed states of subsystem A and subsystem B respectively
$\beta_0(m)/\beta_0(n)$	To denote service (repair) rate for completely failed states of subsystem C and subsystem D respectively
$P_i(t)$	For i =0 to 12, being probability of the system sojourn in S_i at any given time
$\overline{P}(s)$	To denote the Laplace transformation of the state probability $p(t)$
P_i (x, t)	Being probability that the system sojourn in S_i and is receiving repair, and the elapsed time to service (repair) is (x, t)
P_i (y, t)	Being probability that the system sojourn in S_i and is receiving repair, and the elapsed time to service (repair) is (y, t)
P_i (m, t)	Being probability that the system sojourn in S_i and is receiving repair, and the elapsed time to service (repair) and the system is under repair and the elapsed time to service (repair) is (m, t)
P_i (n, t)	Being probability that the system sojourn in S_i and is receiving repair, and the elapsed time to service (repair) and the system is under repair and the elapsed time to service (repair) is (n, t)
$E_p(t)$	To denote Profit expected in the interval [0, t)
C_1, C_2	To denote the expected revenue mobilized and service (repair)
$\mu_0(x)$	Joint probability is expressed as: $c_\theta(u_1(x), u_2(x)) = exp\left(x^\theta + \{log\phi(x)^\theta\}^{\frac{1}{\theta}}\right)$, $1 \leq \theta \leq \infty$. Where $\mu_1 = \phi(x)$ and $u_2 = e^x$

2.2 Assumptions

a. To begin with, all subsystems are presumed to be functional.
b. Two units from systems A and B, one unit from C, and two out of two from subsystem D are required for system operation.
c. Any unit failure leads to adequate machine performance.
d. Any subsystem device that fails is repairable while it is in use or in the failure state.
e. Restoration of system complete failure is done by Copula while restoration of partial failure is done by general distribution.
f. The required machine unit should function as new, and the repair operation should cause no damage.
g. The load is ready for the system's effective performance as soon as the failed unit is fixed.

2.3 Materials and Methods

The model/system considered in this chapter is made up of three distinct subsystems:

a. Subsystem A is comprised of two similar components, A 1 and A 2, each of which has three active clients. Two clients from A 1 and A 2 must be operational for the system to function. When one of the A 1 and A 2 clients fails, the system's capacity is reduced.
b. Subsystem B: This subsystem is made up of two active servers running in parallel. When one of the subsystem's two active servers fails, the system operates at a reduced capacity. While the failure of the two servers causes the system to fail completely.
c. Subsystem C: this subsystem comprises of two parallel active units/servers. A minimum of one unit/server must be operational. When both of the subsystem's units/servers fail, the system as a whole fail.

All system's failure rates are constant and obeys exponential distribution, and their repair is either obeying general repair and copula repair. When we have a situation where we need to fix failed states quickly and urgently, we can use Copula repair to provide the essential input. So, in the event of total failure, Copula can be used to correct all systems simultaneously, whilst General repair can be used to restore partially failed states. To create differential equations for high reliability physiognomies viz; availability, profit expected function, MTTF and reliability, MTTF, we use supplementary variable technique, Gumbel-Hougaard family Copula, and integral transformation (Fig. 1).

3 Model Formulation

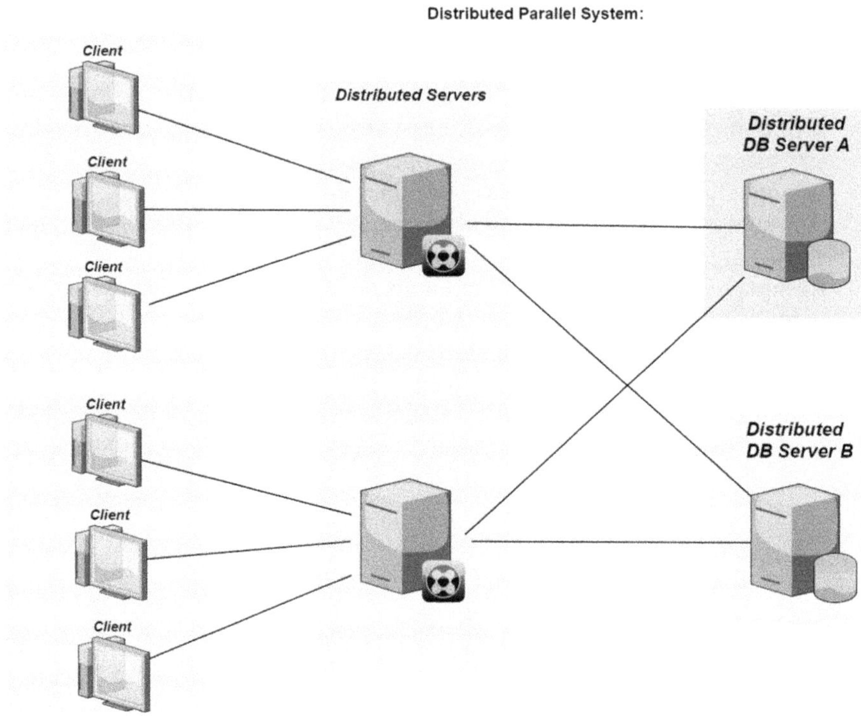

Fig. 1 Distributed parallel system

Using the concept presented in Singh and Hamisu [35], Singh et al. [37], Gulati et al. [8], Gahlot et al. [7], Chopra and Ram [4] and Abubakar and Singh [1] and Fig. 2, the following of equations are derived:

$$\left\{\frac{\partial}{\partial t} + 2\lambda_1 + 2\lambda_2 + 2\lambda_3 + 2\lambda_4\right\} P_0(t) = \int_0^\infty h(x) P_1(x,t) dx + \int_0^\infty h(y) P_2(y,t) dy + \int_0^\infty \alpha_0(y) P_9(y,t) dy + \int_0^\infty \alpha_0(x) P_{10}(x,t) dx + \int_0^\infty \beta_0(m) P_{11}(m,t) dm + \int_0^\infty \beta_0(n) P_{12}(n,t) dn, \quad (1)$$

$$\left\{\frac{\partial}{\partial t} + \frac{\partial}{\partial w} + 2\lambda_1 + 2\lambda_2 + h(w)\right\} P_i(w,t) = 0, w = (x, y) \text{ } and \text{ } i = (1, 2), \quad (2)$$

$$\left\{\frac{\partial}{\partial t} + \frac{\partial}{\partial y} + 2\lambda_k + h(y)\right\} P_i(y,t) = 0, k = (2, 2, 1) \text{ } and \text{ } i = (3, 6, 7), \quad (3)$$

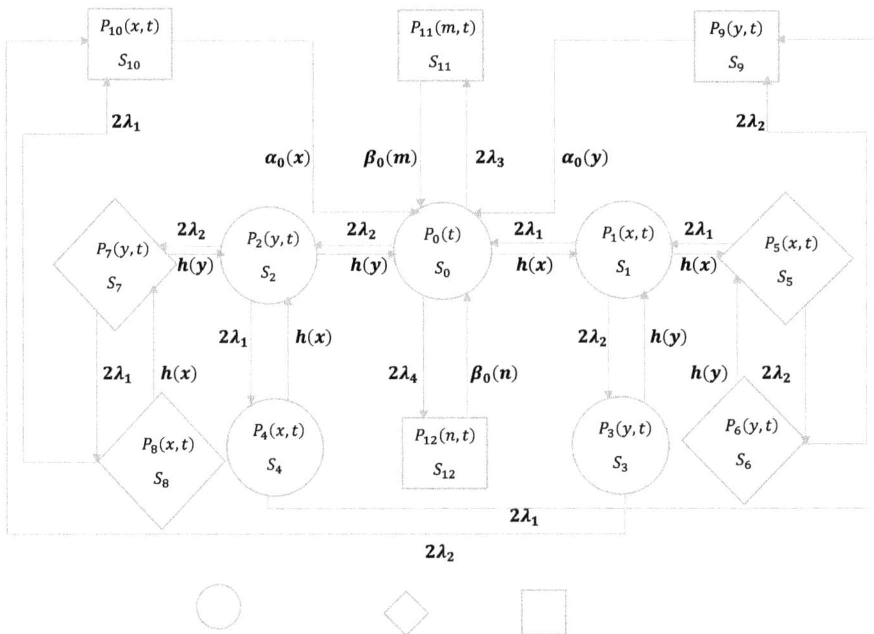

Fig. 2 Transition diagram of the model

$$\left\{\frac{\partial}{\partial t} + \frac{\partial}{\partial x} + 2\lambda_k + h(x)\right\} P_i(x,t) = 0, k = (1, 2, 1)\ and\ i = (4, 5, 8), \tag{4}$$

$$\left\{\frac{\partial}{\partial t} + \frac{\partial}{\partial r} + j_0(r)\right\} P_i(r,t) = 0,$$
$$r = (y, x, m, n),$$
$$j_0 = (\alpha_0, \alpha_0, \beta_0, \beta_0)\ and\ r = (9, 10, 11, 12). \tag{5}$$

Boundary conditions

$$P_i(0,t) = 2\lambda_k P_0(t), k = (1,2)\, and\, i = (1,2), \tag{6}$$

$$P_3(0,t) = 4\lambda_1\lambda_2 P_0(t), \tag{7}$$

$$P_4(0,t) = 4\lambda_1\lambda_2 P_0(t), \tag{8}$$

$$P_i(0,t) = 4\lambda_k^2 P_0(t), k = (1,2)\ and\ i = (5,7), \tag{9}$$

$$P_6(0,t) = 8\lambda_1^2\lambda_2 P_0(t), \tag{10}$$

$$P_8(0,t) = 8\lambda_1\lambda_2^2 P_0(t), \tag{11}$$

$$P_i(0,t) = 16\lambda_1^2\lambda_2^2 P_0(t), i = (9,10), \tag{12}$$

$$P_i(0,t) = 2\lambda_k P_0(t), k = (3,4)\ and\ i = (11,12). \tag{13}$$

3.1 Model Solution

With initial condition $P(0) = 1$ and the Laplace transforms of Eqs. (1)–(5) as:

$$\begin{aligned}\{s + 2\lambda_1 + 2\lambda_2 + 2\lambda_3 + 2\lambda_4\}\overline{P_0}(s) = 1 &+ \int_0^\infty h(x)\overline{P_1}(x,s)dx \\ &+ \int_0^\infty h(y)\overline{P_2}(y,s)dy + \int_0^\infty \alpha_0(y)\overline{P_9}(y,s)dy \\ &+ \int_0^\infty \alpha_0(x)\overline{P_{10}}(x,s)dx + \int_0^\infty \beta_0(m)\overline{P_{11}}(m,s)dm \\ &+ \int_0^\infty \beta_0(n)\overline{P_{12}}(n,s)dn,\end{aligned} \tag{14}$$

$$\left\{s + \frac{\partial}{\partial w} + 2\lambda_1 + 2\lambda_2 + h(w)\right\}\overline{P}_i(w,t) = 0, w = (x,y)\ and\ i = (1,2), \tag{15}$$

$$\left\{s + \frac{\partial}{\partial y} + 2\lambda_k + h(y)\right\}\overline{P}_i(y,t) = 0, k = (2,2,1)\ and\ i = (3,6,7), \tag{16}$$

$$\left\{s + \frac{\partial}{\partial x} + 2\lambda_k + h(x)\right\}\overline{P}_i(x, t) = 0, k = (1, 2, 1)\ and\ i = (4, 5, 8), \tag{17}$$

$$\left\{s + \frac{\partial}{\partial r} + j_0(r)\right\}\overline{P}_i(r, t) = 0,$$

$$r = (y, x, m, n),$$

$$j_0 = (\alpha_0, \alpha_0, \beta_0, \beta_0)\ and\ r = (9, 10, 11, 12). \tag{18}$$

Laplace transforms of the boundary conditions are given by:

$$\overline{\overline{P}}_i\ (0, s) = 2\lambda_k \overline{P_0}(s), k = (1, 2)\ and\ i = (1, 2), \tag{19}$$

$$\overline{P_3}(0, s) = 4\lambda_1\lambda_2\overline{P_0}(s), \tag{20}$$

$$\overline{P_4}(0, s) = 4\lambda_1\lambda_2\overline{P_0}(s), \tag{21}$$

$$\overline{P}_i(0, s) = 4\lambda_k^2\overline{P_0}(s), k = (1, 2)\ and\ i = (5, 7), \tag{22}$$

$$\overline{P_6}(0, s) = 8\lambda_1^2\lambda_2\overline{P_0}(s), \tag{23}$$

$$\overline{P_8}(0, s) = 8\lambda_1\lambda_2^2\overline{P_0}(s), \tag{24}$$

$$\overline{P}_i(0, s) = 16\lambda_1^2\lambda_2^2\overline{P_0}(s), i = (9, 10), \tag{25}$$

$$\overline{P}_i(0, s) = 2\lambda_k\overline{P_0}(s), k = (3, 4)\ and\ i = (11, 12). \tag{26}$$

Equations (14)–(19) can be determined using the Laplace transform of boundary conditions presented in Eqs. (20)–(27).

$$\overline{P_0}(s) = \frac{1}{\Delta(s)}, \tag{27}$$

$$\overline{P}_i(s) = \frac{2\lambda_k}{\Delta(s)}\left\{\frac{1 - \overline{S}_h(s + 2\lambda_1 + 2\lambda_2)}{s + 2\lambda_1 + 2\lambda_2}\right\}, k = (1, 2)\ and\ i = (1, 2), \tag{28}$$

$$\overline{P}_i(s) = \frac{4\lambda_1\lambda_2}{\Delta(s)}\left\{\frac{1 - \overline{S}_h(s + 2\lambda_k)}{s + 2\lambda_k}\right\}, k = (3, 4)\ and\ i = (2, 1), \tag{29}$$

$$\overline{P}_5(s) = \frac{4\lambda_1^2}{\Delta(s)}\left\{\frac{1 - \overline{S}_h(s + 2\lambda_2)}{s + 2\lambda_2}\right\}, \tag{30}$$

$$\overline{P}_6(s) = \frac{8\lambda_1^2\lambda_2}{\Delta(s)}\left\{\frac{1-\overline{S}_h(s+2\lambda_2)}{s+2\lambda_2}\right\}, \tag{31}$$

$$\overline{P}_7(s) = \frac{4\lambda_2^2}{\Delta(s)}\left\{\frac{1-\overline{S}_h(s+2\lambda_1)}{s+2\lambda_1}\right\}, \tag{32}$$

$$\overline{P}_8(s) = \frac{8\lambda_1\lambda_2^2}{\Delta(s)}\left\{\frac{1-\overline{S}_h(s+2\lambda_1)}{s+2\lambda_1}\right\}, \tag{33}$$

$$\overline{P}_i(s) = \frac{16\lambda_1^2\lambda_2^2}{\Delta(s)}\left\{\frac{1-\overline{S}_{\alpha_0}(s)}{s}\right\}, i = (9, 10), \tag{34}$$

$$\overline{P}_i(s) = \frac{2\lambda_k}{\Delta(s)}\left\{\frac{1-\overline{S}_{\beta_0}(s)}{s}\right\}, i = (11, 12)\ and\ K = (3, 4). \tag{35}$$

where:

$$\begin{aligned}\Delta(s) =& (s+2\lambda_1+2\lambda_2+2\lambda_3+2\lambda_4) \\ &- \left\{2\lambda_1\overline{S}_h(s+2\lambda_1+2\lambda_2) + 2\lambda_2\overline{S}_h(s+2\lambda_1+2\lambda_2)\right. \\ &\left.+\ 16\lambda_1^2\lambda_2^2 + 32\lambda_1^2\lambda_2^2\overline{S}_{\alpha_0}(s) + 2\lambda_3\overline{S}_{\beta_0}(s) + 2\lambda_4\overline{S}_{\beta_0}(s)\right\}.\end{aligned}$$

Adding up all the Laplace transformations of the system's state change probabilities that the system is operating. That is,

$$\overline{P_{up}}(s) = \overline{P_0}(s) + \overline{P_1}(s) + \overline{P_2}(s) + \overline{P_3}(s) + \overline{P_4}(s) + \overline{P_5}(s) + \overline{P_6}(s) + \overline{P_7}(s) + \overline{P_8}(s)$$

$$\begin{aligned}&= \frac{1}{\Delta(s)}\left\{2\lambda_1\left(\frac{1-\overline{S}_h(s+2\lambda_1+2\lambda_2)}{s+2\lambda_1+2\lambda_2}\right) + 2\lambda_2\left(\frac{1-\overline{S}_h(s+2\lambda_1+2\lambda_2)}{s+2\lambda_1+2\lambda_2}\right)+\right. \\ &4\lambda_1\lambda_2\left(\frac{1-\overline{S}_h(s+2\lambda_2)}{s+2\lambda_2}\right) + 4\lambda_1\lambda_2\left(\frac{1-\overline{S}_h(s+2\lambda_1)}{s+2\lambda_1}\right) \\ &+4\lambda_1^2\left(\frac{1-\overline{S}_h(s+2\lambda_2)}{s+2\lambda_2}\right) + 8\lambda_1^2\lambda_2\left(\frac{1-\overline{S}_h(s+2\lambda_2)}{s+2\lambda_2}\right)+ \\ &\left.4\lambda_2^2\left(\frac{1-\overline{S}_h(s+2\lambda_1)}{s+2\lambda_1}\right) + 8\lambda_1\lambda_2^2\left(\frac{1-\overline{S}_h(s+2\lambda_1)}{s+2\lambda_1}\right)\right\}\end{aligned} \tag{36}$$

and

$$\overline{P_{down}}(s) = 1 - \overline{P_{up}}(s). \tag{37}$$

4 Investigation of the System in Various Scenarios

a. System's availability analysis

The system performance is termed as availability when regular repairs are supplied to the system's failed units. To distinguish between the two types fixes (repairs) i.e., Copula and General repairs, the system availability was investigated in two ways.

i. When repair follows Copula Distribution

Here, we set $S_{\mu_0}(s) = \overline{S}_{\exp[x^\theta + \{\log\varphi(x)\}^\theta]^{1/\theta}}(s) = \frac{\exp[x^\theta + \{\log\varphi(x)\}^\theta]^{1/\theta}}{s + \exp[x^\theta + \{\log\varphi(x)\}^\theta]^{1/\theta}}$, $\overline{S}_h(s) = \frac{h}{s+h}$, the failure rates are set as λ_j at 0.1, 0.2,0.3 and 0.4,$\varphi = h = x = y = m = n = 1$ and all the repair rates are set equal to 1 i.e. $h(x) = h(y) = \alpha_0(x) = \alpha_0(y) = \beta_0(m) = \beta_0(n) = 1$ in Eq. (37). Applying the inversion of Laplace transform, we can derive the availability equation as follows:

$$P_{up}(s) = \{0.9457494251e^{-0.1181247621t} - 0.1242677829e^{-1.759389402t} + 0.2818807657e^{-4.440785836t} - 0.02409114046e^{-1.400000000t} -0.07927126743e^{-1.200000000t}\}. \tag{38}$$

$$P_{down}(s) = 1 - \{0.9457494251e^{-0.1181247621t} - 0.1242677829e^{-1.759389402t} + 0.2818807657e^{-4.440785836t} - 0.02409114046e^{-1.400000000t} -0.07927126743e^{-1.200000000t}\}. \tag{39}$$

Using varying numbers for time i.e., $t = 0, 1, 2, 3, 4, 5, 6, 7, 8, 9,$ *and* 10. Table 1 and Fig. 3 show the system's availability when Copula distribution is used.

ii. When repair follows general distribution

Setting $\overline{S}_h(s) = \frac{h}{s+h}$ in Eq. (37) and differentiating between parameters by assigning distinct values for λ_j at 0.1, 0.2,0.3 and 0.4, and $\varphi = 1$, $h = 1$, and applying inversion of Laplace transform, one may obtain availability expression as:

Table 1 Availability when Copula distribution is used

Time	0	1	2	3	4	5	6	7	8	9	10
$P_{up}(t)$	1.0000	0.7924	0.7344	0.6603	0.5887	0.5236	0.4654	0.4136	0.3675	0.3266	0.2902
$P_{down}(t)$	0.0000	0.2075	0.2655	0.3396	0.4112	0.4763	0.5345	0.5863	0.6324	0.6733	0.7097

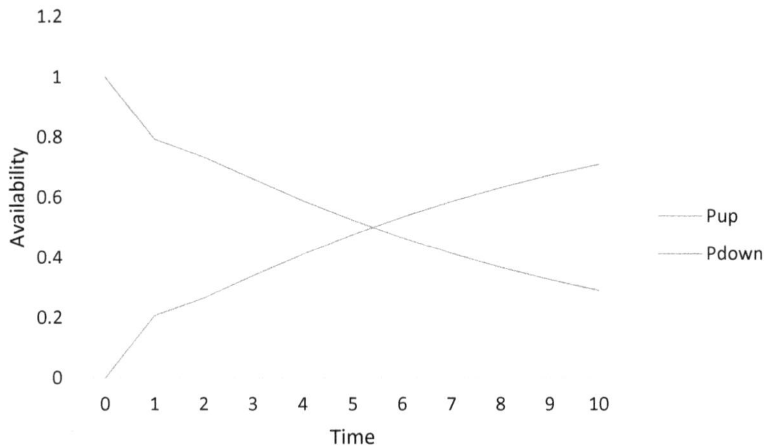

Fig. 3 Availability against time t when Copula distribution is followed by repair

$$P_{up}(s) = \{0.239829654e^{-3.082898540t} + 0.01734502810e^{-1.440657089t} + 0.5895248394e^{-0.07644437064t} + 0.04022627278e^{-1.200000000t} + 0.1130742049e^{-1.400000000t}\}. \quad (40)$$

$$P_{down}(s) = 1 - P_{up}(s). \quad (41)$$

Using time $t = 0, 1, 2, 3$, and so on, one can calculate the system's availability when the general distribution is followed by repair as shown in Table 2 and Fig. 4.

b. System's reliability analysis.

The chance of successful system operation is termed reliability if the repair is not performed. Hence, where there is no repair and the failure rates are taken λ_j at 0.1, 0.2,0.3 and 0.4 in Eq. (37). Then using the inversion Laplace transform, the reliability model is given as:

$$R(t) = \{0.4285714286e^{-0.6000000000t} + 0.3492063492e^{-2.0t} + 0.1422222222e^{-0.2000000000t} + 0.08000000000e^{-0.4000000000t}\}. \quad (42)$$

With the aid of (43) and $t \in [0, 10]$, Table 3 and Fig. 5 are obtained below.

Table 2 Availability when the general distribution is followed by repair

Time	0	1	2	3	4	5	6	7	8	9	10
$P_{up}(t)$	1.0000	0.6012	0.5179	0.4717	0.4350	0.4024	0.3727	0.3452	0.3198	0.2962	0.2744
$P_{down}(t)$	0.0000	0.3987	0.4820	0.5282	0.5649	0.5975	0.6272	0.6547	0.6801	0.7037	0.7255

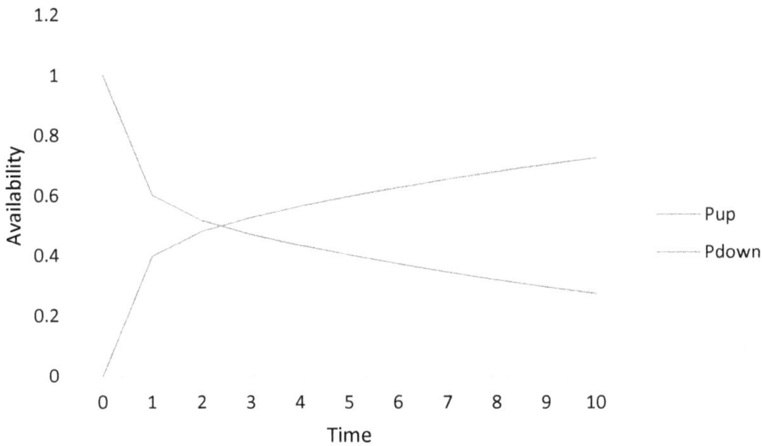

Fig. 4 Availability versus time t when general distribution is followed by repair

iii. Formulation amd analysis of MTTF

Suppose the service facility is unavailable, in which case all repairs in Eq. (37) are zero. Using limit, when s goes to 0, the *MTTF* can be calculated as:

Table 3 Reliability against time t

Time	0	1	2	3	4	5	6	7	8	9	10
Reliability	1.0000	0.4525	0.2667	0.1738	0.1190	0.0845	0.0618	0.0463	0.0355	0.0276	0.0217

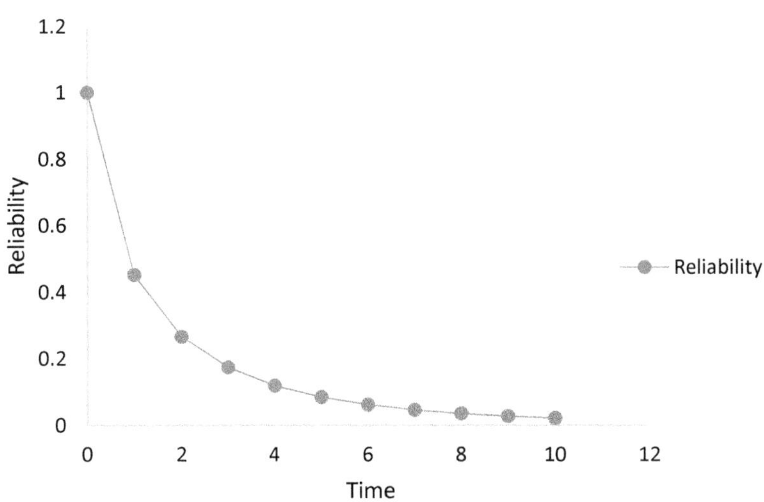

Fig. 5 Reliability versus time t

$$\Delta(s) = \lim_{s\to 0}\Delta(s) = 2\lambda_1 + 2\lambda_2 + 2\lambda_3 + 2\lambda_3.$$

$$MTTF = \lim_{s\to 0}\overline{P_{up}}(s) = \frac{1}{\Delta(s)}$$
$$\left\{1 + \frac{2\lambda_1}{2\lambda_1 + 2\lambda_2} + \frac{2\lambda_2}{2\lambda_1 + 2\lambda_2} + 2\lambda_1 + 2\lambda_2 + \frac{2\lambda_1^2}{\lambda_2} + 4\lambda_1^2 + \frac{2\lambda_2^2}{\lambda_1} + 4\lambda_2^2\right\}. \tag{43}$$

Table 4 and Fig. 6 show the fluctuation of *MTTF* with respect to λ_j for fixed values of some λ_j at 0.1, 0.2,0.3 and 0.4 are used and λ_j is varied between $0.01, 0.02, 0.03, 0.04, 0.05, \ldots, 0.09$. in Eq. (44).

Table 4 MTTF values for different values of λ_j

Failure rate	$MTTF$	$MTTF$	$MTTF$	$MTTF$
	λ_1	λ_2	λ_3	λ_4
0.01	5.8139	2.6311	2.6056	3.0327
0.02	3.5900	2.0058	2.5694	2.9838
0.03	2.8490	1.8001	2.5342	2.9365
0.04	2.4800	1.7014	2.5000	2.8906
0.05	2.2605	1.6470	2.4666	2.8461
0.06	2.1165	1.6161	2.4342	2.8030
0.07	2.0162	1.5996	2.4025	2.7611
0.08	1.9436	1.5929	2.3717	2.7205
0.09	1.8900	1.5916	2.3417	2.6811

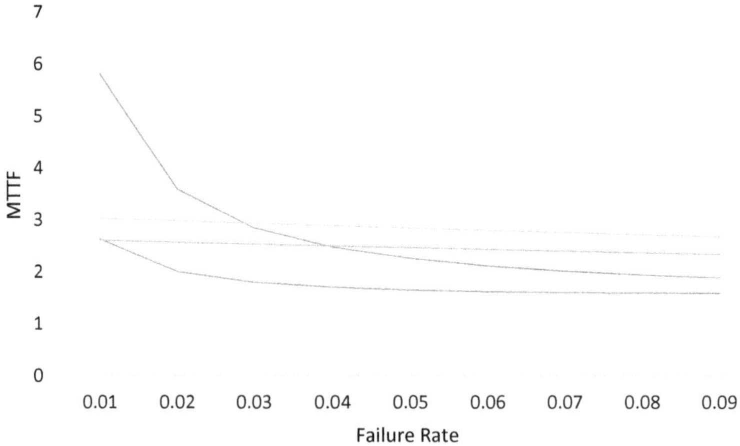

Fig. 6 MTTF against λ_j

iv. Sensitivity analysis

The partial derivative of *MTTF* with respect to failure rates in Eq. (43) can be used to calculate sensitivity. Table 5 and accompanying Fig. 7 show the sensitivity of *MTTF* for fixed values of some λ_j at 0.1, 0.2,0.3 and 0.4 and λ_j is varied between 0.01, 0.02, 0.03, 0.04, 0.05, …, 0.09.

e. Benefit function

The formula below will calculate the profit expected for the interval [0, t)

$$E_p(t) = C_1 \int_0^t P_{up}(t)dt - C_2 t. \tag{44}$$

Table 5 The MTTF's sensitivity to changes in failure rate

Failure rate	$\partial\left(\frac{MTTF}{\lambda_1}\right)$	$\partial\left(\frac{MTTF}{\lambda_2}\right)$	$\partial\left(\frac{MTTF}{\lambda_3}\right)$	$\partial\left(\frac{MTTF}{\lambda_4}\right)$
0.01	−444.6966	−125.1742	−3.6699	−4.9717
0.02	−111.2065	−31.1290	−3.5686	−4.8126
0.03	−49.3263	−13.4834	−3.4715	−4.6611
0.04	−27.5744	−7.1326	−3.3783	−4.5166
0.05	−17.4321	−4.0553	−3.2888	−4.3786
0.06	−11.8621	−2.2719	−3.2029	−4.2470
0.07	−8.4530	−1.1039	−3.1202	−4.1211
0.08	−6.1976	−0.2675	−3.0407	−4.0008
0.09	−4.6144	−0.1675	−2.9642	−3.8857

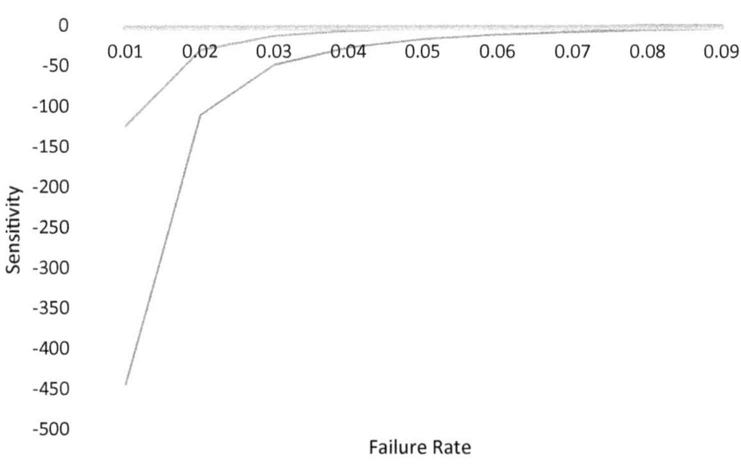

Fig. 7 The MTTF's sensitivity to changes in failure rate

where C_1 revenues to be obtaine and C_2 cost of service in the interval [0, t).

Case 1: Profit expected when the copula distribution is followed by repair.

Assuming that the system's failure rates are as follows: $\lambda_1 = 0.1, \lambda_2 = 0.2, \lambda_3 = 0.3, \lambda_4 = 0.4, \overline{S}_h(s) = \frac{h}{s+h}$ and $\varphi = 1, h = 1$, combining Eqs. (37) and (45), one can obtain Eq. (46) as:

$$E_p(t) = C_1\{0.07063119896e^{-1.759389402t} - 0.06347542442e^{-4.440785836t} - 8.006360464e^{-0.11812476 21t} + 0.01720795747e^{-1.400000000t} + 0.06605938952e^{-1.200000000t} + 7.915937342\} - C_2 t. \quad (45)$$

Table 6 and Fig. 8 can be obtained by using different values of the time variable, such as $t = 0, 1, 2, 3, 4, 5, 6, 7, 8, 9, 10$, and by applying the inverse Laplace

Table 6 Profit expected as a function of time for Copula distribution followed by repair

t	$E_P(t)E_P(t)$					
	C_2					
	0.06	0.05	0.04	0.03	0,02	0.01
0	0	0	0	0	0	0
1	0.7771	0.7871	0.7971	0.8071	0.8171	0.8271
2	1.4833	1.5033	1.5233	1.5433	1.5633	1.5833
3	2.1209	2.1509	2.1809	2.2109	2.2409	2.2709
4	2.6851	2.7251	2.7651	2.8051	2.8451	2.8851
5	3.1807	3.2307	3.2807	3.3307	3.3807	3.4307
6	3.6147	3.6747	3.7347	3.7947	3.8547	3.9147
7	3.9938	4.0638	4.1338	4.2038	4.2738	4.3438
8	4.3240	4.4040	4.4840	4.5640	4.6440	4.7240
9	4.6107	4.7007	4.7907	4.8807	4.9707	5.0607
10	4.8588	4.9588	5.0588	5.1588	5.2588	5.3588

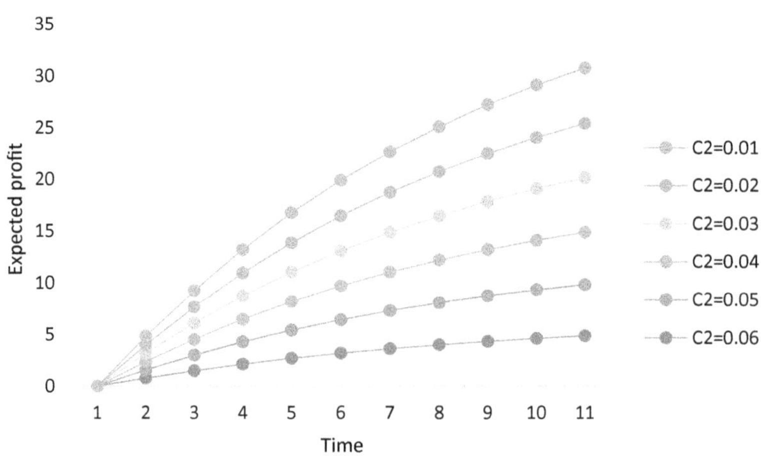

Fig. 8 Profit expected as a function of time for Copula distribution followed by repair

transform to Eq. (46) with $C_1 = 1$ and $C_1 = 0.06, 0.05, 0.04, 0.03, 0.02, 0.01$, respectively.

Case 2: Prpfit expected profit when the general distribution is followed by repair

$$E_p(t) = C_1\{-0.08076728921e^{-1.400000000t} - 0.03352189398e^{-1.200000000t}$$
$$-0.07779356076e^{-3.082898540t} - 0.01203966456e^{-1.440657089t} \quad (46)$$
$$-7.711814938e^{-0.07644437064t} + 7.91593747\} - C_2t.$$

Table 7 Profit expected as a function of time for general distribution followed by repair

t	$E_P(t)E_P(t)$					
	C_2					
	0.06	0.05	0.04	0.03	0,02	0.01
0	0	0	0	0	0	0
1	0.6752	0.6852	0.6952	0.7052	0.7152	0.7252
2	1.1686	1.1886	1.2086	1.2286	1.2486	1.2686
3	1.6022	1.6322	1.6622	1.6922	1.7222	1.7522
4	1.9951	2.0351	2.0751	2.1151	2.1551	2.1951
5	2.3663	2.4036	2.4536	2.5036	2.5536	2.6036
6	2.6810	2.7410	2.8010	2.8610	2.9210	2.9810
7	2.9798	3.0498	3.1198	3.1898	3.2598	3.3298
8	3.2522	3.3322	3.4122	3.4922	3.5722	3.6522
9	3.5001	3.5901	3.6801	3.7701	3.8601	3.9501
10	3.7253	3.8253	3.9253	4.0253	4.1253	4.2253

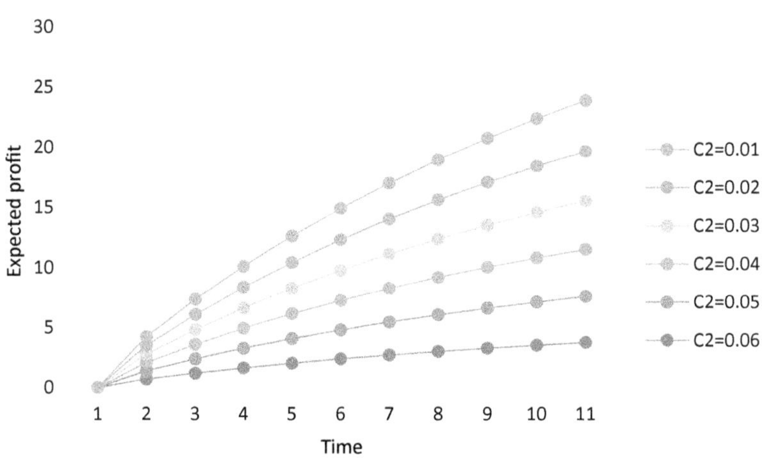

Fig. 9 Profit expected as a function of time for general distribution followed by repair

Table 7 and Fig. 9 can be obtained by varying the time variable's value, such as $t \in [0, 10]$ and by applying the inverse Laplace transform to Eq. (47) with $C_1 = 1$ and $C_1 = 0.06, 0.05, 0.04, 0.03, 0.02, 0.01$, respectively.

5 Result Analysis

This section examines the numerical results presented in the Tables and Figures in order to validate the extracted models and provide quick insight into the system's optimal design.

Table 1 and Fig. 3 depicts the system's availability and chance of failure as function of time for Copula distribution in which failure rates are set to varying levels. When failure rates are reduced, system availability falls until t = 1, at which point the variation slows and the chance of failure increases, eventually stabilizing after a sufficiently long period of time. As a result, the model's graphical analysis demonstrates that the behavior of a complex system can be easily forecasted at any time for any set of parametric values.

Table 2 and Fig. 4 depict the availability and failure probability of the general distribution system over time in which failure rates are varied. Where repair obeys a general distribution, the availability values are lower than when repair obeys a Copula distribution. Figures 3 and 4 show this. According to the findings of this study, Copula repair improves system availability more than General repair.

Table 3 and Fig. 5 provide information on system's reliability in the event that it is not repaired. Table 3 and Fig. 5 indicate the decrement in reliability as time passes for various failure rates. When the value of availability is compared against the value of reliability, it is clear that reliability is declining drastically. Hence, there is a need to keep the maintenance requirements for a good operation to a minimum.

Table 4 and Fig. 6 present the system's MTTF as a function of $\lambda_j,\ j = 1, 2, 3, 4$ when all other parameters remain constant. The MTTF of the system reduces when $\lambda_j,\ j = 1, 2, 3, 4$ change, as shown in Table and Figure. The MTTFs for λ_3 and λ_4 are nearly identical, indicating that both subsystems operate in a comparable manner.

The results of the sensitivity analysis investigated in this study are shown in Table 5 and Fig. 7.

Where C_1 the revenue generated equal to 1 and C_2 the cost of service is assigned values 0.06, 0.05, 0.04, 0.03, 0.02, 0.01, respectively, the expected profit obtained from the system using when the repair is following Copula distribution and General distribution is displayed in Table 6 and its corresponding Table 7 and Fig. 8 and its related Fig. 9, respectively. A close examination of Table 6 and Fig. 8 revealed that as service cost C_2 decreases, predicted profit increases over time. In general, when comparing low service costs ($C_2 = 0.01$) to high service costs ($C_2 = 0.06$), the predicted profit is higher i.e., low service cost gives the maximum profit and high service gives the minimum profit. Same scenario can be observed in Table 7 and Fig. 9 when the repair conforms to general distribution The estimated profit for Copula repair, on the other hand, in Table 6 and Fig. 8 is significantly higher than the expected profit for General repair in Table 7 and Fig. 9. According to this sensitivity study, Copula repair generates greater profit than General repair. This has gone a long way towards demonstrating why Copula repair is preferable to General repair.

6 Conclusion

Failures of systems in various industrial systems can result in a variety of issues, including unsatisfactory usage and a loss in profitability. To prevent these situations, we need to have enough knowledge about system failures as well as certain maintenance strategies. For illustration, we used the Automated Teller Machine, where the repair is performed utilizing Copula and General distributions.

To describe the detail account of the model under consideration, we formulate expressions for system availability, reliability, mean time to failure (MTTF), and cost function related to performance measures. The study present the numerical findings in Tables and Figures to validate the models generated and provide immediate insight for the optimal system design.

Based on numerical data derived in Tables 1, 2, 3, 4, 5, 6, 7 and Figs. 3, 4, 5, 6, 7, 8, 9 for a specific scenario, we note that using Copula repair improves system availability, reliability, and profit function over General repair. We also discovered that the system profit is lowest when the service cost is high and highest when the service cost is low. As a result, system engineers and maintenance managers will benefit more from repairing repairable systems with Copula distribution.

References

1. Abubakar MI, Singh VV (2019) Performance assessment of African textile manufacturers, LTD, in Kano state, Nigeria, through Multi failure and repair using copula. Oper Res Decis 29(4):1–18
2. Berk M, Schubert O, Kroll H-M, Buschardt B, Straub D (2019) Reliability assessment of safety-critical sensor information. IEEE Trans Reliab 68(4):1227–1241
3. Choudhary D, Tripathi M, Shankar R (2019) Reliability, availability and maintainability analysis of a cement plant: a case study. Int J Qual Reliab Manag. https://doi.org/10.1108/IJQRM-10-2017-0215
4. Chopra G, Ram M (2019) Reliability measures of two dissimilar units parallel system using Gumbel–Hougaard family copula. Int J Math, Eng Manag Sci. https://doi.org/10.33889/IJMEMS.2019.4.1-011
5. Cheong MLF, Koo PS, Babu BC (2015) Ad-Hoc automated teller machine failure forecast and field service optimization. automation science and engineering (CASE). In: International Conference on IEEE, pp 1427–1433
6. Dahiya O, Kumar A, Saini M (2019) Mathematical modeling and performance evaluation of A-Pan crystallization system in a sugar industry. SN Appl Sci. https://doi.org/10.1007/s42452-019-0348-0

7. Gahlot M, Singh VV, Ismail Ayagi H, Goel CK (2018) Performance assessment of repairable system in the series configuration under different types of failure and repair policies using Copula Linguistics. Int J Reliab Saf 12(4): 367–374
8. Gulati J, Singh VV, Rawal DK, Goel CK (2016) Performance analysis of complex system in series configuration under different failure and repair discipline using copula. Int J Reliab Qual Saf Eng 23(2):812–832
9. Gupta SK, Choudhari R, Sinha BK (2015) Mathematical modeling of behavior of automatic teller machine with respect to reliability analysis. Int J Math Trends Technol 26(1):29–34
10. Garg H (2017) Performance analysis of an industrial system using soft computing based hybridized technique. J Braz Soc Mech Sci Eng, Springer 39(4):1441–1451
11. Garg H (2016) A novel approach for analyzing the reliability of series-parallel system using credibility theory and different types of intuitionistic fuzzy numbers. J Braz Soc Mech Sci Eng, Springer 38(3):1021–1035. https://doi.org/10.1007/s40430-014-0284-2
12. Garg H (2015) An approach for analyzing the reliability of industrial system using fuzzy Kolmogorov's differential equations. Arab J Sci Eng, Springer 40(3):975–987
13. Ibrahim KH, Singh VV, Lado A (2017) Reliability assessment of complex system having two subsystems arranged in series configuration via Gumbel–Hougaard family copula distribution. J Appl Math Bioinform 7(2):1–27
14. Jain M, Sharma R, Meena RK (2019) Performance modeling of fault-tolerant machining system with working vacation and working breakdown. Arab J Sci Eng 44:2825–2836
15. Kadyan MS, Kumar R (2015) Availability and profit analysis of a feeding system in the sugar industry. Int J Syst Assur Eng Manag 8:301–316
16. Kumar N, Pant S, Singh SB (2017) Availability and cost analysis of an engineering system involving subsystems in series configuration. Int Qual Reliab Manag 34(6):879–894
17. Kumar N, Lather JS (2018) Reliability analysis of a robotic system using hybridized technique. J Ind Eng Int 14:443. https://doi.org/10.1007/s40092-017-0235-5
18. Kumar A (2013) Reliability and cost-benefit analysis of computer systems subject to maximum operation and repair time. PhD thesis, Maharishi Dayanand University, Rohtak, India
19. Lado A, Singh VV (2019) Cost assessment of complex repairable systems in series configuration using Gumbel Hougaard family copula. Int J Qual Reliab Manag 36(10):1683–1698
20. Meena R (2015) Automated teller machine-its benefits and challenges. Int J Commer Bus Manag 4(6):815–821
21. Mehta M, Singh J, Sharma S (2018) Availability analysis of an industrial system using supplementary variable technique. Jordan J Mech Ind Eng 12(4):245–251. ISSN 1995-6665
22. Malik S, Tewari PC (2018) Performance modeling and maintenance priorities decision for water flow system of a coal-based thermal power plant. Int J Qual Reliab Manag 35(4):996–1010
23. Niwas R, Garg H (2018a) An approach for analyzing the reliability and profit of an industrial system based on the cost-free warranty policy. J Braz Soc Mech Sci Eng 40(5)
24. Niwas R, Kadyan MS (2018b) Stochastic analysis of a single-unit system with repairman having multiple vacations. Int J Comput Appl 1(8)
25. Onyessolu MO, Ezeani IM (2012) ATM security using fingerprint biometric identifier: an investitive study. Int J Adv Comput Sci Appl 3(4):68–72
26. Potapov VI, Shafeeva OP, Gritsay AS, Makarov VV, Kuznetsova OP, Kondratukova LK (2019) Reliability in the model of an information system with client-server architecture. Int J Phys: Conf Ser 1260(2): 022007. IOP Publishing
27. Pandey P, Mukhopadhyay AK, Chattopadhyaya S (2018) Reliability analysis and failure rate evaluation for critical subsystems of the dragline. J Braz Soc Mech Sci Eng 40:50. https://doi.org/10.1007/s40430-018-1016-9
28. Pourhassan MR, Raissi S, Hafezalkotob A (2020) A simulation approach on reliability assessment of complex system subject to stochastic degradation and random shock. Eksploatacja i Niezawodnosc—Maintenance Reliab 22(2):370–379. https://doi.org/10.17531/ein.2020.2.20
29. Pourhassan MR, Raissi S, Apornak A (2021) Modeling multi-state system reliability analysis in a power station under fatal and nonfatal shocks: a simulation approach. Int J Qual Reliab Manag. https://doi.org/10.1108/IJQRM-07-2020-0244

30. Ram M, Goyal N (2018) Bi-directional system analysis under copula-coverage approach. Commun Stat-Simul Comput 47(6):1831–1844
31. Ram M, Goyal N (2016) Automated teller machine network inspection under stochastic modelling. J Eng Sci Rev 9(5):1–8
32. Raissi S, Ebadi S (2018) A computer simulation model for reliability estimation of a complex system. Int J Res Ind Eng 7(1):19–31
33. Sanusi A, Yusuf I, Mamuda BY (2020) Performance evaluation of an industrial configured as series-parallel system. J Math Comput Sci 10:692–712
34. Singh VV, Ayagi HI (2018) Stochastic analysis of a complex system under preemptive resume repair policy using Gumbel-Hougaard family copula. Int J Math Oper Res 12(2):273–291. https://doi.org/10.1504/IJMOR.2018.089681
35. Singh VV, Hamishu HI (2018) Stochastic analysis of a complex system under preemptive resume repair policy using Gumbel–Hougaard family copula. Int J Math Oper Res 12(2):273–291
36. Singh VV, Lado Ismail AK, Yusuf I, Abdullahi AH (2021) Probabilistic assessment of computer-based test (CBT) network system consists of four subsystems in series configuration using copula linguistic approach. J Reliab Stat Stud
37. Singh VV, Poonia PK, Rawal DK (2021) Reliability analysis of repairable network system of three computer labs connected with a server under 2- out- of- 3 G configuration. Life Cycle Reliab Saf Eng 10:19–29. https://doi.org/10.1007/s41872-020-00129-w
38. Saini M, Kumar A (2019) Performance analysis of evaporation system in sugar industry using RAMD analysis. J Braz Soc Mech Sci Eng 41:4
39. Tyagi V, Arora R, Ram M, Triantafyllou IS (2021) Copula based measures of repairable parallel system with fault coverage. Int J Math Eng Manag Sci. https://doi.org/10.33889/IJMEMS.2021.6.1.021
40. Yusuf I, Ismail AL, Singh VV, Ali UA, Sufi NA (2020) Performance analysis of multi-computer system consisting of three subsystems in series configuration using copula repair policy. SN Comput Sci 1(5):1–11
41. Yang L, Guo Y, Wang Q (2019) Reliability assessment of a hierarchical system subjected to inconsistent priors and multilevel data. IEEE Trans Reliab 68(4):277–292
42. Zhao B, Xie L, Li H, Zhang S, Wang B, Li C (2020) Reliability analysis of aero-engine compressor rotor system considering cruise characteristics. IEEE Trans Reliab 68(4):245–259

An Efficient Regression Test Cases Selection & Optimization Using Mayfly Optimization Algorithm

Abhishek Singh Verma, Ankur Choudhary, Shailesh Tiwari, and Bhuvan Unhelkar

Abstract Testing has been an inevitable activity in the software development life cycle. In the current scenario, software development has become evolutionary in nature where software is released in cycles, each cycle fulfilling the requirements of the customer on a priority basis. This evolutionary development of software also demands high maintenance in the form of retesting. This re-testing is called regression testing and the literature reveals that it is a proven N-P hard problem that attracts the application of approximation algorithms such as meta-heuristics. In this paper, Mayfly Optimization Algorithm has been adopted to solve the regression test case selection problem to minimize the maintenance cost. The aim is to optimize the number of test cases to re-execute to reduce the execution time and cost. The performance of the adopted approach is further compared with state-of-the-art approaches with the help of statistical tests. The shows that the adopted approach performs well in comparison to state of art approaches.

Keywords Software testing · Mayfly optimization algorithm · SIR · MetaHeuristics · Regression test case selection

---

A. Singh Verma
Dr. APJ Abdul Kalam Technical University, Lucknow, India

A. Singh Verma · A. Choudhary (✉)
School of Engineering & Technology, Sharda University, Greater Noida, India
e-mail: ankur.tomer@gmail.com

S. Tiwari
ABES Engineering College, Ghaziabad, India

B. Unhelkar
Universitu of South Florida, Tampa, USA

V. Kumar and H. Pham (eds.), *Predictive Analytics in System Reliability*,
Springer Series in Reliability Engineering,
https://doi.org/10.1007/978-3-031-05347-4_8

1 Introduction

Regression testing is a significant activitiy of software maintenance phase under SDLC. Today, the growth and success of every software industry is based on full-filing customer needs and delivering good quality products within the stipulated time frame and budget [1]. So, it is very difficult for software industries to fullfill the frequently changing customer requirements and technology upgradation. When an organization wants to modify the existing software during the maintenance phase, the software has to be retest. The process of retesting the modified part of the software is known as Regression testing [2], which helps industries to find out errors in the modified part of the software and ensure its reliability after regression testing. Being a repetitive process the number of test cases will increase due to modification and subsequently the size of the test suite will also increase after every testing cycle. It is very difficult to maintain a large test suite. Retesting of complete test suite consumes more execution time and effort. So, it is necessary to find out obsolete as well as redundant test cases from the existing test suite and remove them to reduce the test suite size. The process of selection of appropriate test cases from the existing test suite as per applicability during regression testing is called as Regression Test Case Selection (RTCS) technique [3, 4]. As evident in the literature RTCS is an N-P complete problem. To solve the NP-complete problem greedy approach, dynamic programming and metaheuristics algorithms are the ways to find the optimal solution. But they are not providing the exact solution. So, to find the solution to RTCS problems various nature-inspired algorithms have been utilized such as Genetic Algorithm [5, 6], ACO [7], PSO [8], Bat Search Algorithm [9], Cuckoo Search Optimization [10], Firefly Optimization [11], Butterfly Optimization [12], Crow Search Algorithm [13] and many more.

This paper utilizes the latest and existing population-based metaheuristic algorithm which combines the properties of swarm intelligence as well as evolutionary algorithms named as Mayfly Optimization Algorithm (MA) [14] to provide the solution of RTCS problem. The mayfly algorithm is inspired by the mating process as well as the flight behavior of mayflies.

The main objective of using the Mayfly Optimization Algorithm (MA) to solve the RTCS problem is to find the maximum fault covered, the execution time of the algorithm as well as unique fault covered [15]. Further, the results have been compared with Bat Search Algorithm (BA) and Improved Grey Wolf Optimization (IGWO) in terms of maximum no. of fault coverage for different experimental objects. The performance of various approaches is evaluated over five benchmarked objects from SIR [16].

The flow of the remaining paper is as follows: The existing related work carried out by previous authors is discussed in Sect. 2. A brief description of the RTCS problem has been discussed in Sect. 3. In Sect. 4, the author briefly discussed the proposed MA approach in detail. Section 5 explained the experimental setup and discusses the results received. Finally, Sect. 6 Concludes the whole paper with discussion.

2 Related Work

Nature always provides sources of learning and inspiration to all of us. According to the literature, this learning has been applied to various domains to solve complex engineering problems and conclude some results. Various researchers performed various studies to solve RTCS problems using a variety of nature-inspired metaheuristic algorithms.

Yoo et al. [17] performed an empirical study and proposed an approach to solve RTCS problem using the concept of Pareto efficient multi-objective optimization and concluded that for multi-objective problems, greedy algorithms are not always Pareto efficient. Maia et al. [18] provided the solution of RTCS problems using a multiobjective algorithm with the help of NSGA-II algorithm and the results showed that the NSGA-II provides optimal solutions to such problems. Singh & Gupta [19] proposed a fusion technique for RTCS using GA and ACO, which identifies and reduces the size of test data. It provides better results in optimum time. Pravin et al. [20] have been proposed an algorithm for prioritizing the test cases which works for both requirement and testing. The algorithm has been testing on limited size data set and validate it by taking large size projects having a large number of test cases. In 2013, Wang et al. [21] proposed weight-based GA's to reduce the test suite size and also achieving maximum fault detection capability and pairwise coverage. Conducted a comparison between three weights-based GA's and evaluated the best performance. De-souza et al. [22] have been developed a hybrid algorithm for structural test case selection by adding a local search approach into binary multiobjective PSO. The algorithm considered both execution cost and branch coverage. Narciso et al. [23] presented a systematic literature review on various test case selection approaches and performed an empirical evaluation on 18 different approaches of 32 papers of test case selection.

In the year 2015, Shi et al. [24] performed a study that compares and combines the two approaches test suite reduction and test case selection, and evaluate the performance of approaches on 17 open source projects and conclude the results. Panichella et al. [25] proposed a novel multiobjective Genetic Algorithm named Diversity-based GA which combined the features of NSGA-II and formulated a diversity preserving method to solve multiobjective test case selection problems. Mondal et al. [26] have been analyzed coverage-based & diversity-based RTCS approaches to solve bi-objective optimization problems where both objectives maximize the coverage/diversity and reduce the test execution time. The study represents that the diversity-based approach is a little more effective than the coverage-based approach. Nagar et al. [27] proposed an algorithm for RTCS using cuckoo search via levy flight algorithm and observed that cuckoo search algorithm reduced approx. 40% test suite size of the problem. Rosero et al. [28] performed a literature survey on various regression testing techniques used in the last 15 years, identified 31 regression testing techniques, and discussed the issues such as identification of new algorithms, use of AI-based techniques, design of new optimization algorithms etc. Srisura et al. [29] proposed a technique to ensure the quality and validity by selecting suitable false test cases, generated during regression testing and conclude that the false TCS technique minimized the test suite size. Kazmi et al. [30] performed a SLR on effective RTCS techniques. This SLR examined 47 empirical studies and categorize the selected studies using various criteria.

Now in the year 2018, Garousi et al. [31] performed a literature review and also proposed a framework called MORTOGA (Multi-objective regression test selection using a genetic algorithm) and found that the proposed framework reduced the cost of regression testing. Bajaj et al. [32] performed a study on various nature-inspired algorithms applied in regression testing & found that Genetic algorithm is performed better than others as well as the use of nature-inspired algorithms provides cost-effective and more accurate results. Agrawal et al. [33] presented the performance comparison of two metaheuristic algorithms: ACO & Hybrid PSO to solve the RTCS problem and consider two performance parameters—fault coverage and execution time. Gupta et al. [34] presented a literature review on various optimization techniques mostly used in the domain of software testing and found that various intelligent and hybrid algorithms are used from 2007 through 2018. Pandey et al. [35] applied a novel approach of a genetic algorithm and hybrid firefly for test data generation and test case selection in regression testing and found that the hybrid approaches provide better results considering various parameters. Staron et al. [36] empirically evaluate the effect of three different feature extraction algorithms on the performance of an existing ML-based selective regression testing technique. Yadav et al. [37] presented a technique for selection and prioritization of regression test cases using UML diagrams & code-based analysis for object-oriented software. Guizzo et al. [38] have performed an empirical study to assess the use of static and dynamic regression test selection techniques with genetic improvement to improve seven real-world programs. MA et al. [39] proposed a method that selects test cases to improve the fault detection rate considering traverse target paths and achieve coverage balance. Chen et al. [40] presented a new approach that evaluates fault detectability of every regression test and proposed two optimization algorithms to optimize a multi-objective function.

The various approaches and techniques discussed above show that both heuristic and metaheuristic algorithms are used to solve regression test case selection problems, but being an NP-complete problem, still, the scope of other optimization algorithms also exists. In this paper, the authors applied the Mayfly optimization algorithm to enhance the performance of the RTCS problem by finding maximum no. of fault coverage in minimum execution time and compare the results with already utilized meta-heuristic algorithms Bat Algorithm (BA) and Improved Grey Wolf Optimization (IGWO).

3 Problem Statement of Regression Test Case Selection

Given: Suppose there exist a software S and a updated software version S'. The test suite (TS) is represented as $TS = (tc_1, tc_2, tc_3, \ldots tc_m)$. Let n $\leq$ m, here n is the number of total test case in the pool and m is the no. of selected test cases in the optimal suite. The objective of the problem is to find the TS' (updated test suite) to provide the maximum coverage in minimum execution time.

4 Proposed Mayfly Optimization Algorithm

Mayflies are the small size of aquatic insects also known as fishflies or up-winged flies. It is estimated that there are approximately 3500 species and 42 families of mayflies worldwide. Despite their name, mayflies are active from May to July.

Mayflies are small to medium-sized insects, belongs the family of Ephemeroptera having a genus called Atalophlebia. The mayfly optimization algorithm is a modified version of PSO and combines the best properties of PSO, GA, and FA. This algorithm was designed and developed based on the behavior of mayflies.

In this algorithm, individual mayflies have been identified as male and female mayflies and explain the different behaviors of mayflies such as movements of male mayflies, movements of female mayflies, and mating of mayflies. These different behaviors of mayflies are mathematically implemented in Sect. 4.1.

4.1 Mathematical Implementation of MA

The inspiration of this algorithm is the social behavior & mating behavior of mayflies. It is assumed that mayflies come as adults after crosshatching from the eggs. Only the fittest mayflies will live. In search space, the mayfly positions have represented a possible solution to the problem. The working procedure of this algorithm is as follows: At the initial stage, two pairs of mayflies are generating randomly in which one male set and the second one female set from the population. Out of these two pairs, each mayfly in search space is randomly placed as an optimal solution denoted by a d-dimensional vector $y = (y_1,\ldots,y_d)$, and the performance of this vector has been assessed with the help of existing objective function $f(y)$. In the search space $w = (w_1,\ldots,w_d)$, the mayfly velocity is defined as the change of mayfly's position. The flying direction of all mayflies is dynamic interaction that reflects the individual and social flying experiences. In particular, every mayfly changes its flight direction towards, either *pbest* position which represents its personal best position, or *gbest* position which represents the global best position of the mayfly.

4.1.1 Movement of Male Mayflies

During iterations, male mayflies have been adopted the exploration or exploitation approach. The position of male mayfly may adjust to follow their own experience and that of their neighbors. Suppose, the current position of ith mayfly in search space at time instance t is represented as y_i^t and the current position has changed by added a velocity factor w_i^{t+1} into the current position. Now, the position of mayfly is denoted as:

$$y_i^{t+1} = y_i^t + w_i^{t+1} \tag{1}$$

While performing the nuptial dance, the male mayflies are maintaining a gap of few meters from the water in the upward direction. It is assumed that they move at a constant speed as they cannot develop the best speed for movement. So, the velocity of ith male mayfly is formulated as:

$$w_{ij}^{t+1} = w_{ij}^t + \alpha_1 e^{-\beta r_p^2}\left(pbest_{ij} - y_{ij}^t\right) + \alpha_2 e^{-\beta r_g^2}(gbest_j - y_{ij}^t) \tag{2}$$

where velocity and position of mayfly i is represented by w_{ij}^t and y_{ij}^t respectively in dimension $j = 1, 2, 3, \ldots, n$ at time instance t, attraction constants are represented by α_1 & α_2. β is the fixed visibility coefficient. The individual best position & global best position of ith mayfly in search space is denoted by $pbest_i$ and $gbest_i$ as well as r_p and r_g represented the cartesian distance between y_i and $pbest_i$,$gbest$ respectively.

The cartesian distances are calculated with the help of equation no. (3):

$$||y_i - Y_i|| = \sqrt{\sum_{j=1}^{n} (y_{ij} - Y_{ij})^2} \tag{3}$$

In Eq. (3), y_{ij} represented the jth element of mayfly i and Y_i is the corresponds to $pbest_i$ and $gbest$.

The best mayflies may perform the nuptial dance in the upward and downward direction as per their best characteristics. The changing velocity of best mayfly is calculated as:

$$w_{ij}^{t+1} = w_{ij}^t + d_n * r_1 \tag{4}$$

where in Eq. (4), r_1 is a random number with values in between $[1, 1]$, and the nuptial dance coefficient is represented by d_n.

4.1.2 Movement of Female Mayflies

While comparing the behavior of male mayflies with females, it has been noticed that female mayflies do not assemble in swarms. The female mayflies have not updated their velocities when they changing movement style. The survival duration of female mayflies is at most one day to one week only. During this period, female mayfly fly towards the males for mate and reproduce themselves. Let us assume, the current position of ith female mayfly is denoted by z_i^t at time instance t, if the velocity w_i^{t+1} added then the current position mentioned as

$$z_i^{t+1} = z_i^t + w_i^{t+1} \tag{5}$$

The process of attraction would be in a random manner as per their fitness function which means that the female mayfly with the best properties should be attracted by the best male mayfly and the other best female attracted by the best male and so on. So, for the ith female mayfly in dimension $j = 1, 2, 3, \ldots, n$ at time instance t, the velocities are formulated as:

$$w_i^{t+1} = \begin{Bmatrix} w_{ij}^t + \alpha_2 e^{-\beta r_{mf}^2}\left(y_{ij}^t - z_{ij}^t\right) \ if \ \ f(z_i) > f(y_i) \\ w_{ij}^t + fl * r_1 \qquad\qquad if \ \ f(z_i) \le f(y_i) \end{Bmatrix} \tag{6}$$

where in Eq. (6), w_i^{t+1} is the female mayfly velocity, z_{ij}^t represented the current position of the ith female mayfly in dimension $j = 1, 2, 3, \ldots, n$ at time instance t, attraction constants are represented by α_2 and β is the fixed visibility coefficient. The cartesian distance is represented by r_{mf} which is calculated by using equation number (3). Whereas r_1 is a random number with values in between [1, 1] and fl is represented a random walk coefficient.

4.1.3 Mating Process of Mayflies

To represent the mating process between male and female mayflies, the crossover operator from the genetic algorithm is utilized in which for mating one male selected one female from the population. All the best half male would be mated with best half female mayflies and the other best male with other best female and so on and produce the pair of children for every one of them. The two offspring as a result of the crossover are generated as follows:

$$offspring1 = l * Male + (1 - l) * Female \tag{7}$$

$$offspring2 = l * Female + (1 - l) * Male \tag{8}$$

where 'l' is a random number with specified values and initial velocities of offspring are considered as zero.

4.1.4 Pseudo Code of Mayfly Optimization Algorithm

The pseudocode of the Mayfly Optimization Algorithm (MA) is explained below:

(1) *calculate objective function f(y) where* $y = (y_1, y_2 \ldots, y_d)^T$
(2) *initialize the population of male mayflies using* mm_i $(i = 1,2,3\ldots,N)$ *&* *velocities* v_{mi}
(3) *initialize the population of female mayflies using* fm_i $(i = 1,2,3\ldots,M)$ *&* *velocities* v_{fi}
(4) *evaluate the possible solutions*
(5) *find global best gbest*
(6) *{*
(7) *Do-while*
(8) *stopping criteria not met*
(9) *{*
(10) *update velocities and solutions of males and females' mayflies*
(11) *assess the solutions*
(12) *ranking of mayflies for mating*
(13) *mating process of mayflies*
(14) *assess the offspring*
(15) *randomly separate offspring to male & female mayflies*
(16) *change worst solutions with the best new solution*
(17) *update pbest & gbest*
(18) *}*
(19) *end of while loop*
(20) *postprocess and visualize the results.*
(21) *}*
(22) *End*

5 Experimental Setup

This section has been described the experimental work & analyzed the performance of the proposed MA approach against BA & IGWO based optimization algorithms for RTCS problem. The subsections have been discussed the research objectives designed, parameters setting of various optimization algorithms, Research Hypotheses, & characteristics of five subject programs utilized to evaluate the performance of proposed approach.

5.1 Research Objectives

The author have designed three research questions to analyze the performance of proposed MA-based approach:

RQ1. Is the performance of MA, BA & IGWO the same?

RQ2. Is there any significant impact on the performance of the various adopted algorithms in terms of execution time?

RQ3. Is there any advantage of selecting different test suite size to evaluate the performance of adopted algorithms?

The parameter setting utilized for MA, BA & IGWO have been shown in Table 1.

5.2 *Research Hypothesis*

To justify the answers to research questions designed in Sect. 5.1, three research hypothesis have been formed:

Ho: MA = BA = IGWO.

Ha: MA ≠ BA ≠ IGWO.

Ho: Execution_Time of MA = Execution_Time of BA = Execution_Time of IGWO.

Ha: Execution_Time of MA ≠ Execution_Time of BA ≠ Execution_Time of IGWO.

Ho: Performance of RTCS_5 = Performance of RTCS_10 = Performance of RTCS_15.

Ha: Performance of RTCS_5 ≠ Performance of RTCS_10 ≠ Performance of RTCS_15.

5.3 *Subject Programs*

The adopted algorithms are evaluated to analyze the performance on a benchmarked dataset consisting of 5 different versions of open-sourced programs written in 'C' and 'JAVA'. These versions are retrieved from benchmarked SIR [16] in regression testing under a controlled experimental setting (Table 2).

6 Result Discussion

The adopted approaches have been executed fifteen times & the faults coverage is considered as performance parameter already mentioned in Table 1. In every run, 500 iterations were performed.

(a) **Response to Research Question RQ 1**
To provide the answer of RQ1, the author have used the mean fault-coverage values collected from experiment executed. Table 3 reflects the fault-coverage capabilities of different approaches on different subjects' programs. The highlighted means in the Table 3 reflects that the MA algorithm performs superior than other adopted optimization algorithms. Figure 1 also confirms the same.

Table 1 Parameter setting values used for BA, IGWO & MA

Parameters values of IGWO algorithm		Parameters values of BA algorithm				Parameters values of mayfly algorithm					
No. of search agents	No of iterations	No. of bats	Frequency	Loudness	Pulse rate	No. of fixed population (N) (20 Males & 20 Females)	Attraction constants α1 & α2 respectively	Visibility coefficient (β)	Nuptial dance (d)	Random flight (fl)	Uniform crossover rate
30	500	30	75	0.75	0.25	40	1 & 1.5	2	0.1	0.1	0.05

Table 2 Subject program's characteristics

Name of objects	Flex_v1	Flex_v2	Flex_v3	Flex_v4	Flex_v5
Total no. of seeded faults	19	20	17	16	9
Total no. of test cases	567	567	567	567	567
Type of test suite	TSL	TSL	TSL	TSL	TSL

Table 3 Tukey HSD of mean fault coverage of adopted algorithms

Name of Algorithms	N	Subsets		
		1	2	3
BA	225	9.85		
IGWO	225		10.54	
MA	225			**10.90**
Sig		1.000	1.000	1.000

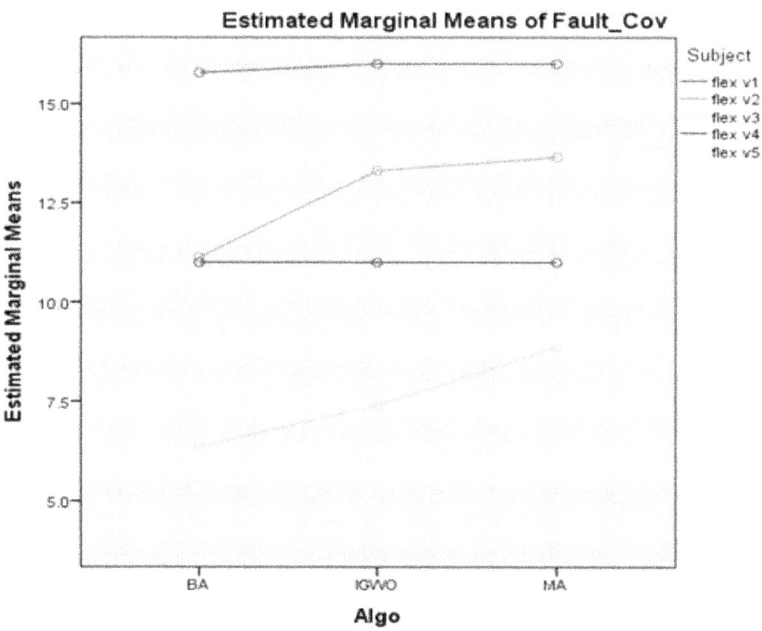

Fig. 1 Performance_Algo. versus Subject programs w.r.t. Fault_Coverage

To further validate the performance a two-way ANOVA test is conducted. The significance value obtained from two-way ANOVA test shown in Table 4 is less than 0.05, which displays that the null hypothesis has been rejected in favor of the alternate hypothesis. So, the fault coverage capabilities of MA algorithm are superior to BA & IGWO optimization algorithms.

Table 4 Results of 2-way ANOVA test conducted on fault coverage

Name of sources	Sum of squares	Difference (df)	Mean square	F value	Significance
Corrected model	10,402.350[a]	44	236.417	1499.423	0.000
Intercept	73,466.317	1	73,466.317	465,944.091	0.000
Algo	127.825	2	63.913	405.352	0.000
Subject	9922.350	4	2480.587	15,732.585	0.000
TS_Size	49.834	2	24.917	158.031	0.000
Algo * Subject	184.264	8	23.033	146.082	0.000
Algo * TS_Size	9.490	4	2.373	15.048	0.000
Subject * TS_Size	70.477	8	8.810	55.873	0.000
Algo * Subject * TS_Size	38.110	16	2.382	15.106	0.000
Error	99.333	630	0.158		
Total	83,968.000	675			
Corrected total	10,501.683	674			

[a]R squared value = 0.991
*Represents multiplication

(b) **Response to Research Question RQ 2**

To provide the answer of RQ2, the author have analyzed the mean of execution time obtained after the experiment conducted. The Table 5 presents the mean of execution time of different utilized algorithms. The highlighted mean values of execution time reflect that MA consumes lesser time as compare to BA and IGWO. The Fig. 2 also confirms the same. To further validate the results a two-way ANOVA test has been performed as shown in Table 6. The significance value obtained from two –way ANOVA test shown in Table 6 is less than 0.05, which shows that null hypothesis has been rejected in favor of alternate hypothesis. Now, it is clear from the evidences that the performance of MA in terms of execution time is superior than BA & IGWO algorithms.

(c) **Response to Research Question RQ 3**

Table 5 Homogenous subsets of mean execution time of algorithms

Name of algorithms	Mean of execution time	Standard error	Confidence interval = 95%	
			Lower bound	Upper bound
BA	1.383	0.023	1.338	1.428
IGWO	1.384	0.023	1.339	1.429
MA	**1.263**	0.023	1.218	1.308

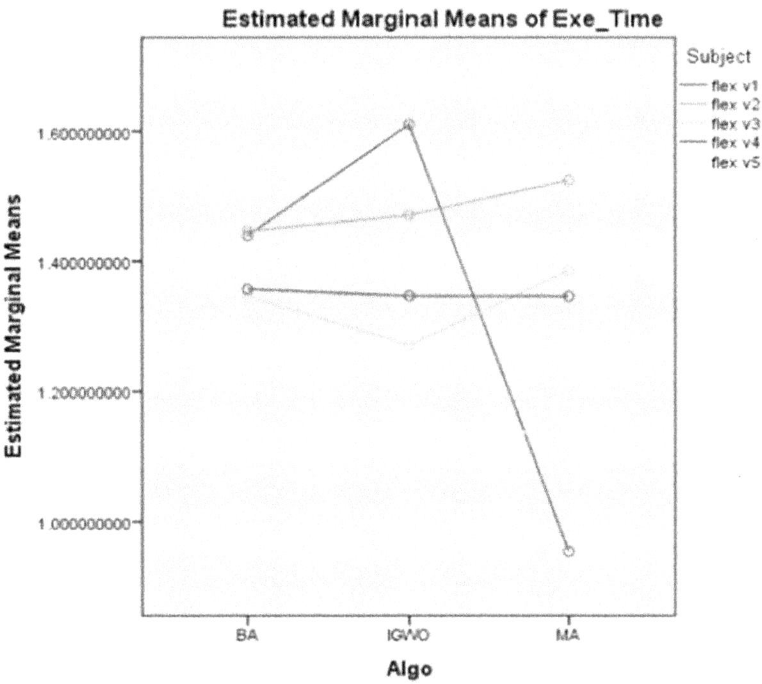

Fig. 2 Performance_Algo. versus Subject programs w.r.t. Execution Time

Table 6 Results of 2-way ANOVA test of variables and their combined effects on execution time

Name of sources	Sum of squares	Difference (df)	Mean square	F value	Significance
Corrected model	104.992[a]	44	2.386	20.261	0.000
Intercept	1217.719	1	1217.719	10,339.506	0.000
Algo	2.172	2	1.086	9.220	0.000
Subject	4.770	4	1.192	10.125	0.000
TS_Size	2.047	2	1.024	8.691	0.000
Algo * Subject	9.805	8	1.226	10.406	0.000
Algo * TS_Size	72.866	4	18.217	154.675	0.000
Subject * TS_Size	6.975	8	0.872	7.403	0.000
Algo * Subject * TS_Size	6.357	16	0.397	3.373	0.000
Error	74.197	630	0.118		
Total	1396.908	675			
Corrected total	179.189	674			

[a]R squared value = 0.586
*Represents multiplication

Table 7 Tukey HSD of mean fault coverage of test-suite size

Test suite size	N	Subsets		
		1	2	3
5	225	10.06		
10	225		10.55	
15	225			**10.69**
Sig.		1.000	1.000	1.000

To answer the research question RQ3, we have selected three different sizes of test cases 5, 10, and 15 from the test suite. To check the impact of these selected different test suites sizes on fault coverage we have collected the mean value of fault coverage as shown in Table 7. Table 7 shows that there is a difference in the mean value of fault coverage for 5 as compared with 10 and 15 test cases (Figs. 3 and 4).

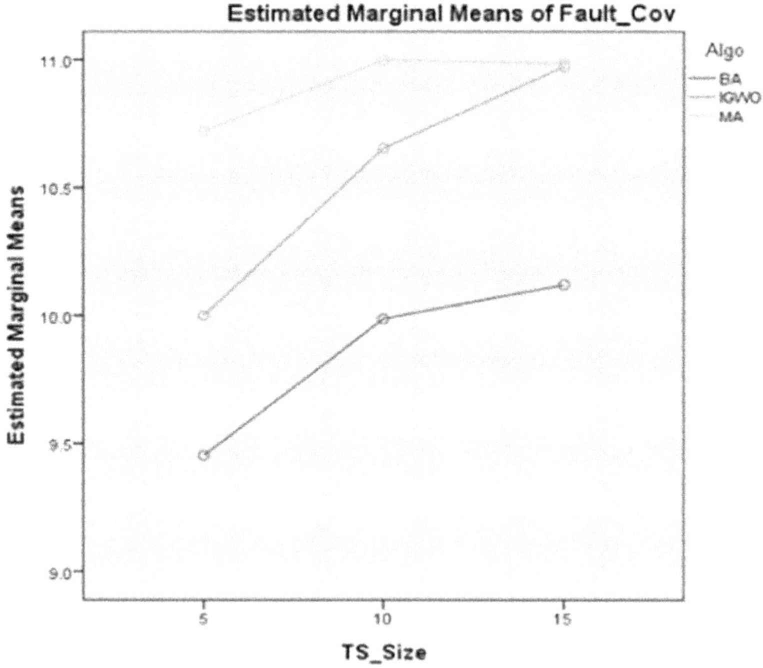

Fig. 3 Test suite sizes versus Fault_Cov w.r.t. Algorithms

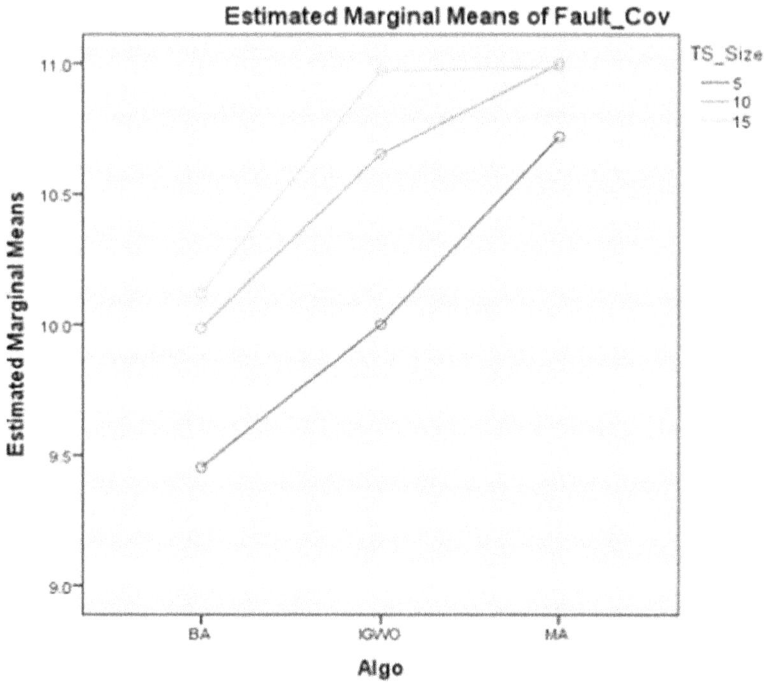

Fig. 4 Performance of algorithms versuss Fault_Cov w.r.t. Test suite size

7 Conclusion & Future Scope

In this paper, the Mayfly Optimization algorithm has been adopted for the selection of test cases during regression testing, and the performance of the adopted algorithm is evaluated against IGWO and Bat search algorithm using five versions of benchmarked subject programs taken from the SIR repository. The answers to three research questions formulated to evaluate the performance of the adopted approach are concluded as follows:

(a) It is clear from the results of Table 3, that the fault detection capabilities of the Mayfly algorithm is superior than the other adopted state-of-the-art algorithms.
(b) From the results reported in Table 5, it is concluded that the execution time of the Mayfly algorithm to find the maximum no. of faults is lesser than the BA as well as IGWO optimization algorithms.
(c) It has been observed from Table 7 that while increasing the size of test suites from 5 to 10 and 15, it will also increase the fault coverage capabilities of adopted optimization algorithms.

So, these concluded point's leads to show the superiority of the adopted approach.

References

1. Kumar V, Khatri SK, Dua H, Sharma M, Mathur P (2014) An assessment of testing cost with effort-dependent FDP and FCP under learning effect: a genetic algorithm approach. Int J Reliab Qual Saf Eng 21(6):1–16
2. Binkley D, Society IC (1997) Semantics guided regression test cost reduction. IEEE Trans Softw Eng. 23(8), 498–516
3. Rothermel G, Harrold MJ (1997) A safe, efficient regression test selection technique. (2), 1–35
4. Harrold MJ et al (2011) Regression test selection for Java software. ACM SIGPLAN Not. 36(11):312–326
5. Goldberg DE, Deb K (1991) A comparative analysis of selection schemes used in genetic algorithms, pp. 69–93
6. Kumar V, Sahni R (2020) Dynamic testing resource allocation modeling for multi-release software using optimal control theory and genetic algorithm. Int J Qual Reliab Manag 37(6–7):1049–1069
7. Dorigo M, Di Caro G (1999) Ant colony optimization: a new meta-heuristic. Proc. 1999 Congr. Evol. Comput. CEC 1999 2:1470–1477
8. Kennedy J, Eberhart R (1995) Particle swarm optimization. Adapt Learn Optim 15:45–82
9. Yang XS (2010) A new metaheuristic bat-inspired algorithm. Stud Comput Intell 284:65–74
10. Yang X, Deb S, Behaviour ACB (2009) Cuckoo Search via L ´ evy Flights. Ieee 210–214
11. Yang XS (2009) Firefly algorithms for multimodal optimization. Lecture Notes in Computer Science (including subseries Lecture Notes in Artificial Intelligence and Lecture Notes in Bioinformatics), vol 5792 LNCS, pp 169–178
12. Arora S, Singh S (2019) Butterfly optimization algorithm: a novel approach for global optimization. Soft Comput 23(3):715–734
13. Askarzadeh A (2016) A novel metaheuristic method for solving constrained engineering optimization problems: crow search algorithm. Comput Struct
14. Zervoudakis K, Tsafarakis S (2020) A mayfly optimization algorithm, vol 145. Elsevier Ltd.
15. Kumar V, Sahni R (2016) An effort allocation model considering different budgetary constraint on fault detection process and fault correction process. Decis Sci Lett 5(1):143–156
16. Do H, Elbaum S, Rothermel G (2005) Supporting controlled experimentation with testing techniques: an infrastructure and its potential impact. Empir Softw Eng 10(4):405–435
17. Yoo S, Harman M (2007) Pareto efficient multi-objective test case selection. In: International Symposium on Software Testing and Analysis, pp 140–150
18. Maia CLB, Do Carmo RAF, De Freitas FG, De Campos GAL, De Souza JT (2009) A multi-objective approach for the regression test case selection problem. XLI Brazilian Symp Oper Res XLI SBPO 2009, pp 1824–1835
19. Singh G, Gupta D (2013) An Integrated Approach to Test Suite Selection Using ACO and Genetic Algorithm. Int. J. Adv. Res. Comput. Sci. Softw. Eng. 3(6):2277–3128
20. Pravin A, Srinivasan S (2013) An efficient algorithm for reducing the test cases which is used for performing regression testing. 2nd Int Conf Comput Tech Artif Intell 119:194–197
21. Wang S, Ali S, Gotlieb A (2013) Minimizing test suites in software product lines using weight-based genetic algorithms. GECCO 2013—Proc 2013 Genet Evol Comput Conf 1493–1500
22. De Souza LS, Prudêncio RBC, De Barros FA (2014) A hybrid binary multi-objective particle swarm optimization with local search for test case selection. In: Proceedings—2014 Brazilian Conference on Intelligent System, BRACIS 2014, pp 414–419
23. Narciso EN, Delamaro ME, De Lourdes Dos Santos Nunes F (2014) Test case selection: a systematic literature review. Int J Softw Eng Knowl Eng 24(4):653–676
24. Shi A, Yung T, Gyori A, Marinov D (2015) Comparing and combining test-suite reduction and regression test selection. In: 2015 10th Joint Meeting European Software Engineering Conference and Symposium on the Foundations of Software Engineering (ESEC/FSE) 2015—Proceedings, pp 237–247
25. Panichella A, Oliveto R, Di Penta M, De Lucia A (2015) Improving multi-objective test case selection by injecting diversity in genetic algorithms. IEEE Trans Softw Eng 41(4):358–383

26. Mondal D, Hemmati H, Durocher S (2015) Exploring test suite diversification and code coverage in multi-objective test case selection. In: IEEE 8th International Conference on Software Testing, Verification and Validation, ICST 2015—Proceedings
27. Nagar R, Kumar A, Singh GP, Kumar S (2015) Test case selection and prioritization using cuckoos search algorithm. In: 2015 1st International conference on futuristic trend in computational analysis and knowledge management ABLAZE 2015, pp 283–288
28. Rosero RH, Gómez OS, Rodríguez G (2016) 15 years of software regression testing techniques—a survey. Int J Softw Eng Knowl Eng 26(5):675–689
29. Srisura B, Lawanna A (2016) False test case selection: Improvement of regression testing approach. In: 2016 13th International conference on futuristic trend in computational analysis and knowledge management ECTI-CON 2016
30. Kazmi R, Jawawi DNA, Mohamad R, Ghani I (2017) Effective regression test case selection: a systematic literature review. ACM Comput Surv 50(2)
31. Garousi V, Özkan R, Betin-Can A (2018) Multi-objective regression test selection in practice: an empirical study in the defense software industry. Inf Softw Technol 103:40–54
32. Bajaj A, Sangwan OP (2018) A survey on regression testing using nature-inspired approaches. In: 2018 4th International Conference on Computing, Communication and Automation ICCCA 2018, pp 1–5
33. Agrawal AP, Kaur A (2018) A comprehensive comparison of ant colony and hybrid particle swarm optimization algorithms through test case selection. Adv Intell Syst Comput 542(August):397–405
34. Gupta N, Sharma A, Pachariya MK (2019) An insight into test case optimization: ideas and trends with future perspectives. IEEE Access 7:22310–22327
35. Pandey A, Banerjee S (2019) Test suite optimization using firefly and genetic algorithm. Int J Softw Sci Comput Intell 11(1):31–46
36. Al-Sabbagh KW, Staron M, Ochodek M, Hebig R, Meding W (2020) Selective regression testing based on big data: comparing feature extraction techniques, pp 322–329
37. Yadav DK, Dutta S (2020) Regression test case selection and prioritization for object oriented software. Microsyst Technol 26(5):1463–1477
38. Guizzo G, Petke J, Sarro F, Harman M (2021) Enhancing genetic improvement of software with regression test selection, pp 1323–1333
39. Ma B, Wan L, Yao N, Fan S, Zhang Y (2021) Evolutionary selection for regression test cases based on diversity. Front Comput Sci 15(2):3–5
40. Chen Y, Chen M (2021) Multi-objective regression test selection 76:105–116

Development of Reliability Block Diagram (RBD) Model for Reliability Analysis of a Steam Boiler System

Suyog S. Patil, Anand K. Bewoor, Ravinder Kumar, and Iliya K. Iliev

Abstract Industrial steam boilers are prone to occurrences such as equipment failures, human errors, and common-cause failures in a context of sophisticated maintenance, inspection, and testing management. These events will have an impact on reliability of safety-related systems as well as the overall risk level. To analyze the impact of item failures on system availability, reliability block diagrams (RBD) are commonly used, taking into account their physical arrangement in the system. In this research, the RBD technique is utilised to estimate the reliability of boiler systems used in Indian textile industries. Furthermore, the boiler system reliability before and after the preventative maintenance PM task is compared.

Keywords Steam boiler · RBD model · Preventive maintenance interval · Reliability

---

S. S. Patil (✉)
Department of Mechanical Engineering, Zeal College of Engineering, SPPU, Pune, Maharashtra, India
e-mail: suyogpatil21@gmail.com

Department of Mechanical Engineering, Sharad Institute of Technology College of Engineering, Yadrav, Maharashtra, India

A. K. Bewoor
Department of Mechanical Engineering, Cummins College of Engineering, Pune, Maharashtra, India

R. Kumar
Department of Mechanical Engineering, Lovely Professional University, Phagwara 144411, Punjab, India

I. K. Iliev
Department of Thermotechnics, Hydraulics and Environmental Engineering, University of Ruse, Ruse, Bulgaria
e-mail: iki@uni-ruse.bg

V. Kumar and H. Pham (eds.), *Predictive Analytics in System Reliability*, Springer Series in Reliability Engineering,
https://doi.org/10.1007/978-3-031-05347-4_9

1 Introduction

The knowledge of operational relationship between the subsystems and components of a system is essential before any system reliability evaluations can be performed. To improve or evaluate a system's reliability, it is vital to understand how each of its components functions and how these functions affect the system. Accurate representations of these interactions are required to create meaningful predictions, allocations, and assessments. Reliability Block Diagram (RBD) can be used to represent this information, as it is clear and easy to comprehend.

A reliability block diagram (RBD) can be used to examine a number of failure-related characteristics of engineering systems such as reliability, availability, and maintainability [1, 2]. A RBD, or graphical structure made up of blocks and connectors, is used to depict the behaviour of a system. During the evaluation of a computational software's reliability, for example, the blocks may represent the computational elements with a given failure rate, and the connectors between them may be used to describe various alternative paths required for a successful computation using the given software [3]. In RBD, individual component failure rates can now be used to assess a system's failure characteristics, whereas the entire system fails when all paths to successful execution fail. The RBD-based analysis has become a popular technique for analysing the trade-offs of various system configurations during the system design stage due to its ability to quantify the impact of component failures on overall system safety and dependability.

RBD-based analysis has typically been carried out using proof methods on paper and pencil, as well as computer simulations. Despite their limitations, these methods cannot be depended on to give absolute accuracy. Formal approaches for dealing with the above-mentioned inaccuracy issues have been proposed for the RBD-based analysis. These solutions do not work for all sorts of complex engineering systems due to their limited scope. This paper provides a brief overview of the aforementioned RBD-based analysis methodologies.

The process industries are either batch or continuous. These can be found all over the world and make a substantial contribution to a country's economy. Steam is frequently used as a heat transfer medium when transferring heat from one process to another in many process industries such as food, beverages, chemicals, pharmaceuticals, petroleum, ceramics, base metals, coal, plastics, rubber, textiles, wood and wood products, paper and paper products, etc. Many process industries rely on the use of industrial steam boilers. The boiler is a complex system that necessitates process integration, modern technology and software interfaces, as well as multidisciplinary tasks. Higher organisational needs, increasing complexity, and lower prices are currently posing challenges to new product development [4]. The availability research classifies boiler systems, subsystems, and components based on reliability and maintainability in order to reduce system failure and safety-related issues. This has switched the emphasis to reliability, maintainability, and lowering product life cycle costs [5]. It is possible to improve the uptime and downtime of boiler subsystems and components in order to increase plant availability. As a result, it is decided to carry out an examination of the reliability, availability, and maintainability of the components of a typical steam boiler used in process industries.

Only a few studies on boiler equipment failures have been undertaken. Visual inspection, microstructural, hardness measurements, and residual stress measurements using X-ray diffraction (XRD) methods [6], scanning electron microscopy, optical microscopy, and energy dispersive spectroscopy methods [7], failure modes and effects analysis (FMEA), stochastic technique [8] are all used to examine boiler tube failures. In the boiler reliability and availability study, numerous studies have also focused on the combustion, ignition, and fuel feeding systems, as well as their reliability and availability. Vandermeer [9] researched and recorded the many causes of a boiler's starting flame failure in order to prevent this type of failure, and a flame loss detector was developed. The coal crusher failures can be caused by a variety of factors, including rotor wear, mill wear, inappropriate hammer usage, and coal bunker issues [10]. Mariajayaprakash and Senthilvelan [11] provides an alternative solution to fuel-feeding system failure using FMEA, the Taguchi technique, and a Cause and Effect Diagram. A probabilistic feedback technique is proposed [12] to construct the fuel feeding system maintenance schedule based on plant maintenance records.

Examining the systems can increase the reliability and maintainability of the existing plant. Some authors have identified the critical components of the boiler system and conducted an analysis. Insufficient inlet air motion is a major cause of FD fan failure in boiler fans, according to Parthiban [10]. According to Rajkumar and Priyaa [13], a low water level in the boiler drum can cause an explosion, while a high water level can cause water particles in the steam. Carazas et al. [14] proposed a reliability and availability evaluation approach based on FMEA. A dynamic programming system is intended to optimise alternative maintenance procedures, determine maintenance methods, and reduce the biomass boiler's operational and cleaning costs Agarwal and Suhane [15]. Kiran et al. [16] established a model for improving plant availability in a process plant using an appropriate maintenance schedule. Arjunwadkar et al. [17] investigated CFB boiler components for agglomeration, gas reflux, and back-sifting, as well as emission control and bed temperature management.

A review of the literature finds that failure analysis of a few boiler components has been fairly rare up until now. A few researchers in the process sector have attempted to reduce boiler failure rates and discover the best maintenance methods. As a result of the current study, steam boilers must be evaluated for reliability, maintainability, and availability (RMA). This book chapter discusses a case study conducted on a typical steam boiler in the process industry to estimate the preventive maintenance plan and increase capacity utilisation using reliability, maintainability, and availability studies. The rest of the chapter is structured as follows: Sect. 2 covers the methodologies used to assess the reliability, maintainability, and availability. Section 3 offers a case study on the boiler system reliability analysis using reliability block diagram (RBD). Finally, Sect. 4 of this chapter finishes with a concluding remarks.

2 Reliability Analysis Model for Boiler System

Barabady and Kumar [18] presented a technique for the selection of TTF and TTR models based on various data trends. This approach is presented in simple way and can be used to examine the maintenance data. On this basis, a modified framework for selecting models is proposed. The model selection framework shown in Fig. 1 is more comprehensive and easier to use for industrial applications. It shows a detailed flow diagram that was used to identify and explore the issues in this case. Various models are available for evaluating reliability data. TTF and TTR data must be used to select the required system analysis model. In the literature, various methods for

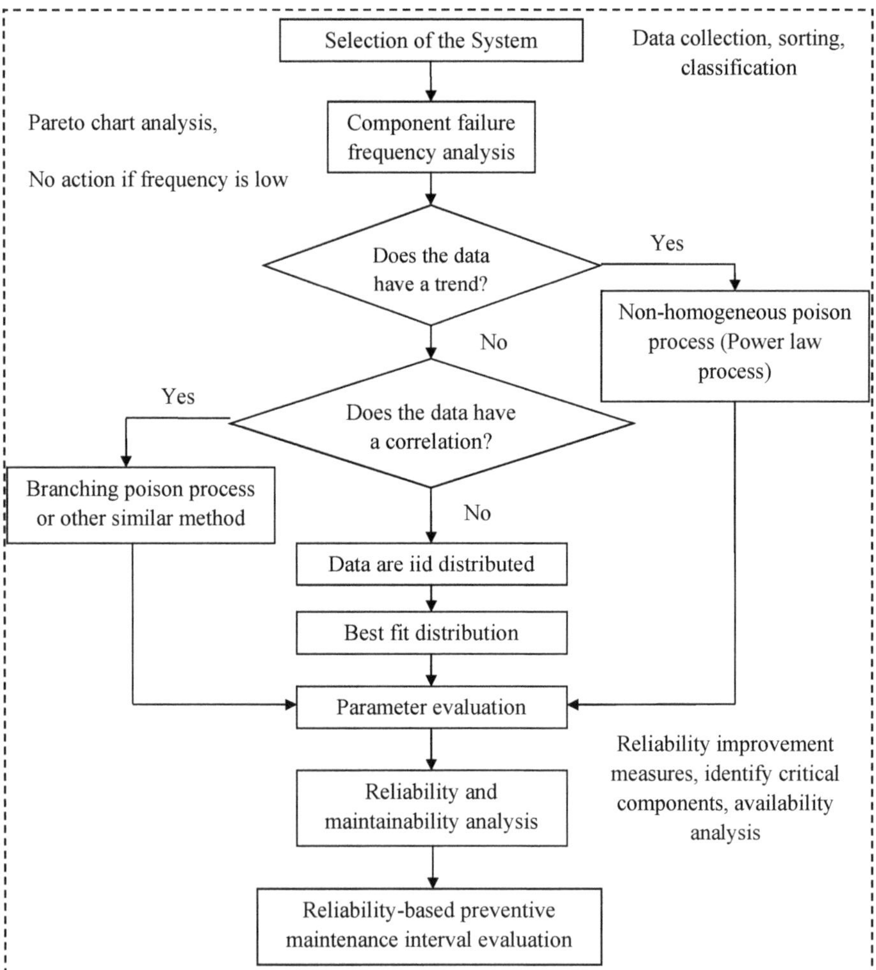

Fig. 1 Framework for the selection of time to failure model [20]

modelling the reliability of repairable and non-repairable systems have been given. To determine the patterns in the data and to assess the goodness of fit, a significant number of studies in many processes must be performed.

Before a steam boiler system is selected for study, it is categorised into numerous levels such as assembly, sub-systems, and component. Data for a RAM study is gathered from a variety of sources, including maintenance history cards, registers, and expert opinions. When there is insufficient data, the Bayesian approach may be used. If enough information is available, a parametric or non-parametric analysis can be performed. Because many system failures are seen as minor, the Pareto chart analysis techniquc is uscful for identifying critical components. This is followed by an examination of data trends using graphical and analytic methods. In order to analyse failure data, graphical techniques including cumulative failure versus time plots, timeline plots, and serial co-relation charts and analytical methods such as Mann test, military handbook tests can be employed [19].

Estimate the "goodness-of-fit" of the data before constructing a failure rate model. The data can be used to fit other distributions, including Weibull, Exponential, Normal, and Lognormal distributions. The most likely distribution is analysed and distribution parameters are determined using the Chi-square, the classic p-value test, or the Kolmogorov–Smirnov (K–S) test. Finally, reliability of the subsystems and components of the boiler system are evaluated. After the reliability features have been discovered and quantified, priority measurement can be used to detect the criticality of each element and the partial failure subsystem. There includes an analysis of the system's weakest points, as well as a discussion of the changes that will improve the system's reliability.

3 Reliability Analysis of the Boiler System by RBD

Table 2 summarises the reliability values of all boiler components computed using an exponential distribution. The Eq. (1) of exponential distribution for reliability estimation is used to estimate reliability values of the components of boiler system.

$$R(t) = e^{[-\lambda t]} \tag{1}$$

Similarly, by doing preventative maintenance, one can increase the reliability values of the boiler components and, as a result, the system's availability. Table 2 also shows the enhanced reliability values of all boiler components.

The reliability of the entire steam boiler system is evaluated by using the reliability block diagram shown in Fig. 2. The boiler RBD is built on the assumption that all of the boiler subsystems are connected in series, and that if one of the subsystems fails, the entire system fails. The different codes used in this reliability block diagram are presented in the Table 1. The Eqs. (2) and (3) are the reliability models of the steam boiler system (Table 2).

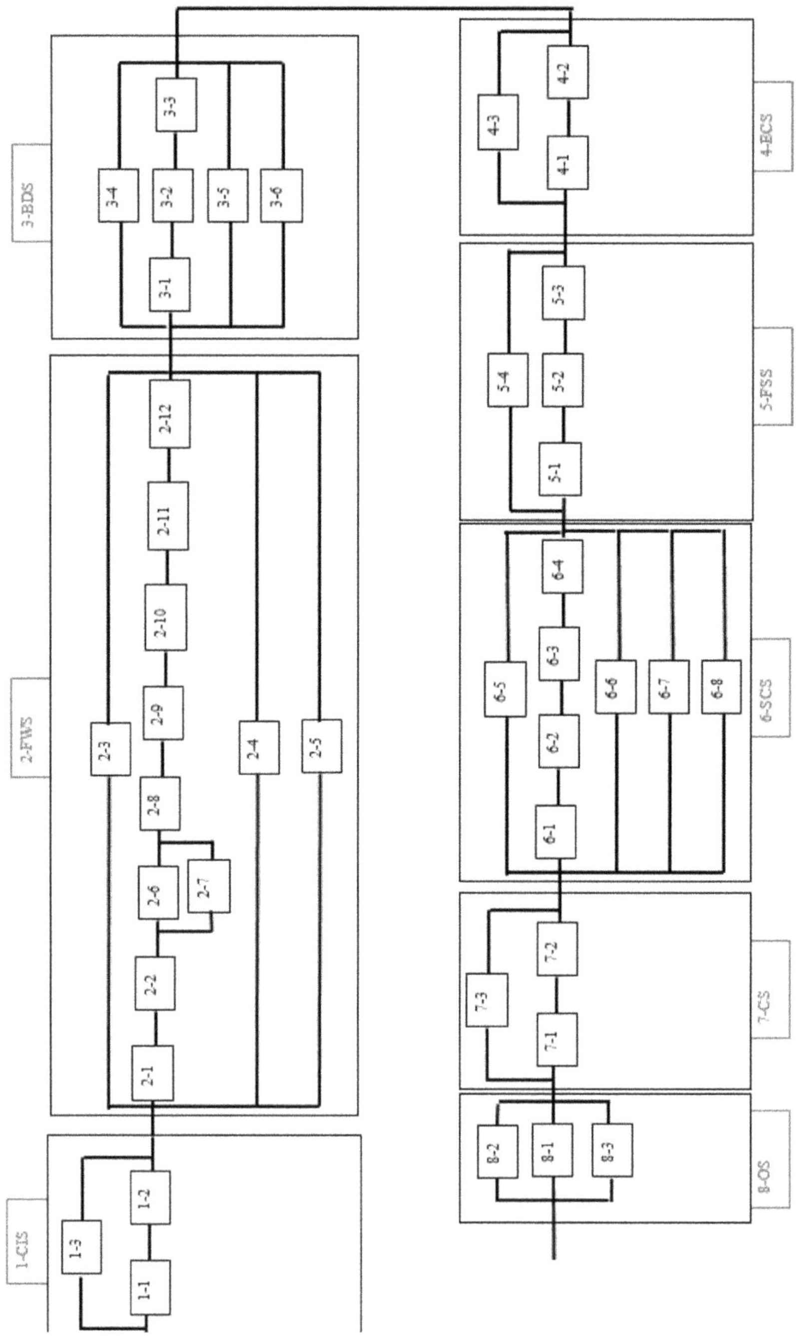

Fig. 2 Reliability block diagram of the boiler system

Table 1 Codes used to develop RBD for boiler system

Sr. No.	Component	Code	Sr. No.	Component	Code
1	Furnace/Combustion chamber	1-1	22	Induced drum (ID) fan	4-1
2	burner	1-2	23	Forced draft (FD) Fan	4-2
3	Temperature regulator	1-3	24	Mechanical dust collector (MDC)	4-3
4	Water tubes	2-1	25	Rack and pinion coal feeding mechanism	5-1
5	Feed water pump	2-2	26	Coal crusher	5-2
6	Back flow preventer valve	2-3	27	Coal crusher motor	5-3
7	Feed water pump-gauge	2-4	28	Coal storage tank	5-4
8	Supply water temperature sensor	2-5	29	Header	6-1
9	Softnar	2-6	30	Steam circulation pipes	6-2
10	Feed water tank	2-7	31	Pressure relief valve (PRV) station	6-3
11	Water level controller (Mobari)	2-8	32	Pressure reducing valve	6-4
12	Feed check valve	2-9	33	Pressure gauge	6-5
13	Feed water hose	2-10	34	Steam water separator	6-6
14	Strainer	2-11	35	By-pass valve	6-7
15	Deaerator	2-12	36	Intake vent/air vent	6-8
16	Drain pump	3-1	37	Safety valves	7-1
17	Condensate filter	3-2	38	Main steam stop valve	7-2
18	Blow-down connections	3-3	39	Fusible plug	7-3
19	Return water temperature sensor	3-4	40	Gate valve	8-1
20	Shut-off valve	3-5	41	Globe valve	8-2
21	Blow down valve	3-6	42	Ball valve	8-3

$$\therefore \mathrm{R_S} = \mathrm{R_1} \times \mathrm{R_2} \times \mathrm{R_3} \times \mathrm{R_4} \times \mathrm{R_5} \times \mathrm{R_6} \times \mathrm{R_7} \times \mathrm{R_8} \tag{2}$$

$$\begin{aligned}\therefore \mathrm{R_S} = \{&[1 - (1 - \mathrm{R_{1-3}})(1 - (\mathrm{R_{1-1}} \times \mathrm{R_{1-2}}))] \times [1 - (1 - \mathrm{R_{2-3}}) \times (1 - \mathrm{R_{2-4}}) \\ &\times (1 - \mathrm{R_{2-5}}) \times (1 - (\mathrm{R_{2-1}} \times \mathrm{R_{2-2}} \times (1 - (1 - \mathrm{R_{2-6}})(1 - \mathrm{R_{2-7}})) \times \mathrm{R_{2-8}} \\ &\times \mathrm{R_{2-9}} \times \mathrm{R_{2-10}} \times \mathrm{R_{2-11}} \times \mathrm{R_{2-12}})] \times [1 - (1 - \mathrm{R_{3-4}}) \times (1 - (\mathrm{R_{3-1}} \times \mathrm{R_{3-2}} \\ &\times \mathrm{R_{3-3}})) \times (1 - \mathrm{R_{3-5}}) \times (1 - \mathrm{R_{3-6}})] \times [1 - (1 - \mathrm{R_{4-3}}) \times (1 - (\mathrm{R_{4-1}} \\ &\times \mathrm{R_{4-2}})] \times [1 - (1 - \mathrm{R_{5-4}}) \times (1 - (\mathrm{R_{5-1}} \times \mathrm{R_{5-2}} \times \mathrm{R_{5-3}}))] \times [1 - (1 - \mathrm{R_{6-5}}) \\ &\times (1 - (\mathrm{R_{6-1}} \times \mathrm{R_{6-2}} \times \mathrm{R_{6-3}} \times \mathrm{R_{6-4}})) \times (1 - \mathrm{R_{6-6}}) \times (1 - \mathrm{R_{6-7}}) \\ &\times (1 - \mathrm{R_{6-8}})] \times [1 - (1 - \mathrm{R_{7-3}}) \times (1 - (\mathrm{R_{7-1}} \times \mathrm{R_{7-2}}))] \\ &\times [1 - (1 - \mathrm{R_{8-1}}) \times (1 - \mathrm{R_{8-2}}) \times (1 - \mathrm{R_{8-3}})]\}\end{aligned} \tag{3}$$

Table 2 Reliability values of boiler components

Sr. No.	Name of component	Earlier reliability values				Improved reliability values			
		MTTF	λ	λt	R(t)	MTTF	λ	λt	R(t)
1	Header	98553.91	1.01467E−05	−0.2630	0.7687	108409.301	9.2243E−06	−0.2391	0.7873
2	Hot gas tubes	98553.91	1.01467E−05	−0.2630	0.7687	108409.301	9.2243E−06	−0.2391	0.7873
3	Furnace/Shell	102,090	9.79527E−06	−0.2539	0.7758	112299.044	8.9048E−06	−0.2308	0.7939
4	Intake vent/air vent	96024.13	1.0414E−05	−0.2699	0.7634	120,000	8.33333E−06	−0.2160	0.8057
5	Water tubes	109681.2	9.11733E−06	−0.2363	0.7895	120649.3	8.28849E−06	−0.2148	0.8067
6	Supply water temperature sensor	6189.33	0.000161568	−4.1879	0.0152	13,000	7.69231E−05	−1.9938	0.1362
7	Back flow preventer valve	75522.4	1.32411E−05	−0.3432	0.7095	90,000	1.11111E−05	−0.2880	0.7498
8	Feed water pump	50071.74	1.99713E−05	−0.5177	0.5959	70,000	1.42857E−05	−0.3703	0.6905
9	Feed water pump-gauge	52497.83	1.90484E−05	−0.4937	0.6103	57747.61	1.73167E−05	−0.4488	0.6384
10	Softnar	77510.29	1.29015E−05	−0.3344	0.7158	82,500	1.21212E−05	−0.3142	0.7304
11	Feed water tank	28767.86	3.4761E−05	−0.9010	0.4062	31644.65	3.16009E−05	−0.8191	0.4408
12	Water level controller (Mobari)	99108.2	1.009E−05	−0.2615	0.7699	120,000	8.33333E−06	−0.2160	0.8057
13	Feed check valve	113,074	8.84377E−06	−0.2292	0.7951	124381.4	8.03979E−06	−0.2084	0.8119
14	Feed water hose	97120.66	1.02965E−05	−0.2669	0.7658	106832.7	9.36043E−06	−0.2426	0.7846
15	Strainer	25283.99	3.95507E−05	−1.0252	0.3587	40,000	0.000025	−0.6480	0.5231
16	Deaerator	104005.2	9.61491E−06	−0.2492	0.7794	114405.665	8.74083E−06	−0.2266	0.7973
17	Return water temperature sensor	9516.97	0.000105075	−2.7236	0.0656	10468.667	9.55231E−05	−2.4760	0.0841
18	Drain pump	43925.53	2.27658E−05	−0.5901	0.5543	48318.083	2.06962E−05	−0.5364	0.5848
19	Condensate filter	5985.82	0.000167061	−4.3302	0.0132	6584.402	0.000151874	−3.9366	0.0195
20	Shut-off valve	105761.8	9.45521E−06	−0.2451	0.7826	116337.958	8.59565E−06	−0.2228	0.8003

(continued)

Table 2 (continued)

Sr. No.	Name of component	Earlier reliability values				Improved reliability values			
		MTTF	λ	λt	R(t)	MTTF	λ	λt	R(t)
21	Blow-down connections	103704.2	9.64281E−06	−0.2499	0.7788	114074.653	8.76619E−06	−0.2272	0.7967
22	Blow down valve	95751.03	1.04438E−05	−0.2707	0.7628	105326.133	9.49432E−06	−0.2461	0.7818
23	Induced drum (ID) fan	94924.99	1.05346E−05	−0.2731	0.7610	104417.489	9.57694E−06	−0.2482	0.7802
24	Forced draft (FD) Fan	102386.8	9.76688E−06	−0.2532	0.7763	112625.469	8.87899E−06	−0.2301	0.7944
25	Secondary air (SA) fan	104256.9	9.59169E−06	−0.2486	0.7799	114682.59	8.71972E−06	−0.2260	0.7977
26	Mechanical dust collector (MDC)	40764.94	2.45309E−05	−0.6358	0.5295	44841.434	2.23008E−05	−0.5780	0.5610
27	Rack and pinion coal feeding mechanism	108341.3	9.23009E−06	−0.2392	0.7872	119175.375	8.391E−06	−0.2175	0.8045
28	Coal crusher	22417.63	4.46077E−05	−1.1562	0.3147	24659.393	4.05525E−05	−1.0511	0.3495
29	Coal crusher motor	100291.8	9.97091E−06	−0.2584	0.7723	130,000	7.69231E−06	−0.1994	0.8192
30	Coal storage tank	80199.83	1.24689E−05	−0.3232	0.7238	88219.81	1.13353E−05	−0.2938	0.7454
31	Pressure gauge	67093.35	1.49046E−05	−0.3863	0.6795	70,000	1.42857E−05	−0.3703	0.6905
32	Steam circulation pipes	105938.8	9.43941E−06	−0.2447	0.7830	116532.7	8.58128E−06	−0.2224	0.8006
33	Pressure relief valve	106050.7	9.42946E−06	−0.2444	0.7832	130,000	7.69231E−06	−0.1994	0.8192
34	Pressure reducing valve (PRV) station	80830.56	1.23716E−05	−0.3207	0.7257	85,000	1.17647E−05	−0.3049	0.7372
35	Strainer	65833.27	1.51899E−05	−0.3937	0.6745	72416.6	1.3809E−05	−0.3579	0.6991
36	Steam water separator	102386.8	9.76688E−06	−0.2532	0.7763	105,000	9.52381E−06	−0.2469	0.7813
37	By-pass valve	104256.9	9.59169E−06	−0.2486	0.7799	114682.6	8.71972E−06	−0.2260	0.7977
38	Safety valves	102228.5	9.78201E−06	−0.2535	0.7760	112451.339	8.89274E−06	−0.2305	0.7941
39	Main steam stop valve	102228.5	9.78201E−06	−0.2535	0.7760	112451.339	8.89274E−06	−0.2305	0.7941

(continued)

Table 2 (continued)

Sr. No.	Name of component	Earlier reliability values				Improved reliability values			
		MTTF	λ	λt	R(t)	MTTF	λ	λt	R(t)
40	Fusible plug	11204.2	8.92522E−05	−2.3134	0.0989	17,500	5.71429E−05	−1.4811	0.2274
41	Gate valve	109447.2	9.13682E−06	−0.2368	0.7891	120391.942	8.3062E−06	−0.2153	0.8063
42	Globe valve	109447.2	9.13682E−06	−0.2368	0.7891	130,000	7.69231E−06	−0.1994	0.8192
43	Ball valve	109447.2	9.13682E−06	−0.2368	0.7891	115,000	8.69565E−06	−0.2254	0.7982

where R_S is the boiler system reliability, and R_1, R_2, R_3, … R_8, are the reliabilities of the boiler subsystems.

The earlier reliability of the steam boiler system after three years is calculated as follows,

$$\therefore R_S = R_1 \times R_2 \times R_3 \times R_4 \times R_5 \times R_6 \times R_7 \times R_8$$
$$\therefore R_S = 0.776 \times 0.894 \times 0.952 \times 0.808 \times 0.777 \times 0.998 \times 0.642 \times 0.991$$
$$\therefore R_{S\ (\text{Earlier})} = 0.2632$$

Similarly, the improved reliability of the steam boiler system after three years is calculated as follows,

$$\therefore R_S = R_1 \times R_2 \times R_3 \times R_4 \times R_5 \times R_6 \times R_7 \times R_8$$
$$\therefore R_S = 0.794 \times 0.930 \times 0.960 \times 0.833 \times 0.804 \times 0.999 \times 0.715 \times 0.993$$
$$\therefore R_{S\ (\text{Improved})} = 0.3367$$

Therefore the Change in system reliability = Improved reliability − Earlier reliability.

$$(\Delta R)3\ \text{year} = 0.3367 - 0.2632 = 0.0735(27.92\%\ \text{increase})$$

The system reliability can be improved up to 30% by performing preventive maintenance.

4 Conclusion

A system, including its subsystems and components, can be represented as a series of blocks using reliability block diagrams (RBDs), allowing equipment failure rates, operating philosophies, and maintenance strategies to be quantitatively assessed in terms of their expected impact on system performance.

The study uses a reliability block diagram (RBD) to offer a qualitative and quantitative reliability analysis of the steam boiler system. This effort may help to create and elaborate the RBD. The reliability values of all boiler system components are evaluated using an exponential distribution. Finally, the RBD technique is used to conduct a system reliability analysis. It is found that the boiler system's reliability may be improved by executing the necessary preventative maintenance at appropriate intervals. This research estimates that system reliability can be improved by about 30%.

References

1. Soszynska S (2010) Reliability and risk evaluation of a port oil pipeline transportation system in variable operation conditions. Int J Press Vessels Pip 87(2–3):81–87
2. Huffman D, Antelme F (2009) Availability analysis of a solar power system with graceful degradation. In: Reliability and Maintainability Symposium IEEE, pp 348–352
3. Abd-Allah A (1997) Extending reliability block diagrams to software architectures, USC-CSE-97-501, Department of Computer Science, University of Southern California, USA
4. Ebling C (2000) An introduction to reliability and maintainability engineering. University of Dayton, Tata McGraw Hill Education Private Limited.
5. Ascher HE, Feingold H (1984) Repairable system reliability: modeling, Interface, misconception and their causes. Marcel Dekker, New York
6. Duarte C, Espejo E, Martinez JC (2017) Failure analysis of the wall tubes of a water-tube boiler. Eng Fail Anal 79:704–713
7. Liu S, Wang W, Liu C (2017) Failure analysis of the boiler water-wall tube. Case Stud Eng Failure Anal 9:35–39
8. Moghanlou L, Pourgol-Mohammad M (2017) Assessment of the pitting corrosion degradation lifetime: a case study of boiler tubes. ASCE-ASME J Risk Uncertainty Eng Syst Part B Mech Eng 3(4)
9. Vandermeer W (1998) Flame safeguard controls multi-burner environments, pp 1–33
10. Parthiban K (2006) Fans at work in boilers. Venus Energy Audit System Report
11. Mariajayaprakash A, Senthilvelan T (2013) Failure detection and optimization of sugar mill boiler using FMEA and Taguchi method. Eng Fail Anal 30:17–26
12. Barry D, Hudson M (1986) Reliability modelling for the scheduling of plant work in majority vote mode. Int J Qual Reliab Manag 3(2):12–20
13. Rajkumar T, Priyaa V (2013) Boiler drum level control by using wide open control with three element control system. Int J Sci Eng Res 4(5):204–210
14. Carazas F, Salazar C, Souza C (2011) Availability analysis of heat recovery steam generators used in thermal power plants. Energy 36:3855–3870
15. Agarwal S, Suhane A (2017) ScienceDirect study of boiler maintenance for enhanced reliability of system a review. Mater Today Proc 4(2):1542–1549
16. Kiran S, Kumar KP, Sreejith B, Muralidharan M (2016) Reliability evaluation and risk based maintenance in a process plant. Procedia Technol 24:576–583
17. Arjunwadkar A, Basu P, Acharya B (2016) A review of some operation and maintenance issues of CFBC boilers. Appl Therm Eng 102:672–694
18. Barabady J, Kumar U (2008) Reliability analysis of mining equipment: a case study of a crushing plant at Jajarm bauxite mine in Iran. Reliab Eng Syst Saf 93:647–653
19. Patil SS, Bewoor AK, Patil RB (2020) Availability analysis of a steam boiler in textile process industries using failure and repair data: a case study. ASCE-ASME J Risk Uncertainty Eng Syst Part B Mech Eng 2020
20. Patil SS, Bewoor AK (2020) Reliability analysis of a steam boiler system by expert judgment method and best-fit failure model method: a new approach. Int. J. Qual. Reliab. Manag. 38(1):389–409

Computation Signature Reliability of Computer Numerical Control System Using Universal Generating Function

Tripty Pandey, Arpita Batra, Mansi Chaudhary, Anjali Ranakoti, Akshay Kumar, and Mangey Ram

Abstract The aim of this research is to deal with a complex manufacturing system using the Computer Numerical Control as the bottom case manufacturing system, where the arrangement of various complex sub-systems is in series, parallel or in both the configurations. The system reliability with several other factors like signature, tail signature, expected time as well as expected cost and sensitivity have been obtained with the assistant of universal generating function technique. The purpose of this chapter is to comparison of the systems on the basis of signature and its measures Further, a numerical example demonstrates the proposed system and technique for a better understanding.

Keywords Signature · Reliability function · Computer numerical control · Tail signature · Expected time · Universal generating function

1 Introduction

In the recent past, Reliability theory has played a key role in the history of engineering fields. Researchers have studied and applied the signature reliability theories in day to day life to solve the real-life problems. In the last few decades, reliability design of the Computer Numerical Control (CNC) machines has been extensively used in the manufacturing field. Ghare and Taylor [9] determined that the finest possible answer to the corresponding issue was equivalent to the optimal solution for the

T. Pandey · A. Batra · M. Chaudhary · A. Ranakoti · A. Kumar (✉)
Department of Mathematics, Graphic Era Hill University, Dehradun, Uttarakhand, India
e-mail: akshaykr1001@gmail.com

M. Ram
Department of Mathematics, Computer Science & Engineering, Graphic Era (Deemed to be University), Dehradun, Uttarakhand, India

Institute of Advanced Manufacturing Technologies, Peter the Great St. Petersburg Polytechnic University, 195251 Saint Petersburg, Russia

V. Kumar and H. Pham (eds.), *Predictive Analytics in System Reliability*, Springer Series in Reliability Engineering,
https://doi.org/10.1007/978-3-031-05347-4_10

optimized redundant problem through a procedure called branch and bound procedure on a zero–one programming problem. Li and Lumb [21] proposed a technique to determine the approximate reliability of an engineering system using methods like curve fitting and numerical integration. This technique could be applied to both non-Gaussian and Gaussian variables having non-linear or linear failure limits and also effective for implicit evaluation tasks. Iyer et al. [12] demonstrated a practice-based approach to study the behaviour of breakdown of computerized systems particularly to examine everlasting failures. A variety of significant methods, which might be generally implemented in failure as well as workload examination, are gathered. A conventional combination was proposed by Shatz and Wang [33] derived a quantitative allocation model, applied its theory to introduce, talk about systems and algorithmic rule for second level or third level redundancy models. The consequences provided a substitute to performance-oriented techniques and contributed to the frame of knowledge on task allotment. Heimann et al. [10] collectively discussed concepts like reliability, maintainability, availability etc. by addressing computer system dependability analysis. Further, model verification and validation along with the decision of the parameters, is also discussed. Enevoldsen and Sorensen [7] considered a reliability based design of structural system, developed direct and consequent optimized processes to solve the optimization problems and hence included a new and efficient technique called bounds iteration method which turned out to become highly efficacious in reliability grounded maximisation of complicated systems. Dugan and Van Buren [6] presented a combined scanning of the flight control model's portion by combining the fault trees with Markov models techniques, to further determine the dependability of every single system and a system is considered reliable if it is giving reasonable outcomes. Coit and Smith [4] designed and demonstrated a complication particular generic algorithm technique to resolve the redundancy assigned issue for a complex system with enough element options at hand for various k-out-of-n subsystems. Barlow and Proschan [1] estimated system reliability at a constituent level, assuming that the rate of failure is constant. Palisano et al. [28] discussed the necessity for the standardized system for the classification of a gross motor function system with cerebral palsy and developed a five-level classification system used in medicine and determined the inter-rater reliability of the classification system. Boland and Samaniego [3] discussed the signature of binary and k-out-of-n system. Authors compared the various systems on the basis of signature analysis. Ding et al. [5] evolved an inclusive structure to approximate the reliability of multi-state weighted k-out-of-n systems for which they defined a pair of multi-state weighted k-out-of-n system models based on fuzzy. The fuzzy universal generating function methods and fuzzy recursive techniques were formed for computing such systems. The curve fitting and clustering technique were accustomed to discover the probabilities of states in the models, and fuzzy weights. Pang et al. [29] presented a novel AHP based on the method of ELECTRE I of reliability design scheme decision for CNC machine. The method of AHP was used in order to quantify the weights of reliability design factors from decision model. The method of ELECTRE I was then constructed for the ranking of reliability design scheme according to the preference of decision maker. Sahner et al. [30] considered an approach based on the SHARPE Software

Package which provided a range of probabilistic and discrete-state models to assess the reliability, communication systems and conduct of the computer. Huang et al. [11] examined the electrical system of a CNC machine tool for the reliability having Bayesian network and the composition of the BNs was framed from fault trees too. Kumar et al. [16] identified the complexity of different sub-systems by the functioning of a multi-state repairable system having hot redundancy. The responsive study of the system (the stochastic model) was carried out using Markov process (probabilistic approach) which derived the first-order differential equations using mnemonic rule affiliated with the stochastic model supposing that the repair rate as well as failure criterion of each sub-system is exponentially distributed and constant.

In the context of signature reliability, Levitin [18] reflected upon a redundancy optimization question for a multi-state system to lessen the amount of investment costs with satisfying demand as well, from genetic and UGF algorithm in which the working of the system is determined if the specified work is larger as compared to the demand, presented by a cumulative demand curve while having the appointed probability. Levitin [19] introduced a new model termed as linear multi-state sliding window system that generalized the successive k-out-of-r-from-n:F system to multi-state case where every single component could have unalike states due to absolute failure up to best functioning and calculated the reliability of the system using genetic algorithm method and UGF. Levitin [20] discussed the UGF technique with its implementation in order to optimize and analyse the several types of multi-state and binary system in a comprehensive up-to-date form. Samaniego [31] addressed the need of system signatures in building reliability into systems and provided the guidance on how reliability problems can be structured, modelled and solved and further compared the actual system lifetimes using or omitting the tool. Samaniego et al. [32] adapted the thesis of system signatures as explained in [31], to type of signatures apt in dynamic reliability settings. The concept of dynamic signature was introduced where a system is considered which is examined at the time (let's say t) and is detected to be running with floundered elements (let's say k). Navarro and Rychlik [25] studied the correlation and limits for the expected lifetime of mixed or rational systems with separate elements that had indefinite distributions based on the elements expected lifetimes, moreover they approximated the lifetime of independent identically distributed components in the upper and lower pattern. Marichal et al. [23] derived a decomposition based on signature of the joint reliability of a pair of system established on the concept of joint structure signature of two systems. In order to evaluate the joint structure signature of two or more systems, they provided an explicit formula and in order for the joint reliability of the systems to acquire a decomposition set up on signature they also discussed the necessary and sufficient condition on this distribution. Kumar and Singh [13] discussed about the complex k-out-of-n coherent system (acquiring independent and identically distributed components) and its signature reliability using structure and reliability functions as well as intended to evaluate the expected lifetime, Barlow-Proschan index, expected cost rate and signature reliability of the proposed systems. Kumar and Singh [14] suggested the evaluation of the signature, Barlow-Proschan index and mean time to failure in the binary and multi-state sliding window system to calculate the cost and reliability while using the UGF technique. Kumar and Singh [15] proposed to study a structure of the sliding window coherent system that consisted of G linearly needed multi-state components and parallel components (G in number) in A-within-B from-D/G for every multi-state and used Owen's method and UGF for approximating the various attributes like Barlow-Proschan index, tail signature, signature, sensitivity, and expected lifetime owning structure or reliability function. Kumar and Singh [17] discussed the signature of consecutive k-out-of-n:F system having two states such as working and failed with given allowable weight and authors also find various parameters of the consider system.

2 Evaluation of Signature, Tail Signature and Expected Cost

The signature of i.i.d. element like order statistics and reliability function methods [2, 26, 27] is defined as

$$S_l = \frac{1}{\binom{n}{n-l+1}} \sum_{\substack{\overline{H}\subseteq[n] \\ |\overline{H}|=n-l+1}} \varphi(\overline{H}) - \frac{1}{\binom{n}{n-1}} \sum_{\substack{\overline{H}\subseteq[n] \\ |\overline{H}|=n-1}} \varphi(\overline{H}) \tag{1}$$

which are some of the various coherent systems. And the polynomial form of structure functions of the arrangement with independent and identically distributed (i.i.d.) elements is $\overline{H}(p) = \sum_{e=1}^{m} C_j \binom{m}{e} p^e q^{n-e}$ and $C_e = \sum_{i=n-e+1}^{n} s_i$, $e = 1, 2, \ldots, n$. For finding the tail signature we'll change the signature of the system $S_l = \sum_{i=l+1}^{n} s_i = \frac{1}{\binom{n}{n-1}} \sum_{|\overline{H}|=n-l} \varphi(\overline{H})$ (given polynomial function) having n-tuples set function like $S = (S_0, \ldots, S_n)$, into $P(X) = X^n \overline{H}\left(\frac{1}{X}\right)$ using Taylor expansion then the signature.

$$S_l = \frac{n-1}{n!} D^l P(1), \; l = 0, 1, \ldots, n \tag{2}$$

of the designated method [22] is

$$s = S_{l-1} - S_l, \; l = 1, \ldots, n \tag{3}$$

Now computing the expected lifetime and cost of system from reliability function [8, 24] defined as $E(X) = \sum_{i=1}^{n} i.s_i,\ i = 1, 2, \ldots, n$ and $E(T) = \mu \sum_{i=1}^{n} \frac{C_i}{i}$ formulated on the quantity of failed elements and minimal signature of the given system possessing mean value is one.

3 Model Description: Computer Numerical Control

Here, using CNC used in the bottom case manufacturing system (the shock absorber manufacturing plant). This system consists of various complicated sub-systems functioning in parallel, series or in both the configurations. The procedure of this system begins with the blank casting connected in series with oil seal machining (which are two in numbers connected in parallel, and if the couple units fails at the same time it results in absolute collapse of the arrangement) which is next attached in series with another sub-system BTA process (which is a sole unit that causes the whole system failure if this unit fails). Then the BTA process is in series with axle hole machining (which has three units attached in parallel and if these units fails at a time it results in the absolute failure of the system) which is then connected in series with the fender milling (It's a single unit so the whole system fails if this unit fails). And finally mounting hole drilling and tapping is executed, both connected in series (with zero chances of failure).

4 Numerical Example

A BCM (bottom case manufacturing) system is a complex system comprising of 10 components. These 10 components can be abated in a binary system acquiring 2 components in series configuration where 8, 9 and 10 are connected in series and the rest of them are attached in a combination of series–parallel configuration. Its system structure function can be described as min(X_1, max(X_2, X_3), X_4, max(X_5, X_6, X_7) X_8, X_9, X_{10}) and the rate of performance of working and non-working components is given as 1, 0 which is based on working or failure of each unit shown as Fig. 1.

From using Fig. 1, UGF of above ten elements can be defined as

$$U_1(z) = p_1 z^1 + (1 - p_1) z^0$$

$$U_2(z) = p_2 z^1 + (1 - p_2) z^0$$

$$U_3(z) = p_3 z^1 + (1 - p_3) z^0$$

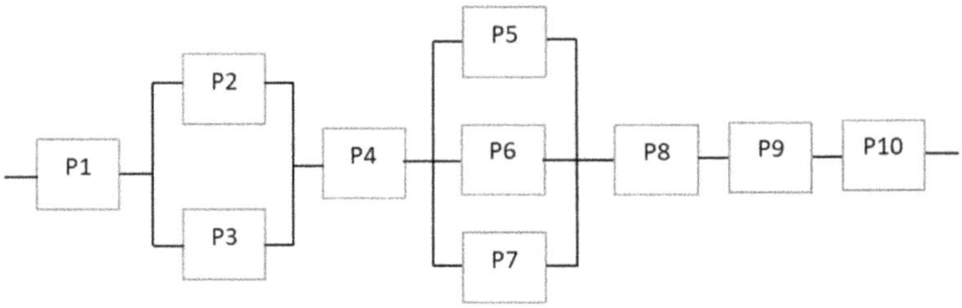

Fig. 1 Block diagram of computer numerical control

$$U_4(z) = p_4 z^1 + (1 - p_4)z^0$$

$$U_5(z) = p_5 z^1 + (1 - p_5)z^0$$

$$U_6(z) = p_6 z^1 + (1 - p_6)z^0$$

$$U_7(z) = p_7 z^1 + (1 - p_7)z^0$$

$$U_8(z) = p_8 z^1 + (1 - p_8)z^0$$

$$U_9(z) = p_9 z^1 + (1 - p_9)z^0$$

$$U_{10}(z) = p_{10} z^1 + (1 - p_{10})z^0.$$

Now operating the UGF of above elements we get,

$$U_{11} = \underset{par}{\varphi}(U_2, U_3) = U_2 \underset{\max}{\otimes} U_3 = [p_2 + (1 - p_2)p_3]z^1 + [(1 - p_2)(1 - p_3)]z^0.$$

$$U_{12} = \underset{par}{\varphi}(U_5, U_6, U_7) = U_5 \underset{\max}{\otimes} U_6 \underset{\max}{\otimes} U_7$$
$$[(p_5 + (1 - p_5)p_6 + (1 - p_5)(1 - p_6)p_7]z^1 + [(1 - p_5)(1 - p_6)(1 - p_7)]z^0.$$

$$U_{13} = \underset{ser}{\varphi}(U_8, U_9, U_{10}) = U_8 \underset{\min}{\otimes} U_9 \underset{\min}{\otimes} U_{10} = [p_8 p_9 p_{10}(1 - p_8 p_9 p_{10})z^0].$$

$$U_{14} = \underset{ser}{\varphi}(U_1, U_{11}) = U_1 \underset{\min}{\otimes} U_{11}$$
$$= [p_2 + (1 - p_2)p_3)p_1]z^1 + [(1 - p_2)(1 - p_3) + (p_2 + (1 - p_2)p_3(1 - p_1)]z^0.$$

$$U_{15} = \underset{ser}{\varphi}(U_4, U_{12}) = U_4 \underset{\min}{\otimes} U_{12}$$
$$= [p_5 + (1 - p_5)p_6)p_4 + [(1 - p_5)(1 - p_6)p_7 p_4]z^1 +$$
$$[p_5 + (1 - p_5)p_6 + (1 - p_5)(1 - p_6)p_7(1 - p_4) + (1 - p_5)(1 - p_6)(1 - p_7)]z^0$$

$$\begin{aligned}U_{16} &= \underset{ser}{\varphi}(U_{14}, U_{15}) = U_{14} \underset{\min}{\otimes} U_{15}\\ &= [p_1p_2 + (1-p_2)p_1p_3)(p_4p_5 + (1-p_5)p_4p_6\\ &\quad + (1-p_5)(1-p_6)p_4p_7)]z^1 + [((1-p_2)(1-p_3)\\ &\quad + p_2 + (1-p_2)p_3)(1-p_7)(1-p_5)(1-p_6) + (p_5\\ &\quad + (1-p_5)p_6 + (1-p_5)(1-p_6)p_7(1-p_4))\\ &\quad + ((1-p_2)(1-p_3) + (p_2 + (1-p_2)p_3)(1-p_1)(p_4p_5)\\ &\quad + (1-p_5)p_4p_6 + (1-p_5)(1-p_6)p_4p_7)]z^0.\end{aligned}$$

Now using the all u-functions connected in series manner such as

$$\begin{aligned}U(z) &= \underset{ser}{\varphi}(U_{16}, U_{13}) = U_{16} \underset{\min}{\otimes} U_{13}\\ &= [p_1p_2 + (1-p_2)p_1p_3)(p_4p_5 + (1-p_5)p_4p_6 + (1-p_5)(1-p_6)p_4p_7)(p_8p_9p_{10}]z^1 +\\ &[((1-p_2)(1-p_3) + p_2 + (1-p_2)p_3)((1-p_7)(1-p_5)(1-p_6) + (p_5 + (1-p_5)p_6 +\\ &(1-p_5)(1-p_6)p_7(1-p_4)) + ((1-p_2)(1-p_3) + (p_2 + (1-p_2)p_3(1-p_2)(1-p_1))\\ &(p_4p_5 + (1-p_5)p_4p_6 + (1-p_5)(1-p_6)p_4p_7 +\\ &(p_1p_2) + (1-p_2)p_1p_3(p_4p_5 + ((1-p_5)p_4p_6 + (1-p_5)(1-p_6)p_4p_7(1-p_8p_9p_{10})z^0.\end{aligned} \tag{4}$$

So, the system reliability can be obtained [20] by

$$\begin{aligned}R &= (p_1p_2 + (1-p_2)p_1p_3)(p_4p_5\\ &\quad + (1-p_5)p_4p_6 + (1-p_5)(1-p_6)p_4p_7(p_8p_9p_{10}).\end{aligned} \tag{5}$$

Let us suppose that all the probabilities are independent identically distributed elements from each other, so let all the probabilities.

$$p_1 = p_2 = \cdots = p_{10} = p$$

$$R = 6p^7 - 9p^8 + 5p^9 - p^{10} \tag{6}$$

Now, obtained tail signature using reliability function from Eqs. (6) and (1) is

$$S = (1, 1/2, 2/5, 1/20, 0, 0, 0, 0, 0, 0, 0). \tag{7}$$

After getting tail signature, calculate the system signature by using Eqs. (2) and (7), we have

$$s = (1/2, 1/10, 7/20, 1/20, 0, 0, 0, 0, 0, 0). \tag{8}$$

Using structure function, we have evaluated the system's minimal signature which is

$$\text{Min. signature} = (0, 0, 0, 0, 0, 0, 6, -7, 5, 1). \tag{9}$$

Hence, from Eq. (11) and $E(T) = \mu \sum_{i=1}^{n} \frac{C_i}{i}$ Expected time is

$$E(T) = 0.188. \tag{10}$$

Expected X and Expected cost rate can be calculated by using $E(X) = \sum_{i=1}^{n} i s_i,\ i = 1, 2, \ldots, 10$ & signature of the system is

$$E(X) = 1.95. \tag{11}$$

From Eqs. (10) and (11), we have.
Cost rate = $E(X)/E(T) = 10.372$.

5 Conclusion

In the present chapter calculate reliability of the bottom case manufacturing system CNC using the UGF technique as it is one of the most compelling methods as compared to others. Signature analysis basically used for comparison system either binary or complex using units of the proposed system. The system CNC having signature (1/2, 1/10, 7/20, 1/20, 0, 0, 0, 0, 0, 0), tail signature (1, 1/2, 2/5, 1/20, 0, 0, 0, 0, 0, 0, 0), minimal signature (0, 0, 0, 0, 0, 0, 6, -7, 5, 1), and expected cost 10.372 have also been quantified to find the model's efficiency. And since the mean value of expected lifetime is 0.188 and cost of system is one the performance of the system will be prominent. CNC types system is a various type of machine tools, decision making, industrial robots, Computer systems for planning, data collection.

References

1. Barlow RE, Proschan F (1996) Mathematical theory of reliability. Soc Ind Appl Math
2. Boland PJ (2001) Signatures of indirect majority systems. J Appl Probab 38(2):597–603
3. Boland PJ, Samaniego FJ (2004) The signature of a coherent system and its applications in reliability. In: Mathematical reliability: an expository perspective, pp 3–30. Springer, Boston, MA
4. Coit DW, Smith AE (1996) Reliability optimization of series-parallel systems using a genetic algorithm. IEEE Trans Reliab 45(2):254–260
5. Ding Y, Zuo MJ, Lisnianski A, Li W (2010) A framework for reliability approximation of multi-state weighted k-out-of-n systems. IEEE Trans Reliab 59(2):297–308

6. Dugan JB, Van Buren R (1994) Reliability evaluation of fly-by-wire computer systems. J Syst Softw 25(1):109–120
7. Enevoldsen I, Sørensen JD (1993) Reliability-based optimization of series systems of parallel systems. J Struct Eng 119(4):1069–1084
8 Eryilmaz S (2012) The number of failed elements in a coherent system with exchangeable elements. IEEE Trans Reliab 61(1):203–207
9. Ghare PM, Taylor RE (1969) Optimal redundancy for reliability in series systems. Oper Res 17(5):838–847
10. Heimann DI, Mittal N, Trivedi KS (1990) Availability and reliability modeling for computer systems. In: Advances in computers, vol 31, pp 175–233. Elsevier
11. Huang T, Yan J, Jiang M, Peng W, Huang H (2016) Reliability analysis of electrical system of computer numerical control machine tool based on bayesian networks. J Shanghai Jiaotong Univ (Sci) 21(5):635–640
12. Iyer RK, Rossetti DJ, Hsueh MC (1986) Measurement and modeling of computer reliability as affected by system activity. ACM Trans Comput Syst (TOCS) 4(3):214–237
13. Kumar A, Singh SB (2017) Computations of the signature reliability of the coherent system. Int J Qual Reliab Manage 34(6):785–797
14. Kumar A, Singh SB (2018) Signature reliability of linear multi-state sliding window system. Int J Qual Reliab Manage 35(10):2403–2413
15. Kumar A, Singh SB (2019) Signature *A*-within-from *B*- *-D/G* sliding window system. Int J Math Eng Manage Sci 4(1): 95–107
16. Kumar A, Kumar V, Modgil V (2019) Behavioral study and availability optimization of a multi-state repairable system with hot redundancy. Int J Qual Reliab Manage 36(3):314–330
17. Kumar A, Singh SB (2021) Signature reliability of consecutive *k*-out-of-*n*: F system using universal generating function. In: Reliability and risk modeling of engineering systems, pp 27–39. Springer, Cham
18. Levitin G (2001) Redundancy optimization for multi-state system with fixed resource-requirements and unreliable sources. IEEE Trans Reliab 50(1):52–59
19. Levitin G (2002) Optimal allocation of elements in a linear multi-state sliding window system. Reliab Eng Syst Saf 76(3):245–254
20. Levitin G (2005) The universal generating function in reliability analysis and optimization, p 442. Springer, London. https://doi.org/10.1007/1-84628-245-4
21. Li KS, Lumb P (1985) Reliability analysis by numerical integration and curve fitting. Struct Saf 3(1):29–36
22. Marichal JL, Mathonet P (2013) Computing system signatures through reliability functions. Statist Probab Lett 83(3):710–717
23. Marichal JL, Mathonet P, Navarro J, Paroissin C (2017) Joint signature of two or more systems with applications to multistate systems made up of two-state components. Eur J Oper Res 263(2):559–570
24. Navarro J, Rubio R (2009) Computations of signatures of coherent systems with five components. Commun Stat Simul Comput 39(1):68–84
25. Navarro J, Rychlik T (2010) Comparisons and bounds for expected lifetimes of reliability systems. Eur J Oper Res 207(1):309–317
26. Navarro J, Ruiz JM, Sandoval CJ (2007a) Properties of coherent systems with dependent components. Commun Stat Theory Methods 36(1):175–191
27. Navarro J, Rychlik T, Shaked M (2007b) Are the order statistics ordered? a survey of recent results. Commun Stat Theory Methods 36(7):1273–1290
28. Palisano R, Rosenbaum P, Walter S, Russell D, Wood E, Galuppi B (1997) Development and reliability of a system to classify gross motor function in children with cerebral palsy. Dev Med Child Neurol 39(4):214–223
29. Pang J, Zhang G, Chen G (2011) ELECTRE I decision model of reliability design scheme for computer numerical control machine. JSW 6(5):894–900
30. Sahner RA, Trivedi K, Puliafito A(2012) Performance and reliability analysis of computer systems: an example-based approach using the SHARPE software package. Springer Science & Business Media

31. Samaniego FJ (2007) System signatures and their applications in engineering reliability, vol. 110. Springer Science & Business Media. ISBN 978-0-387-71796-8
32. Samaniego FJ, Balakrishnan N, Navarro J (2009) Dynamic signatures and their use in comparing the reliability of new and used systems. Naval Res Logist (NRL) 56(6):577–591
33. Shatz SM, Wang JP (1989) Models and algorithms for reliability-oriented task-allocation in redundant distributed-computer systems. IEEE Trans Reliab 38(1):16–27

Evaluate and Measure Agile Software Efficiency by the Integrated Strategy of Fuzzy MOORA and AHP

Abhishek Srivastava, P. K. Kapur, Alakkshendra Rawat, Aditya Mittal, and Vidhyashree Nagaraju

Abstract Now a day's every business stakeholder wants that they will get best output from their software development teams. There are many software developing methodologies present in the market right now and every methodology has its own advantages and disadvantages, even Agile has certain disadvantages, but agile always try to give the best output product to their client and always welcomes the spontaneous market changes, which most of the software developing methodologies are unable to deliver. Now the question arises that how to measure the efficiency of this software development or testing methodology. In this paper, we plan to use an approach or propose an integrated strategy of Fuzzified Multi Objective Optimization on the bases of Ratio Analysis (MOORA) and Analytic Hierarchy Process (AHP) to determine the efficiency of Agile software. In this we first extracted some important factors used in agile software development through correlational research and with the help of agile certified professionals, then we compared those factors with help of the above-mentioned tools Fuzzy based MOORA and AHP. The methodology utilized in this certain research study article acts as very vital toolkit for the researchers or professionals who want to measure and evaluate the efficiency of such software development technology.

Keywords Agile methodologies · Fuzzy MOORA · AHP · Efficiency

---

A. Srivastava (✉) · A. Rawat · A. Mittal
Amity School of Engineering and Technology, Amity University, Noida, Uttar-Pradesh, India
e-mail: abhishek.sri13@gmail.com

P. K. Kapur
Amity Center for Inter-Disciplinary Research, Amity University, Noida, Uttar-Pradesh, India

V. Nagaraju
Tandy School of Computer Science, The University of Tulsa, Tulsa, OK 74104, USA

V. Kumar and H. Pham (eds.), *Predictive Analytics in System Reliability*,
Springer Series in Reliability Engineering,
https://doi.org/10.1007/978-3-031-05347-4_11

1 Introduction

In this mean time of twenty-first century everyone wants to be their work done in hurry and as soon as possibly completed. As we can see the marketing environment is changing extra ordinarily fast and it is really necessary to do both think out-of-the-box and to focus on customer needs in productive and cost-efficient way, also we have to focus on business value of the product. According to market requirements we must be adaptable and flexible for any kind of usual and unusual market changes, here 'Agile' comes in possession. Agile has the ability to tackle mostly any kind of market requirement in minimum given time and giving the impressive outcome to the clients. If we develop culture of agile then it can make a great impact on the long-term success of company, Agile's principles which are commonly utilised in the field of software development can be applied to every other part of business, inclusive of Human Resources (HR). Agile not only satisfy clients with their progressive results but according to the survey the employees who are doing their work in agile environment feel a great overall sense of satisfaction and pride in their work. They feel more empowered as a result of having a clearer understanding of how their function affects the business and working in a more collaborative atmosphere.

The methods employed in the research is briefly detailed here. The system's present design and architecture factors were first evaluated. Measurement of priority weights was used to quantify design and architecture qualities. We estimated the design's utility metrics after successfully implementing two phase assessments. We can take suitable remedial measures to improve the system's efficiency based on the results received.

In this research study we presented a framework to evaluate and measure agile software's efficiency by using an integrated strategy of Fuzzy MOORA and AHP. The multi-objective optimization on the basis of ratio analysis which is collectively referred to as MOORA, this approach was built by Brauers and Zavadskas in 2006 [1]. Fuzzy set theory helps to address the complication in tackling the obscurity in information and the fuzziness in human perception, fuzzy set theory was first applied by Zadeh in 1965. Multi-Objective Optimization on the basis of Ratio Analysis (MOORA) also termed as multi-criteria or multi attribute optimization. It's a framework for synchronously enhancing more than two competing objectives (factors) subject to certain limitations. It has overall two components, i.e., the reference point approach and the ratio system approach. In MOORA various criteria and factors can have different units. The priorities of conflicting criteria are considered by MOORA technique, which were evaluated using AHP. Here AHP stands for Analytic Hierarchy Process, it is a process which solves a problem in three hierarchal steps, the initial part is the problem that it going to resolve, the secondary part is to find the different solutions for one problem because a problem can we solve by alternate methods, the method utilized to evaluate alternative approaches is the tertiary and the most crucial part of the AHP method.

2 Literature Review

In the past, the principles relevant to data distribution methods have been well known and discussed.

Brauers et al. in 2012 [2] done a study on robustness of MULTIMOORA, which states that different objectives have been taken care of by Multi-Objective Optimization with the objectives keeping their own units. A cost-effective technique is developed which process on the principle to balance and manage the criterions that are conflicting, this process is named as multi-Objective optimization on the basis of ratio Analysis (MOORA) [2].

Görener et al. in 2013 acknowledged in their research that using the multi-objective optimization on the basis of ratio analysis (MOORA) and AHP they can rank the locations of alternative branches of banks and target the locations with the profitable client's needs [3].

Kamariah et al. did a study in 2014 about the applications of hybrid MCDM (Multi-Criteria Decision Analysis) for evaluating entrepreneurial intensity among the SMEs they used Fuzzy Analytic Hierarchy Process (FAHP) for discovering the weights of sub criteria and criteria [4].

Ashraf et al. (2013) analysed the effect on data retrieval time of fragmentation and distribution and select the most suitable strategy of fragmentation focused mostly on design and selection characteristics of both the database. It has been concluded that as the technique is altered from centralised to distributed databases, the response time decreases. Mazilu [5] in 2010 explored different approaches to replication of databases and presented many benefits provided by them [6]. Various techniques are also presented by the author in the form of cases in which data replication can be applied. Goel and Buyya [7] proposed the algorithms of replication utilised for various distribution systems and content management system through a survey. Distributed DBMS, P2P networks, Data Grid and WWW were the replication algorithms considered in this paper. Features such as performance, reliability, autonomy of the site, control of the data and heterogeneity were used to evaluate these algorithms [7].

Srivastava et al. in 2012. Highlights in their article the fundamental principles behind distributed database systems, including the management of transactions and access control. The suggested approach to implementing the homogenous distributed database systems showed that contact traffic was decreased, and efficiencies were improved. In order to achieve high data recall efficiency, Chen and others stated in 2015 which was focused upon the use of clustering but explored usage genetically based clustering algorithm for data partitioning. Three new Genetic algorithm operators were suggested by the authors [8].

AHP is a hypothesis that allows to scale or measure the absolute judgements that how much one criteria or attribute is dominating in comparison to other criterions or factors, this was stated by Saaty (2008). For implementing AHP firstly priority scales must be derived then those are synthesized by simply multiplying with the present parent node priorities [9].

Priority weights were calculated by AHP, based on a expert opinion of stakeholders, for the code smells of a business. Kapur et al. have put forward a framework to explore the implementation of ERP systems using AHP through the Analytical Hierarchy Process. In the implementation of ERP, the authors investigated 10 main success factors. A contrast was made based on the opinion of experts and the priorities were decided in relation to the essential success factors. Authors could improve the implementation of the ERP using the proposed AHP-based Methodology [10].

After a comprehensive literature review of MOORA, AHP and data distribution strategies, the authors concluded that a little work is being made on the selection of the data distribution strategy in the research community. There is no work, however, which can rank and choose the majority of effective data distribution strategy in quantitative terms. This led authors to welcome the latest measurement research and choose the best data distribution approach. This research explored a total of five methods for data distribution and five parameters.

3 Agile

Agile software development has differential reaches to development of a software in which solutions and requirements upgrade through the collaborative exertion of cross-functional and self- organizing teams and their clients. Agile team prefer to hand over the work in meagre, but adaptable, increments, instead of giving everything to a "big bang" launch. Agile teams have legitimate structure to give a counter fir a quick change because its results, requirements and plans are continuously evaluating. Agile isn't characterised by a lot of instructions or specific advancement strategies. Or maybe, agile is a gathering of procedures that exhibit a guarantee to tight input cycles and ceaseless.

Agile Manifesto

The approach of Agile towards development of software is explained by its promise to develop the software in regular stages or incrementally. This strategy presents clients with new releases, or versions, of software following brief intervals of work. The brief intervals of work are frequently called sprints.

According to the Agile Manifesto, the four core values of Agile software development are:

- Individuals and communication over process and tools.
- Prioritize working software instead of rigorous paperwork.
- Participation of customer instead contract disputes.
- Instead of following plan focuses on responding to changes.

The 12 principles that are stated in Agile Manifesto are:

1. Fulfilling clients' needs through early and constant delivery of important work.

2. Dividing enormous works in small assignments so that it can be completed rapidly.
3. Observing that the finest work rises up out of self-coordinated teams.
4. Start by giving empowered team members with the bolster & environment required by them, as well as believe them to get the entire job done.
5. Making procedure that support viable exercise.
6. Keeping up a consistent pace for finished work.
7. Always ready to welcome requirement to change anything in project, even it is late.
8. Gathering and meeting with the project team and business professionals consistently throughout the project.
9. Having a team to think about at normal interims on how to turn out to be increasingly powerful, at that point tuning and modifying conduct in like manner.
10. Estimating development by the amount of work done.
11. Consistently looking for greatness.
12. Saddling change for a competitive upper hand.

The purpose of Agile Manifesto is that the component of Agile methodologies explains that the for values which is defines under the Agile manifesto sponsors the process of software development which focuses on making quality products that fulfil customer's expectations and needs. The main motive to make those 12 principles is to support and create a working climate which is totally concentrated on the consumer, that coordinate business intentions and if the market focus and user needs changes that can respond and pivot as quickly as possible.

4 Factors that Affect Agile Software Process

In this section some criterions are discussed which are very necessary for agile software development. So after studying agile software development methodology, some factors are extracted which are important for agile software development procedure, after conveying a survey with the agile certified professionals to give 5 most important factors from the list of many factors, then 5 important criterion are explained as follows.

- **Daily Scrum**: In Agile when team member held meeting to discuss everyday's gaols that what they did yesterday? what will they going to do today? and are there any flaw coming in your way? These meetings are termed as daily scrum meeting which is also ''Daily Scrum''. These meetings are strictly held on same place each day at same time. Daily scrum held in the morning which helps in being clear about the day's goals, daily scrum is typically has time-box of 15 min. The leader of these meetings is addressed as Scrum Master.
- **Iterative Development**: in this method testers divide the whole software into several modules and those modules then further go through the phase of unit testing

independently after the unit testing phase these modules will integrated incrementally and tested to ensure smooth interaction and interface between modules. End time for the testing phase is not fixed in this method. In Iterative development procedure is the process in which the development cycle goes on repeating itself and keeps adding new features and functionalities in a cyclic repetitive manner. Agile methodology joins the way of thinking of iterative and incremental software advancement that is demonstrated around a continuous increment in features additions and a cyclic release and redesigned upgraded pattern. The result of the subsequent iteration is an improved working augmentation of the product. This is rehashed until the product achieves the necessary functionalities.

- **Agile Testing**: We have discussed above about Agile testing where we elaborated properly about each component of agile software testing. Basically agile has some principles to run its testing, agile team focuses on continuous testing of every iteration in the end of it because it is the only to monitor the progress of the project, agile testing gives us proper and regular feedback which helps us to make product met properly with the business requirements, acc to SDLC team members has the approval to run test on the application while in agile BA's and developers are also included to test the application, in agile business teams are involved in tests of every iteration because it give gives continuous feedback which lessens the response time and expenses involved in fixing it, testing runs with the implementation going on, agile does not support heavy documentation, reusable checklists are used by the testers.
- **Retrospective**: In one of the 12 principles of Agile Manifesto, it is mentioned that team must think about on how to become more effective, then adjust and tune according to the behaviour. This principle fused in the agile groups in the face of retrospective meetings. Retrospective meetings have the motive to reflect on the most recent projects/sprints/milestones and to address the points which needs to be improvised and then celebrate team wins. These meetings are to be held at the end of the sprints and before the starting of next sprint, during retrospective meetings teams survey the particular risky situations.
- **User Story**: A user story is a casual language portrayal of one or more aspects of a software system used in product management and software development. A user story is a tool used in agile development to capture a depiction of a software feature from the perspective of the end-user. A user story depicts the user's point of view, what they require, and why they require it. Making a reorganized depiction of requirement is easier with the help of a user story.

5 Approach Used to Measure Efficiency of Agile Software

We used an integrated strategy of Fuzzy MOORA and AHP to evaluate the efficiency of an agile software. We are going to discuss now that what techniques and formulas are used to find the values using the proposed layout of FUZZY MOORA and AHP.

The norms and criterions have been evaluated and assigned by the priority weight by the utilization of AHP, later on Fuzzy Based MOORA will use those priority weights. It is an analytic tool utilized to resolve the problems with complex decision-makings which gives us the quantitative values from the qualitative values.

Zadeh [11] developed the FUZZY set theory in 1965. This hypothesis helps in modelling the framework quantitatively and qualitatively when their unclearness, vulnerability and equivocalness exists. Fuzzy numbers can be of two types knows as trapezoidal fuzzy numbers and triangular fuzzy numbers. We are going to use triangular fuzzy numbers in this paper. Let $\mu_X(a)$ be represent the fuzzy sub-set X, which maps element x in X into the interval of real numbers [0, 1].

$$\begin{aligned} \mu_x(a) &= \frac{\mathrm{a} - \mathrm{y}}{a - z'} \qquad y \leq a \geq z \\ &= \frac{\mathrm{a} - \mathrm{c}}{z - c'} \qquad z \leq a \leq c \\ &= 0, \qquad Others \end{aligned} \tag{1}$$

Above we mentioned the membership function $\mu_X(a)$ which is comprised of triangular fuzzy numbers (y, z, c).

Now we are going to discuss about the steps which are adopted to conquer purpose of this study.

5.1 *Fuzzy Matrix of Decision*

Making of the Fuzzy matrix of decision is the very initial step to use Fuzzy based MOORA technique, which is made on the basis of expert responses. These experts are those who have experience in depth regarding to the topic.

$$A = \begin{bmatrix} [y_{11}^a, y_{11}^b, y_{11}^c] & [y_{12}^a, y_{12}^a, y_{12}^a] & [y_{13}^a, y_{13}^b, y_{13}^c] & \cdots & [y_{1n}^a, y_{1n}^b, y_{1n}^c] \\ [y_{21}^a, y_{21}^b, y_{22}^c] & [y_{22}^a, y_{22}^b, y_{22}^c] & [y_{23}^a, y_{23}^b, y_{23}^c] & \cdots & [y_{2n}^a, y_{2n}^b, y_{2n}^c] \\ [y_{31}^a, y_{31}^b, y_{31}^c] & [y_{32}^a, y_{32}^b, y_{32}^c] & [y_{33}^a, y_{33}^b, y_{33}^c] & \cdots & [y_{3n}^a, y_{3n}^b, y_{3n}^c] \\ \cdots & \cdots & \cdots & \cdots & \cdots \\ \cdots & \cdots & \cdots & \cdots & \cdots \\ [y_{m1}^a, y_{m1}^b, y_{m1}^c] & [y_{m2}^a, y_{m2}^b, y_{m2}^c] & [y_{m3}^a, y_{m3}^b, y_{m3}^c] & \cdots & [y_{mn}^a, y_{mn}^b, y_{mn}^c] \end{bmatrix} \tag{2}$$

In the above matrix the lower, middle, and higher criteria is denoted by $y_{ij}^a, y_{ij}^b, y_{ij}^c$.

5.2 *Fuzzy Decision Matrix's Normalization*

$$T_{ij}^{a} = \frac{y_{ij}^{a}}{\sqrt{\sum_{1=1}^{m}\left[\left(y_{ij}^{a}\right)^{2} + \left(y_{ij}^{b}\right)^{2} + \left(y_{ij}^{c}\right)^{2}\right]}} \tag{3}$$

$$T_{ij}^{b} = \frac{y_{ij}^{b}}{\sqrt{\sum_{1=1}^{m}\left[\left(y_{ij}^{a}\right)^{2} + \left(y_{ij}^{b}\right)^{2} + \left(y_{ij}^{c}\right)^{2}\right]}} \tag{4}$$

$$T_{ij}^{c} = \frac{y_{ij}^{c}}{\sqrt{\sum_{1=1}^{m}\left[\left(y_{ij}^{a}\right)^{2} + \left(y_{ij}^{b}\right)^{2} + \left(y_{ij}^{c}\right)^{2}\right]}} \tag{5}$$

Now in this step we have to normalize the decision matrix which is made by the responses of experts by using the vector normalizing formula. By using above formula, the normalized values for the elements of matrix of decision are evaluated.

5.3 *Priority Weights Determination for Criteria*

AHP is utilised to decide the specific weights for the priority in this step. In this strategy we must conduct the pair-by-pair correlation of the criterions chosen by the experts. All the criteria and factors are compared pair wise with each other (two at a time) in this following technique. By using Saaty's [9] scale 1–9 the qualitative judgements of the comparison of two factors done by professionals, can be converted into quantitative measure. The matrix structure captures the outcome of the comparison which is named as matrix of judgement. When the values are allocated in the matrix of judgement, priority weight (w) is evaluated after the normalization of the initial matrix. The score of b in in the judgement matrix speaks to the overall significance of the component in the row (i) over the component in the column (j).

$$X = \begin{bmatrix} 1 & b_{12} & b_{13} & \cdots & b_{1n} \\ b_{21} & 1 & b_{23} & \ldots & b_{2n} \\ b_{31} & b_{32} & 1 & \ldots & b_{3n} \\ b_{41} & b_{42} & b_{43} & \cdots & b_{4n} \\ \ldots & \ldots & \ldots & \ldots & \ldots \\ b_{n1} & b_{n2} & b_{n3} & \ldots & 1 \end{bmatrix}$$

Sample of the judgement matrix is shown above.

$$W_i = \frac{\sum_{i=1}^{I}\left(\frac{b_{ij}}{\sum_{j=1}^{J} b_{ij}}\right)}{J} \tag{6}$$

The algorithm mentioned above is the algorithm which we going to use in this following report to evaluate the weight of the priorities (W). In the above-mentioned formula "j" denotes the column number while "i" denote the row number.

5.4 Weighted Normalized Fuzzy Matrix of Decision

$$v_{ij}^{a} = W_i T_{ij}^{a} \tag{7}$$

$$v_{ij}^{b} = W_i T_{ij}^{b} \tag{8}$$

$$v_{ij}^{c} = W_i T_{ij}^{c} \tag{9}$$

Above mentioned Eqs. (7), (8), (9) of the algorithm shoes that the weights obtained by the AHP in preceding step is multiplied by each element present in the normalized fuzzy matrix of decision for the extraction of weighted normalized fuzzy matrix of decision.

5.5 Overall Rating for the Non-beneficial and Beneficial Criteria

$$s_i^{+a} = \sum_{j=1}^{n} v_{ij}^{a}, Where\ j\ belongs\ to\ the\ beneficial\ criteria \tag{10}$$

$$s_i^{+b} = \sum_{j=1}^{n} v_{ij}^{b}, Where\ j\ belongs\ to\ the\ beneficial\ criteria \tag{11}$$

$$s_i^{+c} = \sum_{j=1}^{n} v_{ij}^{c}, Where\ j\ belongs\ to\ the\ beneficial\ criteria \tag{12}$$

$$s_i^{-a} = \sum_{j=1}^{n} v_{ij}^{a}, Where\ j\ belongs\ to\ the\ non-beneficial\ criteria \tag{13}$$

$$s_i^{-b} = \sum_{j=1}^{n} v_{ij}^{b}, Where\ j\ belongs\ to\ the\ non-beneficial\ criteria \tag{14}$$

$$s_i^{-c} = \sum_{j=1}^{n} v_{ij}^{c}, Where\ j\ belongs\ to\ the\ non-beneficial\ criteria \tag{15}$$

All the equations mentioned above will assist us to rate non beneficial and beneficial criteria for every strategy. These equations will evaluate the overall ratings for the strategy of data distribution and the beneficial and non-beneficial criterions for the lower, middle, and upper values for triangular function.

5.6 *Performance Index for Every Alternative*

To determine the de-fuzzified values we will use the equation mentioned below in this step, which will evaluate the overall index of performance (S_i) for each and every alternative.

$$S_i\left(s_i^+, s_i^-\right) = \sqrt{\frac{1}{3}\left[\left(s_i^{+a} - s_i^{-a}\right)^2 + \left(s_i^{+b} - s_i^{-b}\right)^2 + \left(s_i^{+c} - s_i^{-c}\right)^2\right]}. \quad (16)$$

6 Discussions

In this paper the following data distribution sheet which contains AHP table and fuzzy based MOORA calculations by keeping in mind the designing phase for the local software like a video media player.

Table 1 is actually known as the cross matrix which is made for the comparison purpose. These values are actually the quantitatively converted qualitative values which are given by professionals/experts of the following field. It is basically calculated on the base of priority means what we want to choose over something and how much priority we assign to something over something which are interrelated to each other. In AHP we compare two criterions at a time, for e.g. Accessibility is compared to A(Accessibility) first then it is compared to K(Availability) and so on, basically every vertical criterion has been compared with every horizontal criterion.

Above in Table 2 we have calculated priority weights for each individual criterion. Table 8.2 is the normalized form of the Table 1 which is evaluated using Eq. (6).

Table 1 Cross matrix—through AHP

Table 1—goal criteria	A	K	M	C
Accessibility	1	2	2	3
Availability	1/2	1	1	2
Manageability	1/2	1	1	2
Costs	1/3	1/2	1/2	1
Total	2.3333	4.5000	4.5000	8.0000

Note A(Accessibility), K(Availability), M(Manageability), C(Cost)

Table 2 Criteria weights—through AHP

Table 1—goal criteria	A	K	M	C	P.E.V
Accessibility	0.43	0.4444	0.44	0.3750	0.42
Availability	0.21	0.2222	0.22	0.2500	0.23
Manageability	0.21	0.2222	0.22	0.2500	0.23
Costs (H/S)	0.14	0.1111	0.11	0.1250	0.12
Total	1	1.0000	1.00	1.0000	1.00

Table 3 Fuzzy matrix of decision

	A			K			M			C		
	a	b	c	a	b	c	a	b	c	a	b	c
User story	0.6	0.8	0.6	0.6	0.6	0.4	0.8	0.8	0.8	0.8	1	1
Daily scrum	0.4	0.6	0.6	0.2	0.2	0.2	0.2	0.4	0.6	0.6	0.6	0.6
Iterative Development	0.8	0.6	0.8	0.8	0.8	0.6	0.6	0.6	0.4	0.8	1	0.6
Agile testing	0.6	0.6	0.6	0.8	0.8	0.8	0.4	0.4	0.4	0.6	0.8	1
Retrospective	0.4	0.6	0.4	0.6	0.4	0.2	0.2	0.4	0.6	0.4	0.6	0.8

As we can see in the Table 3 above it is a matrix which contains values which are filled by the three different experts who are professionals in their fields. The triangular fuzzy number approach for making this matrix is used. The values in the above matrix are the values which denotes the equivocalness on the scale of 0–1 between the two elements in which one is an attribute, and one is a condition which is necessary for our software.

The contents of the Table 3 are normalized using the Eq. (3), (4), and (5) then as the result we get the Table 4, which is normalized fuzzy Matrix of Decision.

The weighted normalized fuzzy matrix of decision mentioned in Table 5 is made with the help of Analytic Hierarchy Process (AHP). The weights are 'd' in the Table 2. The values in the normalized fuzzy matrix of decision are multiplied by the weights which are evaluated in the Table 2 then we get the resultant weighted normalized fuzzy matrix of decision as exhibited above in Table 5.

The performance rating for the Table 6 is evaluated using the Eqs. (10) to (15). Accessibility, Availability, and Manageability comes in the beneficial criteria that means the values of the corresponding elements needs to be added and maximized for the evaluation of performance rating. Now the Cost comes in the non-beneficial criteria which values is needs to be subtracted and minimized using Eqs. (13) to (14).

In Table 7 the ranks of the overall performance rating are calculated by defuzzifying the Beneficial and Non Beneficial matrix by the help of Eq. (16). Hence by using the integrated strategy of AHP and Fuzzy MOORA, we evaluated the ranks of the 5 criteria with the great success, that we extracted through correlational research which are used in agile software development with help of some experts.

Table 4 Normalized fuzzy decision matrix

	A			K			M			C		
	a	b	c	a	b	c	a	b	c	a	b	c
User story	0.252	0.336	0.252	0.265	0.265	0.176	0.381	0.381	0.381	0.268	0.335	0.335
Daily scrum	0.168	0.252	0.252	0.088	0.088	0.088	0.095	0.190	0.286	0.201	0.201	0.201
Iterative Development	0.336	0.252	0.336	0.353	0.353	0.265	0.286	0.286	0.190	0.268	0.335	0.201
Agile testing	0.252	0.252	0.252	0.353	0.353	0.353	0.190	0.190	0.190	0.201	0.268	0.335
Retrospective	0.168	0.252	0.168	0.265	0.176	0.088	0.095	0.190	0.286	0.134	0.201	0.268

Table 5 Normalized weighted fuzzy decision matrix

	A			K			M			C		
	a	b	c	a	b	c	a	b	c	a	b	c
User story	0.106	0.142	0.106	0.060	0.060	0.040	0.086	0.086	0.086	0.032	0.041	0.041
Daily scrum	0.071	0.106	0.106	0.020	0.020	0.020	0.021	0.043	0.064	0.024	0.024	0.024
Iterative development	0.142	0.106	0.142	0.080	0.080	0.060	0.064	0.064	0.043	0.032	0.041	0.024
Agile testing	0.106	0.106	0.106	0.080	0.080	0.080	0.043	0.043	0.043	0.024	0.032	0.041
Retrospective	0.071	0.106	0.071	0.060	0.040	0.020	0.021	0.043	0.064	0.016	0.024	0.032

Table 6 Performance rating

	Beneficial (S +)			Non-beneficial (S−)		
	a	b	c	a	b	c
User story	0.253783	0.289416	0.233703	0.032892	0.041115	0.041115
Daily scrum	0.113007	0.1703	0.191961	0.24669	0.024669	0.024669
iterative Development	0.287835	0.252202	0.246094	0.32892	0.041115	0.024669
Agile testing	0.230541	0.230541	0.230541	0.24669	0.032892	0.041115
Retrospective	0.153167	0.190381	0.156329	0.016446	0.024669	0.032892

Table 7 Ranking of overall performance rating

Name	Rank
User story	3
Daily scrum	1
Iterative development	2
Agile testing	4
Retrospective	5

7 Conclusion

Agile is the mostly used and most advanced approach in software development now a days, but the question is how much efficient this methodology is quantitatively and qualitatively, for solving this problem we used the integrated strategy of Fuzzy based MOORA and Analytic Hierarchy Process (AHP). We extracted 5 factors by doing correlational research i.e., User Story, Daily Scrum, Iterative development, Agile Testing, and Retrospective, we took quantitative values too for 5 factors from certified agile experts those values were converted into fuzzy decision matrix using the integrated strategy of Fuzzy MOORA and then that fuzzy matrix was normalized by some calculations and using Eqs. (3), (4), and (5) that are mentioned in the Sect. 5 of this paper. The above-mentioned strategies then placed using Horizontal and Vertical Partitioning. We used AHP for determining priority weights for the 4 criterions for software quality i.e., Accessibility, Availability, Cost, Manageability, and after using the Eqs. (7), (8), and (9) which are mentioned again in Sect. 5, it is observed through the calculation that Accessibility got maximum weight than other criterions. After that these weights were used in Fuzzy based MOORA for the determination of the Ranks among the 5 Factors. The reason we calculated these ranks is that to determine the best software development approach in Agile methodology. After the calculations Daily Scrum gained the 1st Rank. For saving time and money this method is useful for industry experts and researchers who wish to define factors and criterions quantitively. So, this paper concludes that Daily Scrum is an important entity to consider for better agility and growth.

References

1. Brauers WKM, Zavadskas EK (2006) The MOORA method and its application to privatization in a transition economy. Control Cybern
2. Brauers WKM, Zavadskas EK (2012) Robustness of Multimoora: a method for multi-objective optimization. Informatica
3. Görener A, Dinçer H, Hacıoğlu U (2013) Application of multi-objective optimization on the basis of ratio analysis (MOORA) method for bank branch location selection. Int J Finance Banking Stud
4. Rostamzadeh R, Ismail K, Bodaghi Khajeh Noubar H (2014) An application of a hybrid MCDM method for the evaluation of entrepreneurial intensity among the SMEs: a case study
5. Mazilu MC (2010) Database replication. Database Syst J
6. Ashraf I, Khokhar AS, Lundberg L (2013) Impact of data distribution on response time in telecom databases. Int J Eng Res Appl
7. Goel S, Buyya R (2006) Data replication strategies in wide area distributed systems. Enterprise service computing: from concept to deployment
8. Srivastava A, Shankar U, Tiwari SK (2012) Transaction management in homogenous distributed real-time replicated database systems. Int J Adv Res Comput Sci Softw Eng
9. Saaty TL (2008) Decision making with the Analytic hierarchy process
10. Gupta V, Kapur PK, Kumar D (2016) Modelling and measuring code smells in enterprise applications using TISM and two-way assessment. Int J Syst Assur Eng Manage
11. Zadeh LA (1965) Fuzzy sets. Int J Inform Control 8

Software Reliability Models and Multi-attribute Utility Function Based Strategic Decision for Release Time Optimization

Vishal Pradhan, Joydip Dhar, and Ajay Kumar

Abstract The software industry is working hard to keep up with these rapid changes by devising methods to increase the pace of their work without compromising software quality and reliability. Various factors, such as the testing environment, testing strategy, and resource allocation, can influence the optimal release time. The choice of whether or not to release a software product would become much more complicated and significant. When a software developer, clients, or end-users face significant potential financial losses, a decision has strategic significance. A software release decision is a trade-off between early release to take advantage of an earlier market launch and product release deferral to ensure reliability. If a software product is released too soon, the software developer must pay for post-release costs to correct bugs. To decide the best software release time, two attributes, reliability and cost, must be combined. This study discusses a realistic approach to determining when to stop software testing that considers reliability and cost. A multi-attribute utility theory-based proposed decision model is analyzed on various separate weighted combinations of utility functions.

Keywords Software reliability growth model · Multi-attribute utility function · Optimal release time · Non-homogeneous poisson process

1 Introduction

Software technologies are the most prevalent human-made technology that impacts our daily lives due to the importance of software applications in recent years. In the last two decades, the penetration of software-based technologies into people's everyday lives has been remarkable. Everything we see around us is dependent on software or has some connection to software systems. There is a requirement for

V. Pradhan (✉) · J. Dhar · A. Kumar
ABV-Indian Institute of Information Technology and Management Gwalior,
Gwalior 474015, India
e-mail: vishal.iiitmg@gmail.com; vishalp@iiitm.ac.in

V. Kumar and H. Pham (eds.), *Predictive Analytics in System Reliability*,
Springer Series in Reliability Engineering,
https://doi.org/10.1007/978-3-031-05347-4_12

highly dependable, secure, and high-quality software development since our social structure has become increasingly dependent on software-based technologies [31]. Reliability can only be accomplished by thoroughly testing the program before it is made available to the public. Program errors are found, identified, and fixed throughout the testing process, improving software reliability [13]. Reliability is an essential statistic for evaluating commercial software quality in the testing and operational phases. Software reliability may be defined as the likelihood of error-free software execution in a particular environment over a predetermined duration [17, 23]. Non-homogeneous Poisson process (NHPP) based growth models are frequently used in software systems to describe stochastic failure behavior and measure growth reliability [3, 5, 7, 9, 33].

NHPP models have also been extensively used in the cost-control analysis, software time-to-market analysis, and resource allocation issues [1, 10, 12, 26, 32]. The correctness and security of a software system can only be improved with sufficient testing time and effort, such as CPU hours and qualified testing specialists [10, 29]. In general, software testing uses around half of the resources for software development. Continuous software testing for a more extended period may obstruct the timely delivery of the software system. Furthermore, it will quickly result in significant development expenses. Simultaneously, shorter testing combined with an insufficient debugging procedure would cause customer disappointment, potentially affecting the growth of the software as well as the software firm's goodwill. In today's market, a software testing budget should be prioritized over its development budget [11, 20]. As a result, software reliability engineering provides a cost-effective compromise between client needs for dependability, accessibility, delivery time, and life cycle [16, 19]. SRGMs are used to optimize testing techniques for increased organizational competitiveness, estimate the amount of required resources, and calculate the overall cost of the development process [22, 34, 39, 40].

The software reliability may be predicted using appropriate software reliability growth models (SRGMs) based on the fault count data obtained during the testing process [21, 36]. The testing phase is the most significant since it is at this step that the fault detection and removal procedure takes place, which is critical for the dependability and quality of any software system. A critical decision point for management is when to end testing and release the software system to the user [30]. This is referred to as the "Software Release Time Problem". Before being released, the software is subjected to a rigorous testing procedure in order to identify flaws that might have devastating effects if not corrected. Several methods of software testing are now in use with the goal of eliminating faults. It's possible that many bugs went undiscovered because of the short testing time and the sudden release [37]. The choice to release software is a complicated one, and there are significant dangers involved with a release agreement that is either too rapid or too delayed [18, 24]. One of the most common applications of SRGMs is to assist developers in determining the optimal timing to deploy software [2, 6, 22, 35].

The main contribution of this work is as follows:

1. Proposed new SRGMs with log-logistic and Burr Type XII distribution as a fault detection rate.
2. This study suggests the multi-attribute utility theory based optimal release time.

2 Software Reliability Modeling

The NHPP based SRGM is used in this work. The NHPP is a method of calculating the total number of faults found throughout the testing procedure. In this technique, SRGMs such as exponential [8], delayed S-Shaped [41], inflected delayed S-shaped [27], and power function have been used to anticipate potential bugs laying latent in the program. Let $N(t)$ be the total number of defects discovered at time t, and $m(t)$ be the expected number of faults. The failure intensity $\lambda(t)$ is therefore linked as follows:

$$m(t) = E[N(t)] = \int_0^t \lambda(s)ds \ , \tag{1}$$

where $N(t)$ has a Poisson probability mass function with parameter m(t), which is as follows:

$$Pr\{N(t) = n\} = \frac{m(t)^n.e^{-m(t)}}{n!} \ , \quad x = 0, 1, 2, ... \tag{2}$$

Various time-dependent models that describe the stochastic failure process of an NHPP have been published in the literature. The failure intensity function $\lambda(t)$ differs across these models, and therefore $m(t)$. In the case of finite failure NHPP models, let "A" indicate the estimated total number of faults that would be identified given infinite testing time.

One of the main goals of testing is to identify software faults to fix them. Once the software code has been written, testing can begin. Before the software is released to the public, the software testing team thoroughly tests it to ensure that the software contains the least number of bugs. Despite the fact is that it is almost impossible to eliminate all the software bugs. As a result, when the testing team tests the software, there's a probability they'll only find a finite number of problems in the code (less than the total number of faults).

2.1 Assumption

i. NHPP models the failure observation/fault removal phenomenon.
ii. The software system is susceptible to failure at any time due to errors that have remained in the system.

iii. There are a finite number of bugs present in the software.
iv. When a failure occurs, it is instantly removed.
v. The severity level of all faults is the same.
vi. The perfect debugging environment is taken into account.
vii. All remaining software faults have an equal impact on the failure rate.
viii. The number of defects discovered throughout the testing process is directly proportional to the number of faults still present in the software.
ix. With a probability distribution function, each occurrence of failure is distributed independently and identically across the software life-cycle.

As a result, finite numbers of bugs are perfectly eradicated, with the mathematical equation. The finite failure NHPP models' differential equation formulated based on the modeling assumption and it expressed as:

$$\frac{dm(t)}{dt} = r(t)[\Lambda - m(t)] \tag{3}$$

When Eq. (3) is solved for the initial condition $m(0) = 0$, the MVF can be calculated as follows:

$$m(t) = \Lambda[1 - e^{-\int_0^t r(v)dv}] = \Lambda[1 - e^{-B(t)}] \tag{4}$$

$$m(t) = \Lambda.F(t)\ . \tag{5}$$

where $F(t)$ is a distribution function.

Various researches assume that fault detection is constant throughout the testing process, but it is not possible in practical behavior. For the detection rate, we know that it is low at the initial stage, and in the mid-stage, it's on the peak; in the later stage, it's again low. So, the FDR is modeled through the specific distribution handling the situation. Therefore, this study proposed the SRGMs with the two most applicable distribution functions for $B(t)$, i.e., Log-logistic and Burr type XII distribution functions.

2.2 *Fault Detection Rate*

2.2.1 Log-Logistic Distribution

The logistic distribution and the log-logistic distribution are closely linked. A probability distribution whose logarithm has a logistic distribution is known as a log-logistic distribution. Log(x) is distributed logistically with mean and standard deviation if x is distributed loglogistically with parameters μ and σ. The log-logistic distribution is a good replacement for the Weibull distribution. It's a hybrid of the Gompertz and Gamma distributions, with the mean and variance values equal to one. The log-logistic distribution has its own status as a life testing model; it is an increasing failure rate model as well as a weighted exponential distribution. The generalized log-logistic distribution refers to several distinct distributions that include the

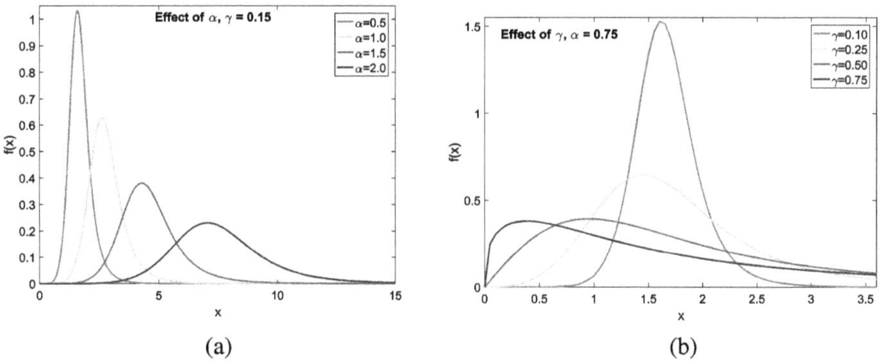

Fig. 1 Log-logistic distributions with effect of **a** shape and **b** scale parameter values

log-logistic as a particular instance. The Burr Type XII distribution and the Dagum distribution, both of which have a second shape parameter, are examples. It's a flexible distribution family that may represent a wide range of distribution types that are shown in Fig. 1. In survival analysis, this distribution is frequently used to simulate events that have an initial rate increase followed by a rate decrease.

The log-logistic distribution with positive scale parameter γ and shape parameter α is described as follows:

$$B(t) = \frac{(\frac{t}{\gamma})^{\alpha}}{1 + (\frac{t}{\gamma})^{\alpha}}, \tag{6}$$

and the density function is:

$$b(t) = \frac{(\frac{\alpha}{\gamma})(\frac{t}{\gamma})^{\alpha-1}}{[1 + (\frac{t}{\gamma})^{\alpha}]^2} \tag{7}$$

2.2.2 Burr Type XII Distribution

The Burr distribution can fit a wide range of empirical data. The parameters' various values span a wide range of skewness and kurtosis. As a result, it is used to represent a range of data types in diverse disciplines such as finance, hydrology, and reliability. The Burr type XII distribution generalizes Burr distribution with additional scale parameters. It is a three-parameter family of positive real-line distributions. It's a versatile distribution family that may represent a variety of different distribution forms that are shown in Fig. 2. Many widely used distributions, such as gamma, log-normal, log-logistic, bell-shaped, and J-shaped beta distributions, are included, overlapped, or have the Burr distribution as a limiting case (but not U-shaped). The Burr distribution is also found in several compound distributions. A Burr distribution is created by compounding a Weibull distribution with a gamma distribution for the scale parameter. There are two asymptotic limiting cases for the Burr distribution: Weibull and Pareto Type I.

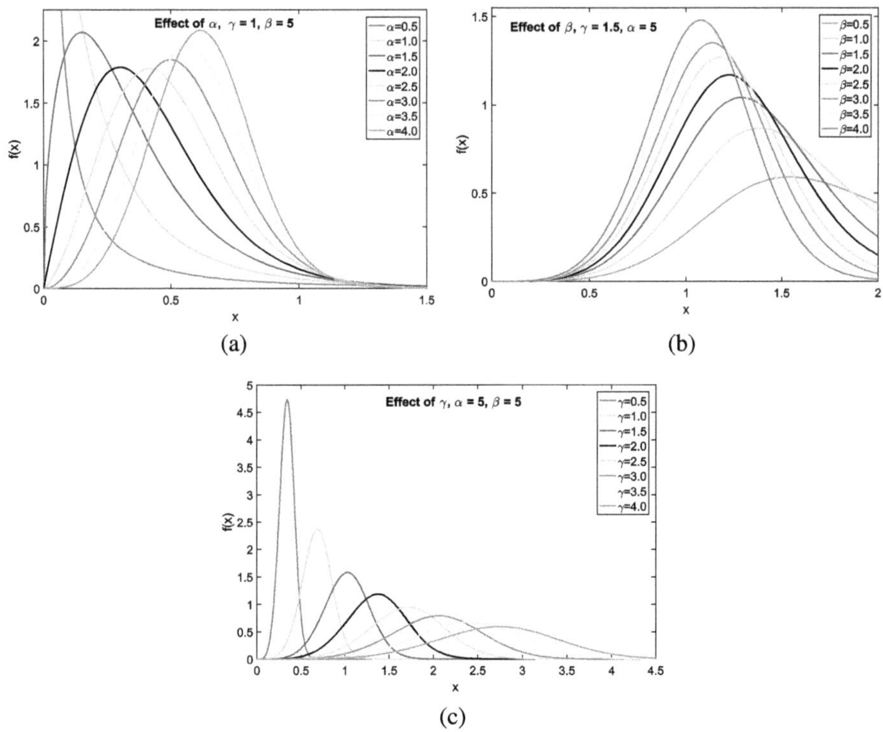

Fig. 2 Burr Type XII distributions with effect of **a** shape, **b** shape and **c** scale parameters

The Burr distribution's cumulative distribution function (cdf) is:

$$B(t) = 1 - \frac{1}{\left[1 + \left(\frac{t}{\gamma}\right)^{\alpha}\right]^{\beta}}, \tag{8}$$

$$b(t) = \frac{\frac{\alpha\beta}{\gamma}(\frac{t}{\gamma})^{\alpha-1}}{[1 - (\frac{t}{\gamma})^{\alpha}]^{\beta+1}} \tag{9}$$

2.3 *Software Reliability Growth Models*

$$m(t) = \Lambda[1 - e^{-B(t)}] \tag{10}$$

2.3.1 Model-1

The MVF of Log-logistic FDR based SRGM is defined as:

$$m(t) = \Lambda \left[1 - e^{-\frac{(\frac{t}{\gamma})^\alpha}{1+(\frac{t}{\gamma})^\alpha}} \right]. \tag{11}$$

2.3.2 Model-2

The MVF of Burr type XII FDR based SRGM is defined as:

$$m(t) = \Lambda \left[1 - e^{-\left(1-\frac{1}{\left[1+\left(\frac{t}{\gamma}\right)^\alpha\right]^\beta}\right)} \right]. \tag{12}$$

One of the most common applications of SRGMs is to assist developers in determining the optimal timing to deploy software. This study formulated a cost model to estimate the best software release timing in the latter portion of this paper. This field of study is strongly connected to the wider software reliability research.

3 Numerical Illustration

The practical applicability of the suggested problem is demonstrated in this section using historical fault discovery data as an example. The fault count data set was used for the numerical illustration. The non-linear least square estimation (LSE) method is used to estimate model parameters. The estimated model parameter findings for detected faults throughout the testing period are shown in Table 1.

The behavior of actual defects data for software release is observed in the graph and most of them are in S-shaped form. This is further supported by the usage of the log-logistic and Burr Type XII FDR function to detect software faults. As evidenced by the values of several comparison criteria, model-1 and model-2 provide a perfect fit. Table 2 present the estimated values of proposed and existing models for DS-1 to DS-6. Table 3 present a comparative analysis of the proposed and existing models. The comparison criteria used here are the sum of square error (SSE), coefficient of determination (R^2) and Adjusted R-square (R^2_{adj}) Fig. 3 illustrate a graphical representation of estimated vs. real cumulative failures over time for a better understanding. Based on these findings, we can conclude that the proposed

Table 1 Datasets from the existing literature

Dataset (DS)	Testing time	Detected faults	Remark
DS-1 [38]	18 Weeks	176	Failure data of large medical record system
DS-2 [38]	17 Weeks	204	Failure data of large medical record system
DS-3 [42]	21 Weeks	43	System test data for a telecommunication system
DS-4 [25]	30 Days	289	Real software project failure data
DS-5 [8]	20 Weeks	100	Computer Programming Center of NTDS data
DS-6 [28]	19 Weeks	328	Reported from Ohba 1984 test data

Table 2 Estimated values of SRGMs parameters for all six datasets

DS	Model-1			Model-2				GO model		DSS model	
	Λ	γ	α	Λ	γ	α	β	Λ	b	Λ	b
DS-1	305.9	9.707	3.082	277.7	171.5	2.672	1805	985.9	0.9243	226.1	0.1741
DS-2	358.9	3.111	0.980	325.2	2.3E+4	0.842	1500	197.4	0.3985	192.5	0.8814
DS-3	106.7	19.63	1.917	82.74	661.8	1.903	987.2	1.6E+4	1.3E+4	62.30	0.1185
DS-4	831.0	35.76	1.880	651.3	171.0	1.841	21.30	6.2E+4	1.4E+4	495.7	0.0645
DS-5	52.85	5.29	1.448	67.19	1.334	4.198	0.083	31.66	0.1906	30.35	0.4601
DS-6	979.8	23.5	1.311	741.9	2768	1.301	626.4	760.5	0.0323	374.1	0.1977

Table 3 Performance comparison of SRGMs for all six datasets

DS	Model-1			Model-2			GO model			DSS model		
	SSE	R^2	R^2_{adj}	SSE	R^2	R^2_{adj}	SSE	R^2	R^2_{adj}	SSE	R^2	R^2_{adj}
DS-1	2544	0.9598	0.9544	2315	0.9634	0.9556	4789	0.9243	0.9196	3246	0.9487	0.9455
DS-2	1034	0.9477	0.9402	910.2	0.9539	0.9433	1210	0.9388	0.9347	3489	0.8234	0.8117
DS-3	59.80	0.9855	0.9839	56.00	0.9864	0.9840	125.8	0.9612	0.9598	62.19	0.9849	0.9841
DS-4	2204	0.9912	0.9905	2194	0.9912	0.9902	9663	0.9612	0.9598	2277	0.9909	0.9905
DS-5	43.80	0.9714	0.9696	26.15	0.9830	0.9810	62.29	0.9596	0.9581	102.4	0.9336	0.9311
DS-6	2111	0.9892	0.9879	2025	0.9897	0.9876	2656	0.9865	0.9857	3205	0.9837	0.9827

model-2 produces good performance and is more realistic when it comes to forecasting the growth behavior of application-based software systems.

In the next section we discuss about the optimal release policy.

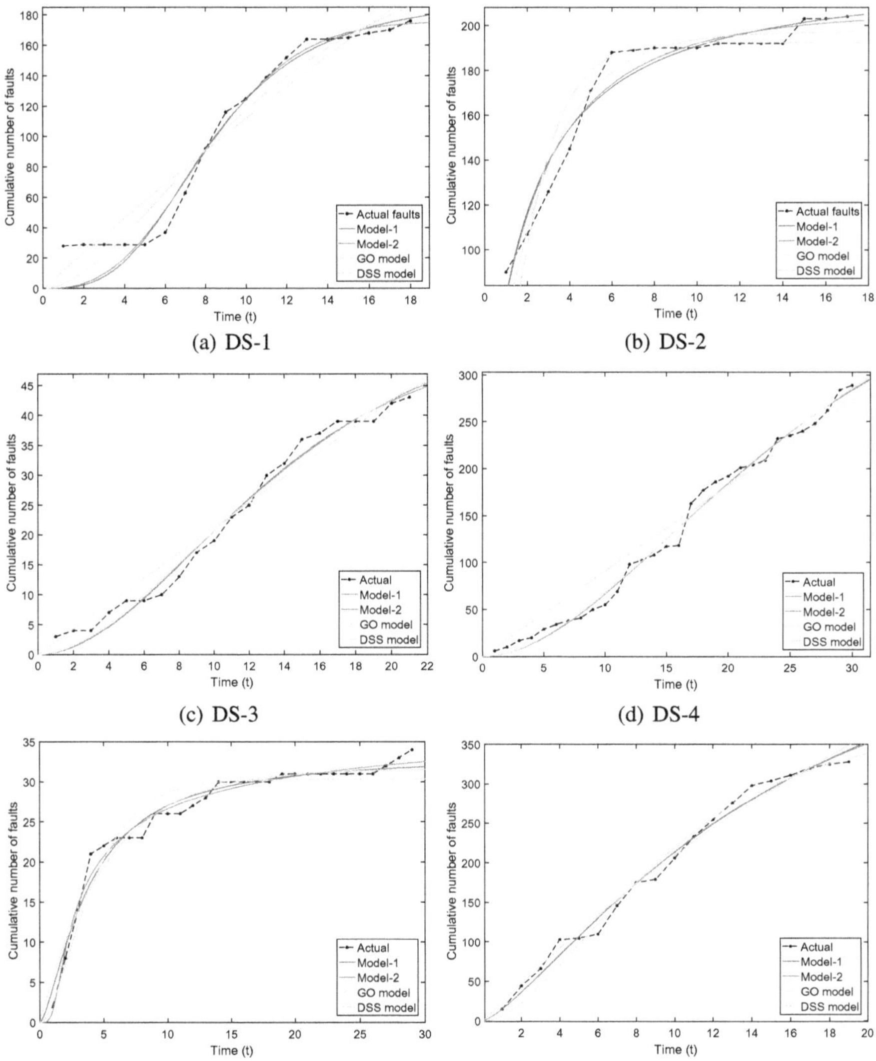

Fig. 3 a–f The fitting results of SRGMs comparison with actual failure data for DS-1-DS-6

4 Optimal Release Policy

With increasing competition in the software industry, continually changing client expectations, and the usual challenges involved with software maintenance, the timing of a new software release has become increasingly critical for a software vendor's success in the market [15]. Given the fierce competition in the market, deploying software on time has become a vital aspect in deciding the software development team's success. The dynamic release problem in software testing processes is discussed in this work [4]. The process of choosing between alternative courses of action in order to achieve goals and objectives is known as decision-making. Software release time, for example, estimating when it should be completed. Other managerial functions rely substantially on decision-making, such as organizing, implementing, and controlling [14].

If the testing period is extended in the software development process, the developed software will presumably be more reliable, but the testing cost will escalate. If we end testing too soon, the program may have too many flaws, resulting in too many failures during operation and significant losses owing to failure penalties or customer discontent. We may incur a considerable testing expense if we spend too much time testing. If the testing period is too short, the software may not be error-free. As a result, software testing and release are mutually exclusive. The testing procedure should determine the release timing dynamically. As a result, our goal is to come up with an appropriate release policy that reduces the cost and time of software testing while increasing the system's reliability. The ideal release time based on the cost-reliability criterion has been described and evaluated.

4.1 Cost and Reliability Modeling

4.1.1 Cost Modeling

1. **Testing cost per unit testing time**: The effort necessary to perform and execute the testing procedure is included in the testing cost. The cost of testing rises linearly with the time of the test. If C_1 is the testing cost per unit time, then the total testing cost is as follows:

$$C_{TCPU} = C_1.T \; . \tag{13}$$

2. **Debugging cost during testing-phase**: This cost includes the testing team's effort to handle failures. The expected number of bugs identified during this time is assumed linearly in software reliability literature. So, in the testing phase, the error-debugging cost is:

$$C_{DCDT} = C_2.m(T) \; . \tag{14}$$

3. **Debugging cost during operational-field**: In the operational phase, it is believed that Debugging cost during operational-field $C_2(T)$ is proportional to the number of software faults that were removed. Thus,

$$C_{DCDO} = C_3.[m(T_{LC}) - m(T)] \; . \tag{15}$$

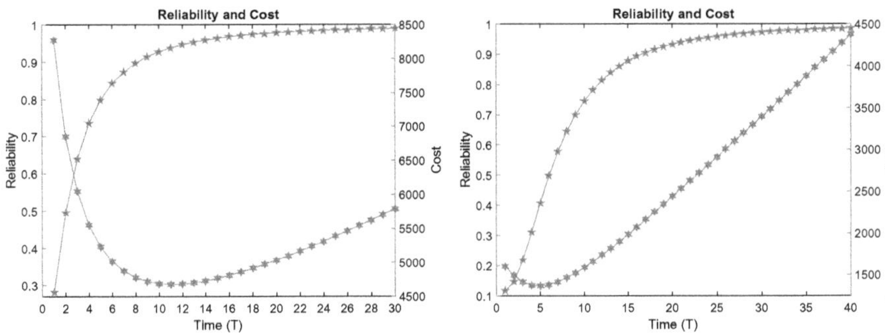

(a) DS-2 (b) DS-5

Fig. 4 Cost versus testing time and reliability vs testing time for DS-2 and DS-5

Because C_2 represents the deterministic cost to remove each fault per unit time during testing, and $C_3(T)$ represents the cost of eliminating a fault during the operational phase, C_3 is typically more than C_2, i.e., $C_3 > C_2$.

They presented a three-part software cost model structure: testing cost per unit time, debugging cost in the testing phase, and debugging cost in the operational phase. The mathematical version of the overall cost model is:

$$C(T) = C_{DCDT} + C_{DCDT} + C_{DCDO} \,. \tag{16}$$

$$= C_1 T + C_2 m(T) + C_3[m(T_{LC}) - m(T)] \,. \tag{17}$$

4.1.2 Reliability Modeling

$$R(\Delta t/T) = e^{-[m(T+\Delta t)-m(T)]} \tag{18}$$

Cost and reliability analysis with time is shown in Fig. 4.

4.1.3 Release Time Problem Using MAUT

When a sequence of possibilities is presented, the goal is to obtain a conjoint measure indicating how desirable one conclusion is in comparison to the others. It is a classical multi-objective optimization technique that addresses the optimization problem by applying weights and utility functions to determine which objectives should be prioritized [30]. The following is the formula for the multi-attribute utility function (MAUF), a weighted sum of single utility functions. It is defined as follows:

$$U(x_1, x_2, ...x_n) = f(u(x_1), u(x_2), ...u(x_n)) = \sum_{i=1}^{n} \theta_i u_i(x_i) \tag{19}$$

This work uses MAUT to construct a new decision model for software release schedule determination that trades off two conflicting objectives at the same time.

The process of determining the utility value consists of four steps.

1. Selection of Attributes.
2. Evaluate the utility function for a single attribute.
3. Allocation of credit and preference for trade-offs.
4. Single attribute to multi attribute utility function transformation.

1. **Attribute selection** Reliability is a necessary attribute that influences optimal software time-to-market and testing length selections. As a result, the proposed optimization problem's first attribute is reliability (R). The second attribute is overall software development cost (C), because no company wants to spend more than it can afford. We take the R and C as two attributes in this study. Our initial goal is to strike a compromise between these two goals by maximizing reliability while minimizing total software development costs:

$$max: \quad R(T) = e^{-[m(T+\Delta t)-m(T)]}\ , \tag{20}$$

$$min: \quad C(T) = \left[\frac{C(T)}{C_b}\right], \quad \frac{C(T)}{C_b} \leq 1\ . \tag{21}$$

   The total budget available to the testing team is denoted by C_b.

2. **Single attribute utility function**
   Each attribute's aim is represented by a utility function applied to each attribute. The single-attribute utility theory (SAUF) expresses the level of satisfaction of management concerning each of the attributes. There are many different functional forms of the utility function, such as linear, exponential, and so on. The utility function of two qualities, namely, reliability and cost function, is used in this study. The linear (additive) form $u(x) = y_1 + y_2 x$ should be employed if they are equivalent to each other because management is risk-neutral. The proposed framework is illustrated as follows:

$$u(R) = l_r R(T) + k_r\ , \tag{22}$$

$$u(C) = l_c C(T) + k_c\ . \tag{23}$$

   where, k_r, l_r, k_c, l_c are constants.

Table 4 Optimal release time by MAUT for DS-2 and DS-5

Attribute weights		Release time (T^*)	
w_r	w_l	DS-2	DS-5
0.4	0.5	14	11
0.5	0.6	15	13
0.6	0.4	16	15
0.7	0.3	18	17
0.8	0.2	20	21
0.9	0.1	25	27

3. **Weight parameter estimation**
   The management decision determines the relative value of each attribute. In this study, we perform various weight combinations values for each attribute. The weight parameter has a value between 0 and 1, with a value closer to 1 denoting greater significance. Furthermore, the sum of the weight parameters must equal 1, i.e.,

$$w_r + w_c = 1 . \tag{24}$$

   where w_r and w_c are weight for the reliability and cost respectively.
4. **Formulation of MAUT**
   The MAUT function is created by multiplying all of the single utility functions by their corresponding weights. The MAUT function with the maximizing objective for the given problem is:

$$Max: \quad U(R, C) = w_r.u(R) - w_c.u(C) . \tag{25}$$

   where

$$u(R) = 2R - 1 , \tag{26}$$

$$u(C) = 2C - 1 . \tag{27}$$

   $U(R, C)$ is a max function that has been written in terms of R and C. From the manager's perspective, R should be maximized while C should be minimized. where, $T_{LC} = 1000$, $C_1 = 100$, $C_2 = 10$, $C_2 = 50$. For DS-2, $C_b = 8500\$$, $\Delta t = 0.025$ and for DS-5, $C_b = 8500\$$, $\Delta t = 0.4$. With the different combination of weights to reliability and cost based optimal release time is shown in Table 4.

5 Conclusions

It is also possible to optimize software release and testing times by maximizing utility. The results show that a corporation should publish software early to achieve a competitive edge. The solution to the problem can also assist software firms design efficient release and testing procedures. In this work, we propose an effort-based optimum decision model that takes into account the cost of detection during testing and operational phases separately using MAUT. SRGMs provide a statistical foundation for determining optimal software testing release time. A decision model based on MAUT is suggested to make wise decisions on optimal test runs before software release. This study optimizes cost and reliability using multi-attribute utility theory and gets optimal release time. These models may help the software industry anticipate software system dependability and release time.

Acknowledgements One of the authors, Mr. Vishal Pradhan, would like to thank to Ministry of Education (earlier MHRD) for financial support provided. The support is gratefully acknowledged.

References

1. Chatterjee S, Chaudhuri B, Bhar C, Shukla A (2020) Optimal release time determination using FMOCCP involving randomized cost budget for FSDE-based software reliability growth model. Int J Reliab Qual Saf Eng 27(01):2050004
2. Chatterjee S, Shukla A (2017) An ideal software release policy for an improved software reliability growth model incorporating imperfect debugging with fault removal efficiency and change point. Asia-Pac J Oper Res 34(03):1740017, e2150
3. Chatterjee S, Shukla A (2019) A unified approach of testing coverage-based software reliability growth modelling with fault detection probability, imperfect debugging, and change point. J Softw Evol Process 31(3):e2150
4. Choudhary C, Kapur PK, Khatri SK, Muthukumar R, Shrivastava AK (2020) Effort based release time of software for detection and correction processes using MAUT. Int J Syst Assur Eng Manage 11(2):367–378
5. Dhaka R, Pachauri B, Jain A (2021) Two-dimensional SRGM with delay in debugging by considering the uncertainty factor and predictive analysis. Reliab Theory Appl SI 2(64):82–94
6. Dhaka R, Pachauri B, Jain A (2022) Two-dimensional software reliability model with considering the uncertainty in operating environment and predictive analysis. In: Data engineering for smart systems, pp 57–69. Springer
7. Goel AL (1983) A guidebook for software reliability assessment. Technical report, SYRACUSE UNIV NY
8. Goel AL, Okumoto K (1979) Time-dependent error-detection rate model for software reliability and other performance measures. IEEE Trans Reliab 28(3):206–211
9. Hsu C-J, Huang C-Y, Chang J-R (2011) Enhancing software reliability modeling and prediction through the introduction of time-variable fault reduction factor. Appl Math Model 35(1):506–521
10. Huang CY, Lo JH, Kuo SY, Lyu MR (2004) Optimal allocation of testing-resource considering cost, reliability, and testing-effort. In: 10th IEEE Pacific Rim international symposium on dependable computing proceedings, pp 103–112. IEEE

11. Huang CY, Lyu MR (2005) Optimal release time for software systems considering cost, testing-effort, and test efficiency. IEEE Trans Reliab 54(4):583–591
12. Jain M, Manjula T, Gulati TR (2014) Prediction of reliability growth and warranty cost of software with fault reduction factor, imperfect debugging and multiple change point. Int J Oper Res 21(2):201–220
13. Kapur PK, Goswami DN, Bardhan A, Singh O (2008) Flexible software reliability growth model with testing effort dependent learning process. Appl Math Model 32(7):1298–1307
14. Kapur PK, Khatri SK, Tickoo A, Shatnawi O (2014) Release time determination depending on number of test runs using multi attribute utility theory. Int J Syst Assur Eng Manage 5(2):186–194
15. Kapur PK, Panwar S, Singh O, Kumar V (2019) Joint release and testing stop time policy with testing-effort and change point. In: Risk based technologies, pp 209–222. Springer
16. Kapur PK, Pham H, Aggarwal AG, Kaur G (2012) Two dimensional multi-release software reliability modeling and optimal release planning. IEEE Trans Reliab 61(3):758–768
17. Kapur PK, Pham H, Gupta A, Jha PC (2011) Software reliability assessment with OR applications. Springer
18. Kapur PK, Shrivastava AK, Singh O (2017) When to release and stop testing of a software. J Ind Soc Probab Stat 18(1):19–37
19. Kapur PK, Singh O, Shrivastava AK (2014) Optimal price and testing time of a software under warranty and two types of imperfect debugging. Int J Syst Assur Engi Manage 5(2):120–126
20. Kumar V, Sahni R, Shrivastava AK (2016) Two-dimensional multi-release software modelling with testing effort, time and two types of imperfect debugging. Int J Reliab Saf 10(4):368–388
21. Kumar V, Saxena P, Garg H (2021) Selection of optimal software reliability growth models using an integrated entropy–technique for order preference by similarity to an ideal solution (topsis) approach. Math Methods Appl Sci
22. Lee DH, Hong Chang I, Pham H, Song KY (2018) A software reliability model considering the syntax error in uncertainty environment, optimal release time, and sensitivity analysis. Appl Sci 8(9):1483
23. Lin C-T, Huang C-Y (2008) Enhancing and measuring the predictive capabilities of testing-effort dependent software reliability models. J Syst Softw 81(6):1025–1038
24. Majumdar R, Shrivastava AK, Kapur PK, Khatri SK (2017) Release and testing stop time of a software using multi-attribute utility theory. Life Cycle Reliab Saf Eng 6(1):47–55
25. Meeker WQ, Hong Y (2014) Reliability meets big data: opportunities and challenges. Qual Eng 26(1):102–116
26. Mishra G, Kapur PK, Shrivastava AK (2017) Multi release cost model : a new perspective. Int J Reliab Qual Saf Eng 24(06):1740007
27. Ohba M, Yamada S (1984) S-shaped software reliability growth models, pp 430–436
28. Ohba M (1984) Software reliability analysis models. IBM J Res Dev 28(4):428–443
29. Pachauri B, Kumar A, Dhar J (2013) Modeling optimal release policy under fuzzy paradigm in imperfect debugging environment. Inf Softw Technol 55(11):1974–1980
30. Pachauri B, Kumar A, Dhar J (2014) Software reliability growth modeling with dynamic faults and release time optimization using GA and MAUT. Appl Math Comput 242:500–509
31. Pham H (2007) System software reliability. Springer Science & Business Media
32. Pham H, Zhang X (2003) NHPP software reliability and cost models with testing coverage. Eur J Oper Res 145(2):443–454
33. Pradhan V, Dhar J, Kumar A (2022) Testing-effort based NHPP software reliability growth model with change-point approach. J Inf Sci Eng 38(2)
34. Pradhan V, Dhar J, Kumar A, Bhargava A (2020) An S-shaped fault detection and correction SRGM subject to gamma-distributed random field environment and release time optimization. Decis Anal Appl Ind 285–300. Springer
35. Pradhan V, Kumar A, Dhar J (2021) Modelling software reliability growth through generalized inflection s-shaped fault reduction factor and optimal release time. In: Proceedings of the institution of mechanical engineers, Part O: journal of risk and reliability, p 1748006X211033713

36. Saxena P, Kumar V, Ram M (2021) Ranking of software reliability growth models: a entropy-electre hybrid approach. Reliab Theory Appl SI 2(64):95–113
37. Shrivastava AK, Kumar V, Kapur PK, Singh O (2020) Software release and testing stop time decision with change point. Int J Syst Assur Eng Manage 11(2):196–207
38. Stringfellow C, Amschler Andrews A (2002) An empirical method for selecting software reliability growth models. Empir Softw Eng 7(4):319–343
39. Tamura Y, Yamada S (2019) Software reliability model selection based on deep learning with application to the optimal release problem. J Ind Eng Manage Sci 2019(1):43–58
40. Tickoo A, Kapur PK, Shrivastava AK, Khatri SK (2016) Testing effort based modeling to determine optimal release and patching time of software. Int J Syst Assur Eng Manage 7(4):427–434
41. Yamada S, Ohba M, Osaki S (1984) S-shaped software reliability growth models and their applications. IEEE Trans Reliab 33(4):289–292
42. Zhang X, Jeske DR, Pham H (2002) Calibrating software reliability models when the test environment does not match the user environment. Appl Stochast Models Bus Ind 18(1):87–99

Reliability Analysis of Centerless Grinding Machine Using Fault Tree Analysis

Rajkumar B. Patil, Sameer Al-Dahidi, Saurabh Newale, and Mohamed Arezki Mellal

Abstract Reliability analysis plays a crucial role in the design and operational process. Fault Tree Analysis (FTA), one of the reliability evaluation techniques, plays a crucial role in the design process. Fault Tree is a graphical representation of major faults or critical failures associated with a system. It uses Boolean logic and low-level event methods to analyze the possible mechanisms of failures and evaluate the expected frequency of their occurrences by describing undesired states of the system. To increase the reliability of a system, analysis of failure data is essential. This Chapter addresses the FTA of the Centerless Grinding Machine (CGM) for safety purposes.

Keywords Reliability analysis · Fault tree analysis · Failure data analysis · Reliability block diagram · Centerless grinding machine

1 Introduction

In 1961, the Fault Tree Analysis (FTA) concept was originated in Bell Telephone Laboratories as a technique to evaluate U.S. Air Force's Minuteman missile launch control System [1, 2]. It was recognized by Boing in 1963 and caught attention in System Safety Conference, held in Seattle, June 1965. Thereafter, the U.S. Federal Aviation Administration (FAA) adopted FTA in 1970. FTA used in nuclear reactor

R. B. Patil (✉) · S. Newale
Department of Mechanical Engineering, Pimpri Chinchwad College of Engineering, Pune 411044, India
e-mail: rajkumar.patil@pccoepune.org; rajkumarpatil2009@gmail.com

S. Al-Dahidi
Department of Mechanical and Maintenance Engineering, School of Applied Technical Sciences, German Jordanian University, Amman 11180, Jordan
e-mail: sameer.aldahidi@gju.edu.jo

M. Arezki Mellal
LMSS, Faculty of Technology, M'Hamed Bougara University, Boumerdes, Algeria
e-mail: mellal.mohamed@univ-boumerdes.dz

V. Kumar and H. Pham (eds.), *Predictive Analytics in System Reliability*, Springer Series in Reliability Engineering,
https://doi.org/10.1007/978-3-031-05347-4_13

safety study presented in WASH-1400 commissioned under Prof. N. Rasmussen [3]. After the Challenger space shuttle accident, FTA is considered as one of the most significant system safety analysis techniques by NASA [1].

FTA is one of the reliability and maintainability analysis tools that can be used to solve complicated problems both proactively and reactively [4, 5]. FTA is often used from the product design stage to identify failure causes or failures and resolve design issues, especially related to risk and performance in a proactive manner. In a reactive approach, FTA provides a mechanism for determining causality in order to support the examination of unwanted or undesired events. FTA is a well-established and widely acknowledged method for assessing the reliability and risk of aeronautical [6], nuclear [7–9], chemical [10–14], mechanical [15], electrical [16, 17], electronics [18], mechatronics [19], renewable energy facilities [5, 20], and communication systems in the engineering domain [21]. This technique also considers how systems interact with their surroundings, human performance at the individual and organizational level, processes such as manufacturing, installation, and commissioning, and procedures such as operation and maintenance. In the system architecture, FTA applies to all levels of indenture.

The Fault Tree (FT) is a deductive methodology that graphically represents the failure events, both fault and normal, that occur in a given system that causes the occurrence of a well-defined outcome, called a top event [22, 23]. The top event is the most undesired state of the system. FT determines, logically, how a lower-level failure mode produces unwanted failures at a higher level. Boolean logic is used to develop the mathematical model that lower-level events. Fault tree has similarities with Reliability Block Diagram (RBD).

The top-down deduction of the fault tree diagram comprises qualitative and quantitative analysis [24]. The qualitative analysis involves defining all the combinations of basic faults that result in the top event, and it is often carried out using Minimal Cut Sets (MCSs) methodology. A given fault typically has more than one cut set that can result in the top event. An MCS is a group or combination of events and occurrences of which causes the top event to occur. The quantitative analysis uses Boolean algebra, probability, and failure rate evaluation. The quantitative methods include developing a mathematical equation to evaluate the FT, identifying Cut Sets (CSs) and MCSs. The application area of the FTA technique includes modeling system dependability, faults, reliability analysis, and the associated application to safety analysis. The discussion will include algorithms to identify MCSs, techniques to measure event importance, and estimating the occurrence probability for the specified top event. FTA is a proven analytical tool for complex systems.

Centerless Grinding Machine (CGM) is one of the essential machine tools widely used in manufacturing industries [25], especially in Western Maharashtra, India, which is one of the largest manufacturing hubs. The reliability of the component largely depends upon manufacturing processes and confirming the design tolerances [26]. The CGM is widely used in grinding operations and is responsible for the quality and reliability of the component. The CGM is also required to maintain high precision as it is used in mass production systems in which failure of a single CGM hamper the manufacturing rate [25, 27]. Furthermore, the degraded state of the CGM can affect the quality of the manufactured components, and reaction rate increases. In both these cases, the manufacturer has to bear the losses due to increased rejection rate, degradation in the quality, and pay penalty as manufacturing targets could not get achieved.

An extensive literature survey states that reliability analysis of CGM can be carried out by using various methods such as Failure Modes and Effects Analysis (FMEA), Reliability Block Diagram (RBD), Fault Tree Analysis (FTA), Markov chains, Petri-nets, and fishbone diagram. Artificial Intelligence (AI), Machine Learning (ML), Multi-Criteria Decision Marking (MCDM), and Internet of Things (IoT) are some of the recent methods that can be integrated with traditional reliability analysis methods at data collection methods, data analysis, system reliability modeling, and estimating reliability metric [28–35]. The first level of reliability analysis requires identifying possible faults, failures, events (external or internal) that cause the system to fail or hamper the system performance. It also provides logical propagation of failure from the lowest level events to the top event. The outcomes of the FTA are generally helpful in carrying out a detailed investigation of the critical components.

In this regard, in this Book Chapter, the FTA is used in identifying the possible failures or faults of the CGM and their effects on the performance and reliability of the CGM are presented. The data collection, analysis, and modeling steps include some of the recent methods mentioned above. This study considers the significant faults observed in the field and is based on the field's experts. The required data has been collected from the maintenance engineers and their records, service engineers, design engineers, and machine operators. An FTA methodology is proposed to construct fault of the CGM, and carry out its qualitative and quantitative analysis. The primary objectives of this study are:

- Review the concept of FT and propose a simple framework for the FTA of the CGM;
- Constructing FT of the CGM considering dominating faults/ failures by resorting to field failure data and expert judgments;
- Developing equivalent RBD, mathematical model, and prioritizing the events based on their occurrence probabilities and reliability.

This Chapter presents the concepts of FT, definitions of terminology, illustrations of mathematics, and discussions of application to a Centerless Grinding Machine (CGM) for safety purposes. The remainder of the Chapter is organized as follows: Sect. 2 presents the basics of Fault Tree Analysis (FTA); Sect. 3 illustrates the terminology and symbols used in FTA; Sect. 4 presents an integrated case study on FTA of a Centerless Grinding Machine (CGM); The insights of the FTA and the interpretation of the case study are summarized in Sect. 5.

2 Basics of Fault Tree Analysis (FTA)

FTA is a top-down method of analysis to identify the faults/failure causes (henceforth called events) or group of events that can lead to the defined top event. The term "undesired event", also called "top event" is used to describe a pivotal event defined for a given situation. The top events are typically defined from either a safety or reliability perspective. The top event is an end for the logic associations of intermediate and basic events that result in its occurrence. The deduction process is performed in steps from the top down in order to identify all basic/incomplete events that contribute to the occurrence of the top event. The result is FTA considers combinatorial events, events that occur in combination to result in a top event, and single failure events that result in a top event. Thus, it is a vital tool for safety, reliability, and root cause analyses. Central to the analysis is the graphical model showing the logical connections between events in relation to the top event. The graphical model uses three basic symbols: events, gates, and transfers. The event symbol is a text box containing a short description of the event. The first event described is the top event. Other events are logically deduced downward from the top event. There is an inherent cause-and-effect relationship between events.

However, the FT should not be confused with other analysis methods, like Failure Modes, Effects, and Criticality Analysis (FMECA), which seeks to identify all possible failures of components in a system [36–41]. The FT does not necessarily contain all possible events of the components of a system or its operation. Only those failure modes that contribute to the existence occurrence of the top event are modeled. Each top event description is specific so that only those events that result in the specific top event apply. If a new top event is described, the FT must be constructed that contains only events that result in the new top event. As a result, systems or subsystems may require multiple FTAs to identify and characterize the failure modes and mechanisms for all undesired events. Gate symbols define the logical requirements of the lower-level events to result in the event immediately above. As the gates are completed, the FT progresses through the gates concluding in the top event after the uppermost gate inputs are satisfied.

The general methodology used for the fault tree analysis of the CGM is summarized in Fig. 13.1. The first step in the FTA is to select a system, including a clear understanding of the system functionality and the need for fault tree analysis. Thereafter, the top event of the fault tree should be defined based on the primary objectives of the FTA. The significant events (intermediate and basic) causing the top event to occur are then to be identified. The ground rules, including symbols and logic gates, need to be defined properly to construct the fault tree and connect the events to intermediate and top events logically. The fault tree is then constructed by connecting the basic events to intermediate and top events. The developed fault is then converted into a mathematical equation (qualitative evaluation) and the probability of occurrence of the basic, intermediate, and top events (quantitative evaluation) are calculated using field failure. Constructing a model and producing an evaluation suitable for practical interpretation to support a correct decision is easy for a clear and well-defined top

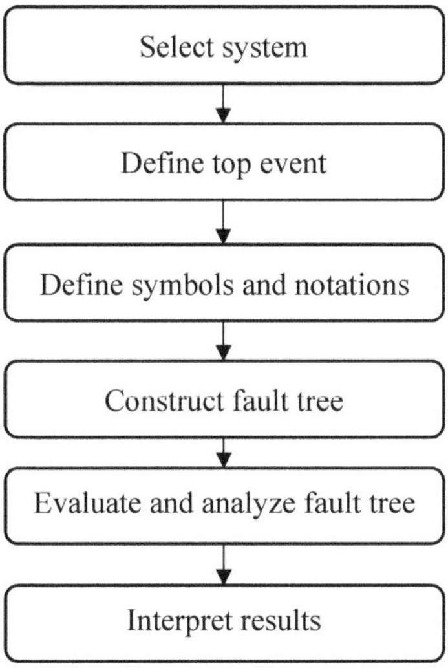

Fig. 13.1 Methodology used in FTA

event, with documented decisions relating to scope and resolution and ground rules governing the conduct of the analysis.

The most critical part of the FTA is interpreting results. It includes identifying the critical events, intermediate events, components, and subsystems, and estimating their probability of occurrence and minimal cut-set. The obtained results need to be used to improve system design, selection of components, deciding maintenance practices, and defining troubleshooting procedures. The logic and math identify the high-risk paths, deficiencies, common-mode, and common cause failures of the system. The qualitative and quantitative interpretation provides better information for better decisions. Better decisions early in design, development, and production enable higher in-service reliability and reduced risk. In-service, FTA is applied in investigative efforts to identify causality and eliminate failure modes that result in undesired events.

3 Terminology and Symbols

The graphical model uses three basic symbols: events, gates, and transfers. Events are shown by a text box (rectangular, circular, etc.) containing a basic description of the event. Gate symbols are the symbols that identify the logical construct for the input to output relationship. Transfer symbols are used to depict internal and external relationships in the organization of the tree. This section describes the terminologies and symbols used in the construction of the fault tree.

Table 13.1 Event symbols

Symbol	Event	Meaning
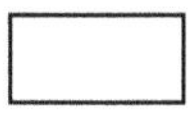	Top or intermediate event	It represents the most unwanted or undesired event associated with the system or sub-system
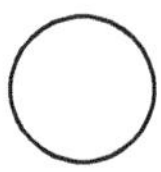	Basic event	It is an independent elementary event representing a basic fault or component. The analysis ends with the basic event
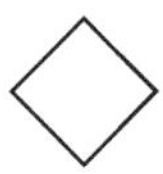	Undeveloped or incomplete event	The events which cannot be developed further either information or knowledge is not available or because it is of insufficient consequence
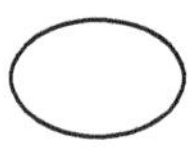	Conditioning event	Specific circumstances or conditions that are applicable to a logic gate. It is primarily used with the "PRIORITY AND" and "INHIBIT" gates
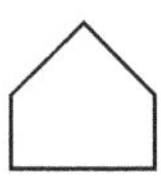	House or external or normal event	An event that is normally expected to occur. A normally occurring event that may not be considered as a fault

3.1 Event Symbols

The creation of any fault tree commences with the definition of the top event and identifying all internal and external events that result in the occurrence of the top event. The lower-level events, intermediate, basic, undeveloped, conditioning, and external events are connected to the top events. The symbols used for showing these event symbols are summarized in Table 13.1. The event symbol for an event is a box containing a basic description of the event. Once the first event description is complete, other events are deduced downward from the top event. The deduction breaks down the higher event into parts with lower events describing all the possible ways to get the event immediately above. The breaking down continues until a point is reached where basic events, which serve as the enabling inputs to the tree, are identified. Basic events are represented by circles, always at the furthest point down any route leading to the top event. Between the top event and the basic events are all the intermediate events. Top events and intermediate events are denoted in rectangle boxes.

3.2 Gate Symbols

Each identified event should be connected to the intermediate and top events using logic gate symbols. The logic gate symbols define the logical requirements of the

Table 13.2 Gate symbols

Symbol	Gate	Meaning
	AND	The higher-level (output) event occurs if and only if all lower-level (input) events occur. Therefore, in Boolean algebra, the output is the intersection of the input events
	OR	The occurrence of one of the lower-level (input) events causes the higher-level (output) event to occur. Therefore, in Boolean algebra, the output is the union of the input
	Priority AND	The higher-level (output) event occurs if and only if all of the lower-level (input) events occur in a definite order. It is a special case of AND gate
	Exclusive OR	The higher-level (output) event occurs if exactly one of the lower-level (input) events occurs. Both events cannot occur at the same time. It is a special case of OR gate
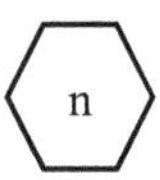	Combination	The higher-level (output) event occurs if 'n' of the lower-level (input) events occurs. It allows a user to specify the number of failures within a group of inputs that will result in output from the gate
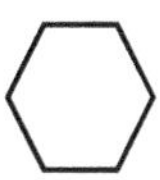	Inhibit	The higher-level (output) event occurs if a single lower-level (input) event occurs in the presence of an enabling condition

lower-level events to result in the event immediately above. The fundamental logic gate symbols are AND and OR. Each of these has at least two lower-level events and one immediate higher-level event. In the higher-level event, the AND gate occurs if and only if all lower-level event occurs. The higher-level event for the OR gate occurs if and only if at least one of the lower-level events occurs. If more complex logical relationships are required, other logical representations such as priority AND, exclusive OR, combination, and inhibit gates are used to describe the relationships. The gate symbols, along with their logical meaning, are summarized in Table 13.2. The gates provide an organized logic for the fault tree and are the analytical foundation of the fault tree analysis.

3.3 *Transfer Symbols and Definitions*

The transfer symbol is typically a triangle, and the transfer-in and transfer-out symbols are shown in Table 13.3. The transfer symbols help to construct and display the fault tree with clarity if there are more and more events that cause the top event

Table 13.3 Transfer symbols

Symbol	Gate	Meaning
	Transfer-in	Transfer-in indicates that the tree is developed further at the occurrence of the corresponding transfer-out
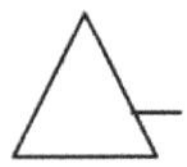	Transfer-out	Transfer-out indicates that this portion of the tree must be attached at the corresponding transfer-in

to occur. It allows the fault tree constructor to insert the sub-fault tree into the main fault tree. Transfer symbols also provide a means to manage time by eradicating the need to redraw tree branches that are duplicated in other locations.

3.4 Fault Tree Construction

Fault tree construction is a top-down approach and starts with defining the top event and connecting all possible events to the top event by using logic gates. Event descriptions, predominantly the top event, are critical elements of the fault tree methodology. Event descriptions, though simple and brief, must comprise not only undesired or failure states of the component but also a description of when the event resulting in that state occurs. The "what" and "when" criteria are essential to all event descriptions.

A sample fault tree for a "Pen is not working" (top event) is shown in Fig. 13.2. The reason failure of a ballpoint pen can be internal or external and therefore these two intermediate events are connected to the top event by OR logic gate as the occurrence of one of the intermediate events leads to the occurrence of the top event. Broken cover and thread failure are taken here as external events and they are connected to the intermediate event by or gate as the occurrence of any event will cause the intermediate to occur. Furthermore, refill failure and pointer failure due to human mistakes are considered internal failures, and they are connected to the intermediate event by the OR gate. Several other events could lead to the occurrence of the top event, intermediate events, and undeveloped events, and can be considered for more detailed analysis.

4 A Case Study: Centerless Grinding Machine

In this section, a detailed case study on a centerless grinding machine is presented. The basic principle of centerless grinding operation is presented in Fig. 13.3, and a

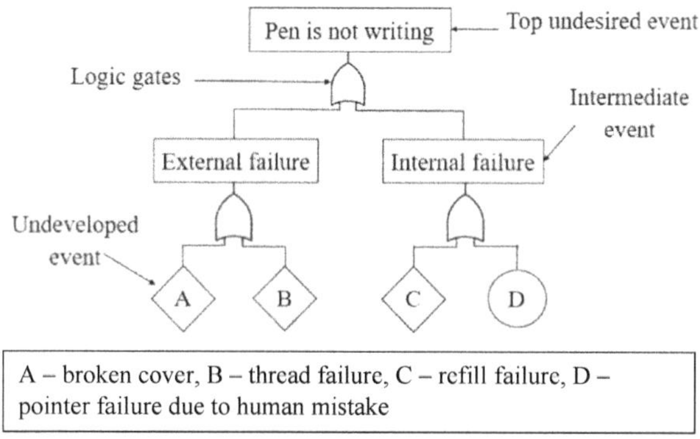

Fig. 13.2 A simple fault for ballpoint pen failure

typical centerless grinding machine is shown in Fig. 13.4. Centerless grinding is a process for uninterruptedly grinding cylindrical surfaces in which the workpiece is supported by rest blades and not by the chuck. The two wheels ground the workpiece. The bigger grinding wheel grinds, while the smaller regulating wheel, which is slanted at an angle, controls the speed of the workpiece's axial movement. External or internal centerless grinding, traverse feed, or plunge grinding are all options. External traverse feed grinding is the most prevalent method of centerless grinding. The centerless grinder has the grinding wheel on the left, the work blade in the center, and the smaller diameter regulating wheel on the right, as seen from the operator's perspective. The centerlines of the grinding wheel and regulating wheel are in the same plane, at identical heights above the machine bed, in most applications. The work blade has to be adjusted so that the centerline of the workpiece is above the centerline of the grinding and regulating wheels to provide rounding action. For successful centerless grinding, this is a crucial relationship.

The contact points form three sides of a square of the workpiece rests on a flat work blade in the center with the regulating and grinding wheels. Any high location on the workpiece will displace the work slightly on the blade as the part is ground in this technique, allowing the grinding wheel to cut a low spot precisely opposite the high spot. This configuration will eventually result in three lobes on

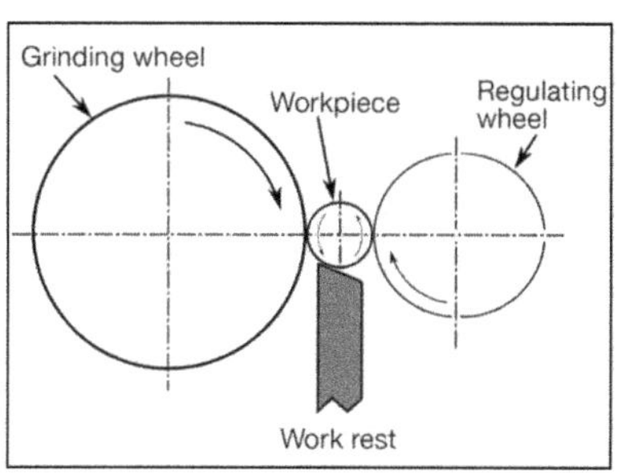

Fig. 13.3 Principle of centerless grinding machine [42]

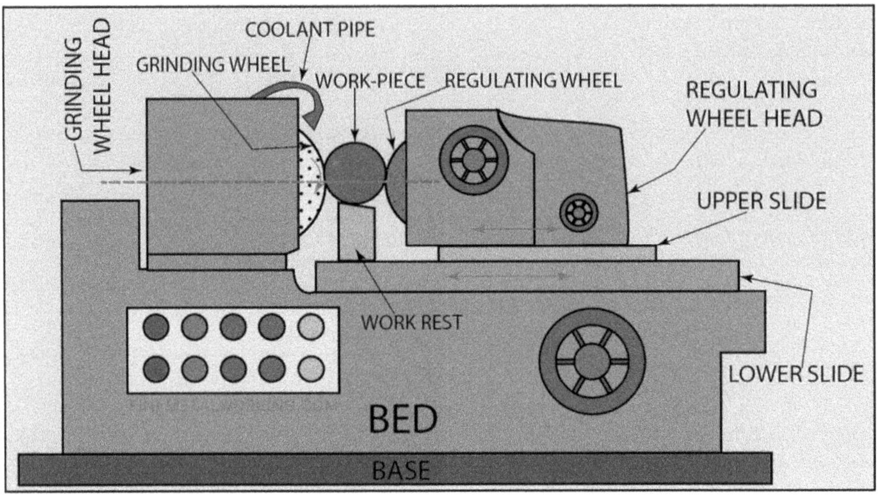

Fig. 13.4 A typical centerless grinding machine (adapted from [43])

the workpiece that are dimensionally exact but not round. The centerless operation generates roundness by using an angled work blade that slopes toward the regulating wheel and supports the workpiece centerline above the centerlines of the regulating and grinding wheels. Because of the angle produced between the centerlines of the wheels and the workpiece, if a high spot comes into contact with either the blade or the regulating wheel, it does not create a directly opposite low spot.

The work blade's angle helps keep the workpiece in contact with and under control of the slower rotating regulating wheel, preventing the workpiece from "spinning up" to the grinding wheel's speed. A spin-up can move a workpiece from 850 rpm to nearly 60,000 rpm in the blink of an eye in some situations. This isn't something you'd like to happen. One-half of the workpiece diameter above the centerline of the grinding and regulating wheels is a good rule of thumb for establishing the correct height for a workpiece up to 1 inch in diameter. For a workpiece with a diameter of one inch, the height should be half an inch above the wheel's centerline.

Blade angles for centerless grinding work range from 0 to 45°. A top blade angle of 30° appears to be ideal for most centerless grinding applications. However, there are limitations. A shallower blade angle is appropriate for greater diameter and longer work. Setting the regulating wheel to a low speed, around 30 rpm, is also a good place to start improving the centerless grinding process. The diameter of the workpiece and the rate of stock removal are factors in regulating wheel rotation speed. A black diagram of the typical centerless grinding machine is shown in Fig. 13.5. Its functionality and performance depend primarily upon the regenerative center function, regenerative function, grinding stiffness, contact compliance of regulating wheel, contact compliance of grinding wheel, and grinding machine.

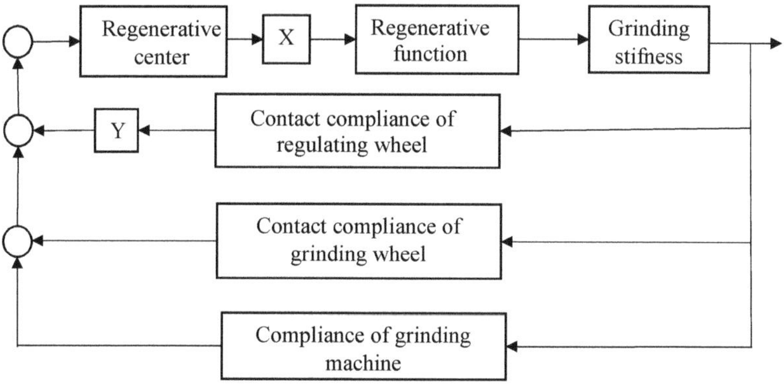

Fig. 13.5 Block diagram of a typical centerless grinding machine

4.1 Construction of Fault Tree

The centerless grinding machine consists of a large number of components, assemblies, and components and several failure causes (failure events), the occurrence of which results in the failure of the whole centerless grinding machine or stops functioning or will not give the required output. The required failure data is collected from the user of the centerless grinding machine. It includes several subsystems, components, failure causes, the occurrence of events, and expert judgments. Records such as maintenance registers, daily, weekly and monthly maintenance predictive and corrective maintenance sheets, and observations of a maintenance engineer, operator, and supervisors are collected for one year. A detailed fault of the centerless grinding machine (it includes major failures, faults, failure causes, and failure modes) has been prepared considering only significant failure events, the occurrence of which led to the failure of the centerless grinding machine. Several components rarely fail, or failure of which does not lead to the failure of the whole system are neglected. The symbols used to show events, and logic gates defined in Sect. 3 are used to construct the fault tree.

4.1.1 Defining Top Event

The centerless grinding machine is expected to manufacture/provide the required surface finish to the components over a given period of time without failure. If the CGM is continuously producing the components which does not the design requirement then it is also considered a failure. therefore, the top event is considered as Failure of CGM (T). It means the occurrence of any event which causes the CGM to manufacture the component does not fulfill the design requirements. The next step in the fault tree construction is to find out the next level intermediate/basic events and occurrence of which lead to the occurrence of the top event. There are such 12 events

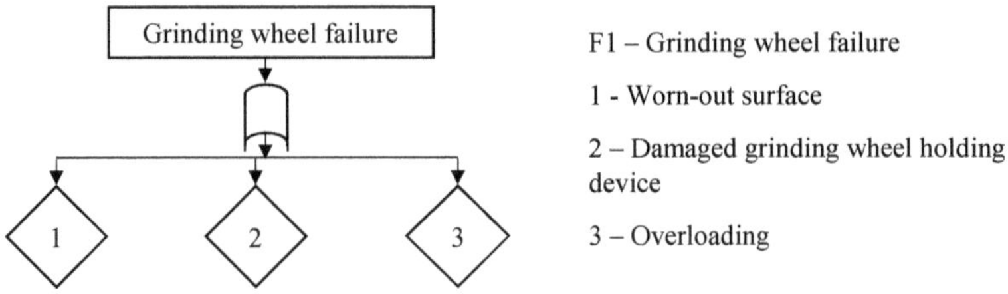

Fig. 13.6 Fault tree of grinding wheel failure

namely grinding wheel failure, work blade failure, swivel plate failure, ball screw assembly failure, cooling system failure, grinding wheel head failure, regulating wheel failure, regulating wheel head failure, main motor failure, lubrication system failure, hydraulic system failure, and human error. The name and terminologies of intermediate events are defined based on the literature review, names used in industries, and referring to the manuals and registers.

4.1.2 First Level Lower-Level Event

The grinding wheel is one of the critical components as it continuously carries out the direct grinding operation and has to provide the required surface finish to the components to be manufactured. The failure of the grinding wheel may result in a burnt workpiece, chatter marks on the workpiece, and poor surface finish. The failure of the grinding wheel (F1) is primarily due to the worn-out surface due to continuous usage (1), damaged grinding wheel holding unit (2), and overloading due to known or unknown reasons (3). These three events (1, 2, and 3) are connected to the F1 by the OR gate, as shown in Fig. 13.6. The failure of the grinding wheel will result in burning out the workpiece, chatter marks on the workpiece, poor surface finish, and increased rejection rate of the components.

Work blade is another component that plays a crucial role in the smooth functioning of the CGM. The failure of the work blade (F2) occurs due to several lower-level events: if it is integrated with the wrong tool (4), fixed at incorrect tension (5), set at incorrect tension (6), and wear and tear (7). Events 4, 5, 6, and 7 are connected to F2 by the OR gate as the occurrence of one of the events leading to the occurrence of the event F2, shown in Fig. 13.7.

Another intermediate event, the occurrence of which leads to the occurrence of the top event, is swivel plate failure (F3). The event F3 occurs impact loading, reduced hardness, and swivel offset and connected to it by OR gate, shown in Fig. 13.8.

The failure of the CGM can also be due to the failure of the ball screw assembly (F4). Misalignment (11), improper lubrication (12), and worn-out balls or screws (13) are the events that cause the ball screws to function improperly. The fault tree of the ball screw assembly failure is shown in Fig. 13.9.

The grinding operation generates a lot of heads and should be dissipated properly to manufacture the workpiece with designed tolerances. Therefore, the failure of the

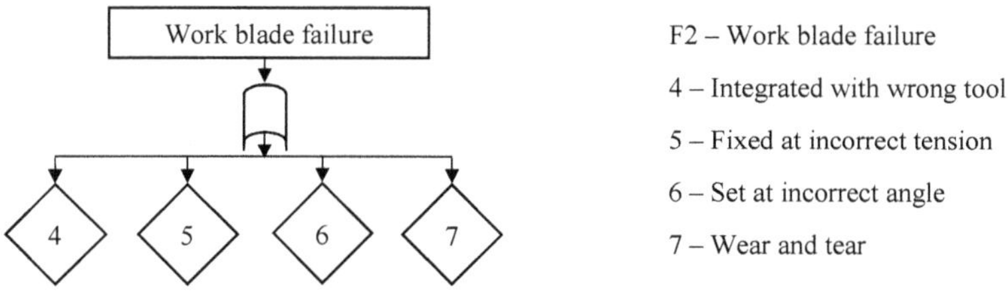

Fig. 13.7 Fault tree of work blade failure

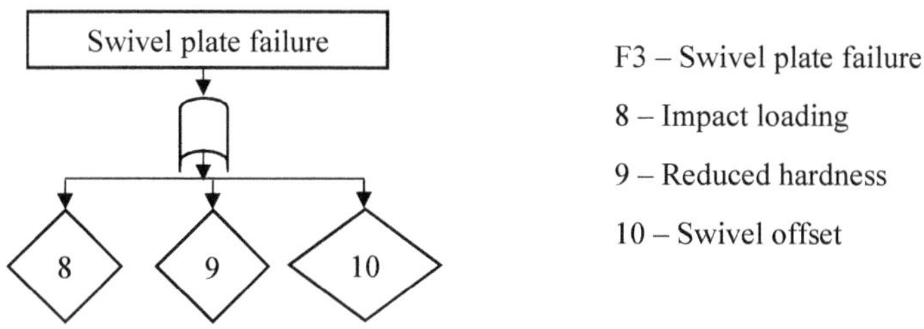

Fig. 13.8 Fault tree of swivel plate

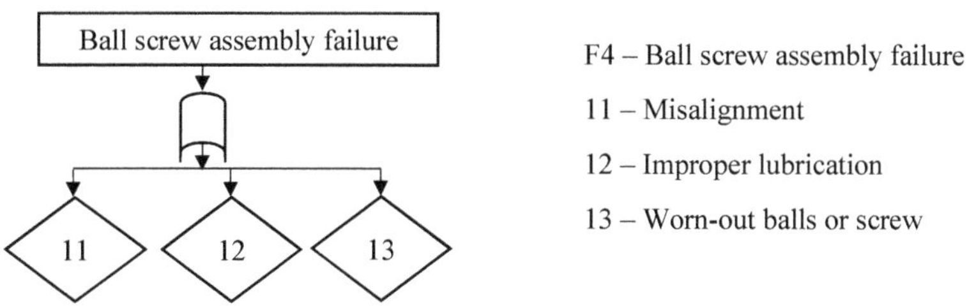

Fig. 13.9 Fault tree of ball screw assembly failure

cooling system led to the occurrence of the top event. The cooling system failure (F5) generally takes place due to contamination of foreign particles (14), coolant pump failure (15), the incorrect combination of coolant and water (16), leaked hose (17), and failure of control system (18). The fault tree of the cooling system failure is shown in Fig. 13.10.

The failure of grinding wheel head (19), regulating wheel (20), regulating wheel head (21), main motor (22), lubrication system (23), hydraulic system (24), and human error (25) are some of the identified first-level intermediate events. There may be many lower-level events that may cause these events (19 to 25) to occur. However, the detailed information was not available and or recorded. Furthermore, these events are connected to the top event by the OR gate and are shown in Fig. 13.11.

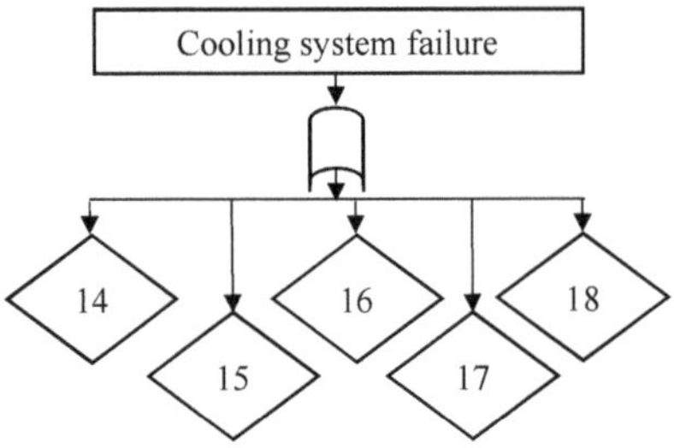

F5 – Cooling system failure
14 – Contamination of foreign particles
15 – Coolant pump failure
16 – Incorrect combination of coolant and water
17 – Leaked hose
18 – Failure of control systems

Fig. 13.10 Fault tree of cooling system failure

4.2 Qualitative Evaluation of Fault Tree

Qualitative analysis of FT of CGM developed in the previous section gives the relationship between undeveloped/basic and intermediate events with the top event. Minimal cut sets i.e., all the combinations of undeveloped events which, when they happen simultaneously, lead to the system failure can also be obtained. For analysis purposes, the developed FT diagram is transformed into an equivalent reliability block diagram (RBD). All the undeveloped events are connected to the top event directly or indirectly by the 'OR' gate and the occurrence of a single event leads to the occurrence of the top event. Therefore, all the basic events are connected in series (series configuration) as shown in Fig. 13.12.

In the case of series configuration, all events are critical events as the failure of a single event led to system failure or causes the top event to occur. Since reliability is a probability, the reliability of centerless grinding machine may be determined from the probability of non-occurrence of the events as:

1 - event 1 does not occur
2 - event 2 does not occur
.
.
.
25 - event 25 does not occur

The probability of survival (reliability) of the centerless grinding machine is estimated using the laws of probability as:

$$R_{CGM} = P(X_1 \cap X_2 \cap X_3 \cap X_4 \cap X_5 \cap X_6 \cap X_7 \cap X_8 \cap X_9 \cap X_{10} \cap X_{11} \cap X_{12} \cap X_{13} \cap X_{14} \cap X_{15} \cap X_{16} \cap X_{17} \cap X_{18} \cap X_{19} \cap X_{20} \cap X_{21} \cap X_{22} \cap X_{23} \cap X_{24} \cap X_{25}$$

$$R_{CGM} = P(X_1) \times P(X_2) \times P(X_3) \times (X_4) \times P(X_5) \times P(X_6) \times P(X_7) \times P(X_8) \times P(X_9) \times P(X_{10}) \times P(X_{11}) \times P(X_{12}) \times P(X_{13}) \times P(X_{14}) \times P(X_{15}) \times P(X_{16}) \times P(X_{17}) \times P(X_{18}) \times P(X_{19}) \times P(X_{20}) \times P(X_{21}) \times P(X_{22}) \times P(X_{23}) \times P(X_{24}) \times P(X_{25})$$

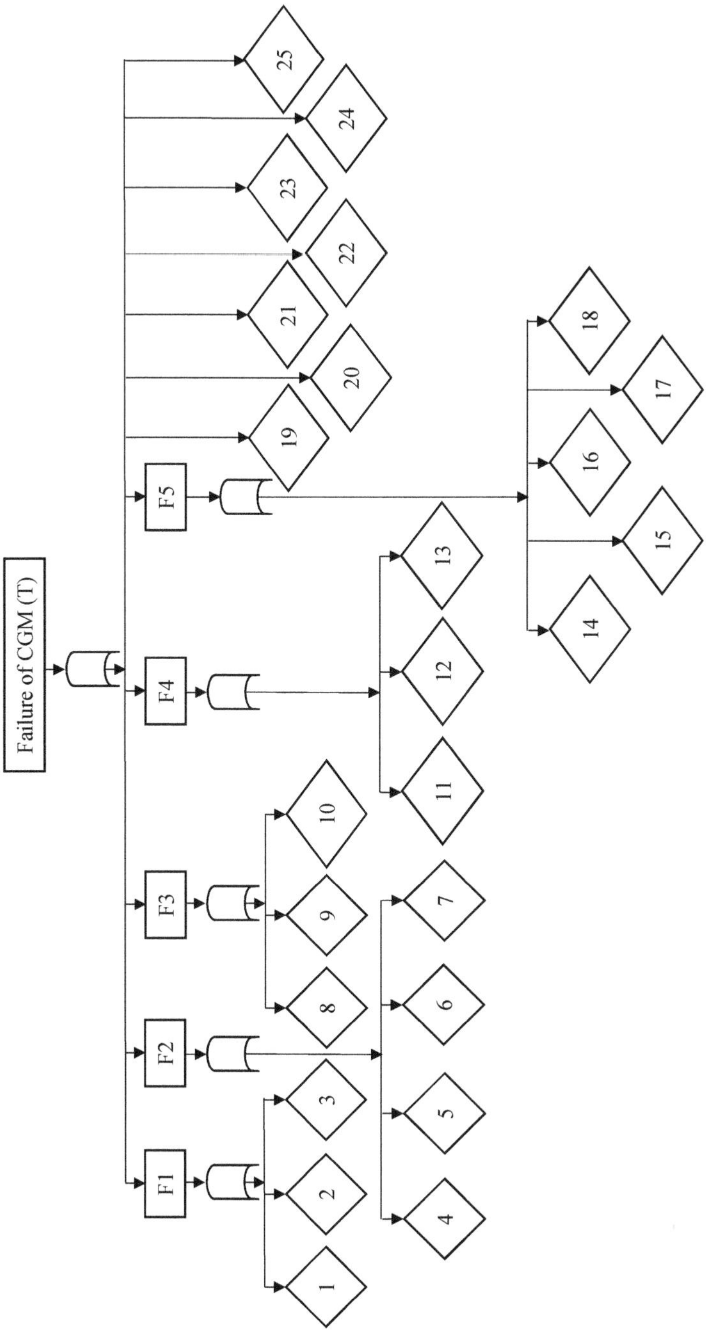

Fig. 13.11 Fault tree of the centerless grinding machine

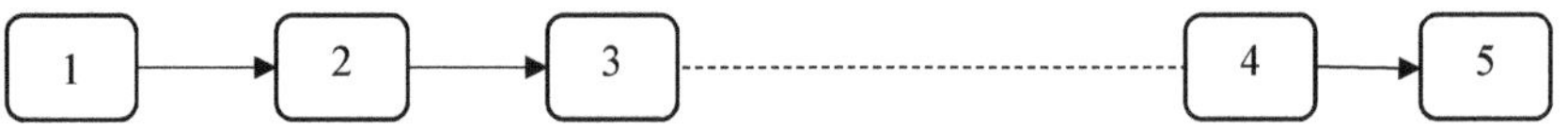

Fig. 13.12 Equivalent reliability block diagram of centerless grinding machine

However,

$$P(X_1) = R_1; P(X_2) = R_2; P(X_3) = R_3; \ldots; P(X_{25}) = R_{25};$$

where, R_1—the reliability or probability of non-occurrence of event 1, R_2—the reliability or probability of non-occurrence of event 2, R_3—the reliability or probability of non-occurrence of event 3, …, and R_{25}—the reliability or probability of non-occurrence of event 25.

$$\begin{aligned} R_{CGM} = {} & R_1 \times R_2 \times R_3 \times R_4 \times R_5 \times R_6 \times R_7 \times R_8 \times R_9 \times R_{10} \\ & \times R_{11} \times R_{12} \times R_{13} \times R_{14} \times R_{15} \times R_{16} \times R_{17} \times R_{18} \times R_{19} \times R_{20} \\ & \times R_{21} \times R_{22} \times R_{23} \times R_{24} \times R_{25} \end{aligned}$$

$$\therefore R_{CGM} = \prod_{i=1}^{25} R_i = R_1 \times R_2 \times R_2 \times \cdots \times R_{25} \quad (1)$$

Equation (1) is the governing reliability model of the centerless grinding machine. In this case, all the events of the centerless grinding machine are assumed to be independent (i.e., the occurrence or non-occurrence of one event component does not affect the occurrence or non-occurrence of the other event). In other words, in order for the system to function, all 25 events should not occur. Furthermore, if the values of probability of occurrence (F_i) for all the 25 events are known, the reliability of the centerless grinding machine (R_{CGM}) is calculated as:

$$\therefore R_{CGM} = \prod_{i=1}^{25} (1 - P_i) = (1 - P_1) \times (1 - P_2) \times (1 - P_3) \times \cdots \times (1 - P_{25}) \quad (2)$$

The reliability (or probability of non-occurrence) of the first level intermediate events can be estimated as:

$$R_{F1} = (1 - P_1) \times (1 - P_2) \times (1 - P_3) \quad (3)$$

$$R_{F2} = (1 - P_4) \times (1 - P_5) \times (1 - P_6) \times (1 - P_7) \quad (4)$$

$$R_{F3} = (1 - P_8) \times (1 - P_9) \times (1 - P_{10}) \quad (5)$$

$$R_{F4} = (1 - P_{11}) \times (1 - P_{12}) \times (1 - P_{13}) \quad (6)$$

$$R_{F5} = (1 - P_{14}) \times (1 - P_{15}) \times (1 - P_{16}) \times (1 - P_{17}) \times (1 - P_{18}) \quad (7)$$

Equations (3) to (7) can be used to estimate the reliability/probability of non-occurrence of intermediate events F1 to F5.

4.3 Quantitative Evaluation of Fault Tree

In quantitative analysis, the probability of occurrence of all 25 events is estimated using the laws of probability. As mentioned in the previous section that the required database for the analysis has been collected from the users of the centerless grinding machine. For some events, the failure probabilities are assumed or estimated using expert judgment. The experts included the employees engaged in maintenance activity, operators, service engineers, and design engineers of the centerless grinding machine. The probability of occurrence of all basic events is also cross-verified using expert judgments. The probability of occurrence of the events is summarized in Table 13.4.

The estimation of the probability of non-occurrence of intermediate events, F1 to F5, using Eqs. (3)-(7) are as given below:

$$R_{F1} = (1 - 0.09) \times (1 - 0.001) \times (1 - 0.002) = 0.91 \times 0.999 \times 0.998 = 0.9072$$

$$R_{F2} = (1 - 0.03) \times (1 - 0.03) \times (1 - 0.06) \times (1 - 0.075) = 0.8181$$

$$R_{F3} = (1 - 0.085) \times (1 - 0.06) \times (1 - 0.009) = 0.8523$$

$$R_{F4} = (1 - 0.08) \times (1 - 0.06) \times (1 - 0.06) = 0.8129$$

$$R_{F5} = (1 - 0.04) \times (1 - 0.08) \times (1 - 0.004) \times (1 - 0.009) \times (1 - 0.004) = 0.8683$$

Furthermore, if the values of probability of non-occurrence i.e., reliability of the centerless grinding machine are calculated as:

$$R_{CGM} = R_{F1} \times R_{F2} \times R_{F3} \times R_{F4} \times R_{F5} = 0.9072 \times 0.8181 \times 0.8523 \times 0.8129 \times 0.8683 = 0.4465$$

Therefore, the system reliability is estimated to be 0.4465. It is observed that the events human error (25), worn-out grinding wheel (1), impact loading on swivel plate

Table 13.4 Probability of occurrence of the events

Event Name	Symbol	Probability of occurrence	Probability of non-occurrence
Grinding wheel failure	F1	0.0928	0.9072
Worn-out surface	1	0.09	0.91
Damaged grinding wheel holding device	2	0.001	0.999
Overloading	3	0.002	0.998
Work blade failure	F2	0.1819	0.8181
Integrated with wrong tool	4	0.03	0.97
Fixed at incorrect tension	5	0.03	0.97
Set at incorrect angle	6	0.06	0.94
Wear and tear	7	0.075	0.925
Swivel plate failure	F3	0.1477	0.8523
Impact loading	8	0.085	0.915
Reduced hardness	9	0.006	0.994
Swivel offset	10	0.009	0.991
Ball screw assembly failure	F4	0.1829	0.8129
Misalignment	11	0.08	0.92
Improper lubrication	12	0.06	0.94
Worn-out balls or screw	13	0.06	0.94
Cooling system failure	F5	0.1317	0.8683
Contamination of foreign particles	14	0.04	0.96
Coolant pump failure	15	0.08	0.92
Incorrect combination of coolant and water	16	0.004	0.996
Leaked hose	17	0.009	0.991
Failure of control systems	18	0.004	0.996
Grinding wheel head	19	0.008	0.992
Regulating wheel	20	0.009	0.991
Regulating wheel head	21	0.001	0.999
Main motor	22	0.083	0.917
Lubrication system	23	0.031	0.969
Hydraulic system	24	0.023	0.977
Human error	25	0.12	0.88

(8), main motor (22), coolant pump (15), and misalignment in the ball screw bearing (11) are the most critical events and needs to be monitored closely.

5 Discussion and Summary

Fault tree analysis is one of the reliability analyses tools that identify internal and external faults/ failure causes/ failure events that lead to the occurrence of the top event, i.e., the most undesirable event associated with the system. It is a graphical tool that includes qualitative and quantitative analysis and therefore it is effective in new system design (design stage) as well as in analyzing systems behavior in operations (operation stage).

The effectiveness of this method is observed throughout the case study on the reliability analysis of the Centerless Grinding Machine (CGM). The qualitative and quantitative analysis of the CGM helped in carrying out the critical events/ components of the CGM where more focus should be given during the operations. Furthermore, this analysis provides insights for design modifications and improvements. However, this case study is based on a limited database. Detailed investigation can be carried out with more data sets that can be collected from a greater number of CGM users, experts in the field, component manufacturers, maintenance, and service engineers. It is also essential to expand the fault tree by identifying all possible faults, components, assemblies, and subsystems occurrence (failure) of which may lead to the occurrence of the top event.

Fault tree analysis is one of the popular analytical methods and needs to be applied with discipline. The probability of occurrence of intermediate (higher level) events is carried by assuming the events occur independently and identically. However, in actual practice several failures may be dependent, can occur simultaneously and the occurrence of one event can trigger the other events to occur or it may accelerate the life of another component. There is a need to integrate these aspects in the fault tree analysis methodology.

One of the critical parameters in systems operation is human and needs to be considered during analysis. It is required to calculate the occurrence of an event/failure due to its inherent failure mechanism and the intervention of the human needs to be investigated. It will reveal how, where and how much the human intervention affects the system reliability. Such study may provide future direction to carry out effective maintenance activity, how artificial intelligence and machine learning (AI & ML) techniques can be effectively integrated with the fault tree analysis and improve system reliability. A process of validation and verification of the results obtained from the fault tree analysis should be developed and integrated.

References

1. Haasl FD (1965) Advanced concepts in fault tree analysis. In: System safety symposium, pp 1–14
2. Kapur KC, Pecht M (2014) Reliability engineering. Wiley & Sons, Inc
3. Rasmussen N et al (1975) Reactor safety study: an assessment of accident risks in U.S. commercial nuclear power plants
4. Signoret JP, Leroy A (2021) Fault tree analysis (FTA). In: Pham H (ed) Springer series in reliability engineering. Springer, pp 209–225
5. Zhang J, Kang J, Sun L, Bai X (2021) Risk assessment of floating offshore wind turbines based on fuzzy fault tree analysis. Ocean Eng 239:109859
6. Boeing Aerospace Company (1968) Fault tree for safety. Seattle, WA
7. Zhao D, Cai L, Gao C, Sun Y (2000) Application of FTA in operation safety design of nuclear waste carrying manipulator. In: Proceedings of the world congress on intelligent control and automation (WCICA) (Cat. No.00EX393), pp 729–732
8. Yoon S, Jo J, Yoo J (2011) A domain-specific safety analysis for digital nuclear plant protection systems. In: 2011 5th international conference on secure software integration and reliability improvement—companion, SSIRI-C 2011, pp 68–75
9. Nasimi E, Gabbar HA (2022) Chapter 11—challenges to probabilistic risk assessment of nuclear power plants. In: Boucau J (ed) Fundamental issues critical to the success of nuclear projects. Woodhead Publishing, pp 333–344
10. Ashraf AM, Imran W, Véchot L (2022) Analysis of the impact of a pandemic on the control of the process safety risk in major hazards industries using a fault tree analysis approach. J Loss Prevent Process Indus 74
11. Markulik S et al (2021) Application of FTA analysis for calculation of the probability of the failure of the pressure leaching process. Appl Sci (Switzerland) 11(15):6731
12. Baig AA, Ruzli R, Buang AB (2013) Reliability analysis using fault tree analysis: a review. Int J Chem Eng Appl 4(3):169–173
13. Kim H, Koh J-S, Kim Y, Theofanous TG (2005) Risk assessment of membrane type LNG storage tanks in Korea-based on fault tree analysis. Korean J Chem Eng 22(1):1–8
14. Zhou K, Huang G, Wang S, Fang K (2022) Research on transportation safety of hazardous chemicals based on fault tree analysis (FTA). In: ICITM 2020—2020 9th international conference on industrial technology and management, pp 206–209
15. Mohammadi H, Fazli Z, Kaleh H, Azimi HR, Moradi Hanifi S, Shafiee N (2021) Risk analysis and reliability assessment of overhead cranes using fault tree analysis integrated with Markov chain and fuzzy Bayesian networks. Math ProblEng 2021:6530541
16. Baek S, Heo G (2021) Application of dynamic fault tree analysis to prioritize electric power systems in nuclear power plants. Energies 14(4):4119
17. Chen Y, He XQ, Lai P (2013) The application of fault tree analysis method in electrical component. In: Proceedings of the 20th IEEE international symposium on the physical and failure analysis of integrated circuits (IPFA), pp 658–661
18. Lakhotia A, Chang R, Santos D, Greene C (2020) Fault tree analysis to understand and improve reliability of memory modules used in data center server racks. Procedia Manuf 51:989–997
19. Shu X, Guo Y, Yang H, Zou H, Wei K (2021) Reliability study of motor controller in electric vehicle by the approach of fault tree analysis. Eng Failure Anal 121:105165
20. Akhtar I, Kirmani S (2020) An application of fuzzy fault tree analysis for reliability evaluation of wind energy system. IETE J Res
21. Ikwan F, Sanders D, Hassan M (2021) Safety evaluation of leak in a storage tank using fault tree analysis and risk matrix analysis. J Loss Prev Process Ind 73:104597
22. Fuentes-Bargues JL, González-Cruz MC, González-Gaya C, Baixauli-Pérez MP (2017) Risk analysis of a fuel storage terminal using HAZOP and FTA. Int J Environ Res Public Health 14(7):705
23. Kabir S (2017) An overview of fault tree analysis and its application in model based dependability analysis. Expert Syst Appl 77:114–135

24. Senol YE, Aydogdu YV, Sahin B, Kilic I (2015) Fault Tree Analysis of chemical cargo contamination by using fuzzy approach. Expert Syst Appl 42(12):5232–5244
25. Hashimoto F et al (2012) Advances in centerless grinding technology. CIRP Ann Manuf Technol 61(2):747–770
26. Patil RB, Mellal MA (2020) Fault tree analysis of a computerized numerical control turning center. In: Kumar V, Ram M (eds) Predictive analytics, 1st ed. CRC Press Taylor & Francis Group, pp 19–30
27. Lee SW, Choi HJ, Nam SH, Choi YJ (2005) Reliability prediction of centerless grinding machine. Key Eng Mater 291–292:151–156
28. Mondal SC, Mandal P (2015) An application of particle swarm optimization technique for optimization of surface roughness in centerless grinding operation. In: ICoRD'15—research into design across boundaries, vol 2. Smart Innovation, Systems and Technologies, pp 687–697
29. Lee ES, Chun YJ, Kim NK (2005) A study on the optimum condition selection of rotary dressing system of ultra-precision centerless grinding machine for ferrule. Key Eng Mater 291–292:189–194
30. Choi HZ, Lee SW, Kim GH, Chol YJ (2004) Reliability prediction of Centerless grinding machine. In: Proceedings of the KSME conference, pp 1105–1108
31. Kim S-I, Cho J-W (2007) Thermal characteristic analysis of a high-precision centerless grinding machine for machining ferrules. Int J Precis Eng Manuf 8(1):32–37
32. Enparantza R, Revilla O, Azkarate A, Zendoia J (2006) A life cycle cost calculation and management system for machine tools. In: Proceedings of the 13th CIRP international conference on life cycle engineering, LCE 2006, pp 717–722
33. Tuan NA (2021) Multi-objective optimization of process parameters to enhance efficiency in the shoe-type centerless grinding operation for internal raceway of ball bearings. Metals 11(6):893
34. Jinfei L, Yinglei L, Xueming M, Liang W, Jielin L (2021) Fault tree analysis using Bayesian optimization: a reliable and effective fault diagnosis approaches. J Fail Anal Prev 21(2):619–630
35. Katielnikov DI, Pustylnik LV, Rotshtein AP (2021) Reliability analysis: from fault tree to catastrophe tree. J Comput Syst Sci Int 60(5):793–801
36. Mohanty JK, Dash PR, Pradhan PK (2020) FMECA analysis and condition monitoring of critical equipments in super thermal power plant. Int J Syst Assur Eng Manage 11(3):583–599
37. Ahmed S, Gu X-C (2020) Accident-based FMECA study of Marine boiler for risk prioritization using fuzzy expert system. Results Eng 6:100123
38. Piumatti D, Sini J, Borlo S, Reorda MS, Bojoi R, Violante M (2020) Multilevel simulation methodology for fmeca study applied to a complex cyber-physical system. Electronics (Switzerland) 9(10):1736
39. Giardina M, Morale M (2015) Safety study of an LNG regasification plant using an FMECA and HAZOP integrated methodology. J Loss Prev Process Ind 35:35–45
40. Catelani M, Ciani L, Galar D, Guidi G, Matucci S, Patrizi G (2021) FMECA assessment for railway safety-critical systems investigating a new risk threshold method. IEEE Access 9:86243–86253
41. Gupta G, Ghasemian H, Janvekar AA (2021) A novel failure mode effect and criticality analysis (FMECA) using fuzzy rule-based method: a case study of industrial centrifugal pump. Eng Fail Anal 123:105305
42. Hanson K (2016) Basics of centerless grinding. [Online]. Available: https://www.ctemag.com/news/articles/basics-centerless-grinding. Accessed: 14 Nov 2021
43. Staff (2021) Centerless grinding. [Online]. Available: https://finemetalworking.com/centerless-grinding

Machine Learning Based Software Defect Categorization Using Crowd Labeling

Sushil Kumar, Meera Sharma, S. K. Muttoo, and V. B. Singh

Abstract Defect categorization is an important task which helps in software maintenance. It also helps in prioritizing the defects, resource allocation, etc. Standard machine learning techniques can be used to automate the categorization of defects. Labeled data is needed for learning models. The expert is required for obtaining the labeled data. Sometimes, it is costly or expert is not available. So, to overcome this dependency, crowd labeled data is used to train a model. Crowd (a set of novices) is asked to assign a category as defined by IBM's Orthogonal Defect Classification (ODC) to the defect reports. Obtaining categories through crowd can be inaccurate or noisy. Inferencing ground truth is a challenge in crowd labeling. Support Vector Machine, k Nearest Neighbor and Gaussian Naive Bayes classifier, are learnt effectively using new methodology from data labeled by a set of novices. In this chapter, we have proposed a learning model which learns effectively to predict the impact category of software defects using the expectation maximization algorithm and shows the better performance according to the various types of metrics by improving the existing technique by 8% and 11% accuracy for Compendium and Mozilla datasets respectively.

Keywords Crowd labeling · Naive Bayes classifier · Categorization · Expectation maximization

---

S. Kumar
Department of Computer Science, Shyam Lal College, University of Delhi, Delhi, India

M. Sharma
Department of Computer Science, Swami Shraddhanand College, University of Delhi, Delhi, India
e-mail: meerasharma@ss.du.ac.in

S. K. Muttoo
Department of Computer Science, University of Delhi, Delhi, India
e-mail: skmuttoo@cs.du.ac.in

V. B. Singh (✉)
School of Computer and Systems Sciences, Jawaharlal Nehru University, Delhi, India
e-mail: vbsingh@mail.jnu.ac.in

V. Kumar and H. Pham (eds.), *Predictive Analytics in System Reliability*,
Springer Series in Reliability Engineering,
https://doi.org/10.1007/978-3-031-05347-4_14

1 Introduction

Defect categorization is an important task that improves the efforts needed for software maintenance during software development process [1]. Defect categorization is a time consuming task which is done manually by the experts and involves high cost. Machine learning techniques like Supervised learning algorithms [2] can be used to categorize the defects as these algorithms need labeled dataset to train a classifier. To get the labelled dataset in case of unavailability of experts is a very challenging task. Crowd labeling can be used to obtain the labeled data. It is a process to get the data labeled by a set of novices or nonexperts. It is possible to learn to classify from this type of labeled data [3]. The number and quality of these novices influence the learning of classifier from this data set. The major concern of learning form crowd labeled data is the reliability. This chapter addresses the realiability issue of non-experts through expectation maximization algorithm.

Data set contains the category that is defined by the Orthogonal Defect Classification (ODC). The ODC was initially specified in [4] for defect classification to improve software development process. The main motive behind ODC was to extract defect information and to know the relation between cause and effect. The ODC permits software developers to distinguish defects based on their impact on customer. Categorization of software defects provides valuable information which is very useful to prioritize and fix the defects. It can also be helpful in prediction of defects and assigning defects to software developers.

This chapter explores the possibility and proposes a methodology to learn an accurate classifier for predicting the impact of software defects from the crowd labeled data using expectation maximization algorithm. The process of defect categorization is shown in Fig. 1. A methodology similar to [21] is used to integrate the labels to find the ground truth to train three classifiers, namely Naïve Bayes, k Nearest Neighbor and Support Vector machine.

The main contribution of this chapter is defect categorization from unstructured text from summary and description and analysis of subjective labeling assigned to the defect report by non-experts using Expectation–Maximization algorithm. Training of classifier has been done by taking into account the reliability of each non-expert. The performance has analysed based on majority voting and Expectation–Maximization algorithm.

This chapter is organized in following sections. Section 2 discusses the related work, ODC and Crowdsourcing. Datasets, Classifier and Expectation Maximization are explained in Sect. 3. Section 4 explains the methodology. The experimental work and metrics are explained in Sect. 5. Results are presented and discussed in Sect. 6.

Fig. 1 Process of defect categorization

Section 7 discusses the threats to validity. Section 8 concludes the chapter with future work.

2 Related Work

Learning from crowd (a set of non-experts) is a new Supervised learning paradigm in which the real or true labels of examples or instances are unavailable. However, each instance is provided with a set of noisy class labels, each indicating the class-membership of the instance according to the subjective opinion of an annotator. Many research works have been carried out in recent years. Most of them focus on labeling techniques and on the quality of the labels. Snow et al. [5] evaluated that the knowledge of four annotators is equal to the one expert. Sheng et al. [6] in his study proposed the idea of relabeling and also compared advantages of it. The idea of weak labeling was proposed by Benaran-Munoz et al. [7] in which every annotator provides more than one label for each instance. GLAD Whitehill et al. [8] proposes different levels of expertise and difficulty of examples. Donmez et al. [9] proposes a novel method of repeated trials to get the knowledge about a label and as well as about a labeller. Welinder and Perona [10] distinguished between a reliable and unreliable labeller. In case of unreliable labeler, more labels need to be asked. On the other hand for a reliable labeler, acquired label is a true label. The probability of getting true category/label follows a Bernoulli distribution by Yan et al. [11]. Gonzalez et al. [12] proposes to learn a classifier using five novices with k Means clustering and EM method. Dermartini et al. [13] proposed the method based on probabilistic reasoning and crowdsourcing. Furthermore, severity prediction of defect reports based on the textual description of defects using machine learning algorithms has been performed. Chaturvedi and Singh [14], proposed a severity prediction method which classifies the severity of the defect reports using supervised machine learning algortihms, namely Multinonmial Naïve Bayes, Support Vector Machine, k-Nearest Neighbor, Naïve Bayes, J48 and RIPPER. To carry out the experimental work, the authors collected the two bug reports data sets from NASA and PROMISE repository. Text mining techniques are applied on bug description to extract the relevant features. Liu et al. [15] present a ranking-based technique to improve the feature selection algorithms and also propose an ensemble feature selection algorithm. To evaluate the performace, the authors collect bug reports from two projects, namely Eclipse and Mozilla. They improve the existing methods by 54.76% in terms of f-measure. In [16], authors present a severity prediction technique using textual features of bug reports from three projects Eclipse, Mozilla and Gnome. They were able to achieve 67% accuracy using adaboost classifer. Yang et al. [17] present a severity prediction approach based on emotion similarity of the reporter by calculating the emotion similarity probability. To validate the proposed approach, they collected the bug reports from five projects: GNU, JBoss, Mozilla, Eclipse and Wireshark.

Table 1 Defect impact category and their definition

Impact	Definition
Capability	The ability of the product/system to perform its intended functions and satisfy the customer's functional requirements
Usability	The ability to use and utilize functions of a system by the user
Performance	The speed and responsiveness of the product/system as perceived by the customer
Reliability	The ability of the product/system to consistently perform its intended functions without unplanned interruption
Installability	The ability to easily install a product
Maintainability	The ease with which a failure can be diagnosed and the product/system can be upgraded to apply corrective fixes without impacting the customer's data and operations
Documentation	The ability of a system to provide user manuals and documentation to its user to understand a system easily
Migration	The ease and degree to which the product/system can be upgraded to the newer release without impacting the customer's data and/or operations
Standards	The degree to which the product/system conforms to established pertinent standards
Integrity/security	The degree to which the product/system is protected from inadvertent or malicious destruction, modification, or disclosure
Capacity	The loss of capability when configured at full capacity
Serviceability	The capacity to diagnose faults and failures easily

2.1 Orthogonal Defect Classification (ODC)

Orthogonal Defect Classification (ODC) is a precise system for Software Defect Classification created by IBM in the mid of 1990s [4]. ODC empowers in-process input to designers by separating marks on the improvement procedure from defects. The 13-classification ODC enables engineers to isolate absconds relying upon their effect. It is especially appropriate for open-source ventures. The impact category and the definition are provided in Table 1. The program structure involved in defect can be indicated by ODC [18].

2.2 Crowdsourcing and Learning from Crowd

The author distributed an article in the wired magazine in 2006 [19]. In this article, He profoundly broke down the effect of a rising miniaturized scale outsourcing through Internet on current business conditions and the term crowdsourcing was first presented. Crowdsourcing has become an important strategy to manage issues at any phase of Software Development Life Cycle (SDLC) from software requirements to maintenance [20, 21]. It is a way of solving a problem with collective efforts [22, 23]. There are various online platforms such as Amazon's MTurk and crowdFlower where a problem can be posted. Crowdsourcing is very helpful in decision making to a software development team. The enthusiasm for the learning from crowd is because of getting large amount of data labeled at very cheap cost through web.

Learning form labeled data by crowd is challenging as each instance of a dataset is assigned a category by non-expert. These non-experts are of obscure trustfulness. The low reliability of these non-experts is another challenge. There are various strategies proposed in past literature. However, in such cases where there is no ground truth and trustworthiness of each non-expert is doubtful, a classifier can be learnt by combining the opinion of each non-expert. Snow et al. [5] estimated the contribution of the non-expert annotators: they recommend that the blend of four non-expert explanations coordinates the information of domain expert.

The following research question has been addressed in this chapter:

Research question: Can we predict more accurately the impact of software defect by estimating the reliability of each non-expert?

The chapter in address to the above question, studied the two datasets Compendium and Mozilla that covers the entire product in both the datasets. To do further analysis, three classifiers Naïve Bayes, Support Vector Machine and k Nearest Neighbor are trained. EM based technique similar to [24, 25] are used. The technique uses the subjective opinion of all the non-expert and estimates the reliability of each non-expert.

3 Datasets and Methods

3.1 Datasets

Two datasets Compendium and Mozilla have been used directly from [12]. The Compendium dataset is taken from http://compendium.open.ac.uk/bugzilla/ which is a software tool. All issues reported in August 2014 are considered. Total 846 defects were obtained. Another dataset Mozilla has 598 defects. Mozilla is an open source application. For both the datasets, two fields summary and description are considered. Figures 2 and 3 show the number of labels assigned by the non-experts (labelers) according to the impact categories defined by ODC for the Compendium and mozilla datasets respectively. Usability, requirement and Capability are the most assigned categories for Compendium dataset.

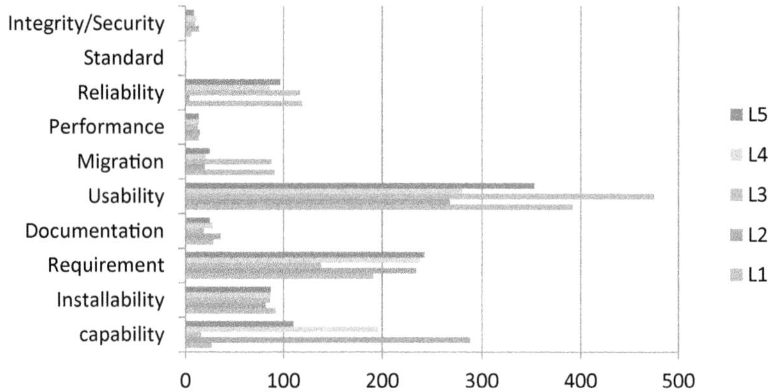

Fig. 2 Number of labels assigned by five labellers for Compendium dataset

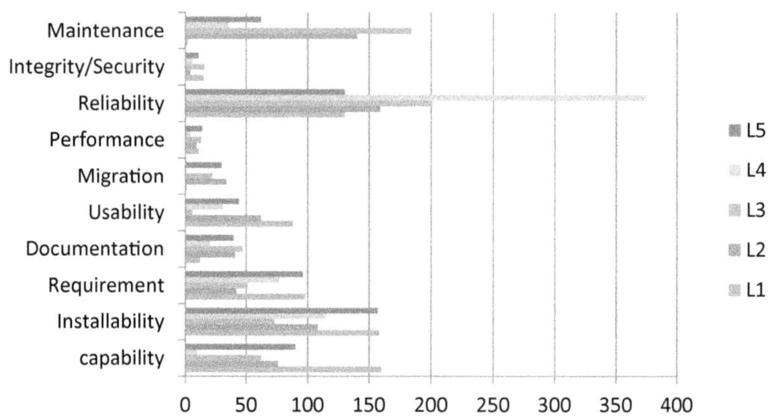

Fig. 3 Number of labels assigned by five labellers for moziila dataset

3.2 Expectation–Maximization

Expectation Maximization algorithm has been widely used [8, 10, 23–25]. The Methods based on expectation maximization is not new in crowd learning methods. The EM based technique proposed similar to [25] for multidimensional learning from crowd labeling is used to categorize the software defects.

Let N be the number of defect reports (instance or examples) in a dataset. Let n_d^l be the number of times, a defect instance d is labeled with label l. Let a function b_i^l is defined as, $b_i^l = 1$, if the assigned label is same as the true label (i.e. $l = l'$) and 0 otherwise. We assume that the labels are assigned independently by the labelers (non-experts). By the definition of multinomial distribution we can define the probability of a observed (assigned) label while the true label l' (most voted label) is known using Eq. (1) as

$$p\left(l|l',d\right) \propto \prod_{l=1}^{L} p(l'|l)^{n_d^l} \tag{1}$$

Let each defect example 'd' of a dataset D is labeled independently, then we can rewrite the above equation as (2) for each category 'c' for all the defect examples.

$$p(l,l_d') \propto \prod_{d=1}^{N}\prod_{c=1}^{C}(p(l)\prod_{l=1}^{L} p\left(l'|l\right)^{n_d^l})^{B_i^l} \tag{2}$$

Algorithm: EM (D, n, ∈)

- $D =< (d_i, l_i) >$ where $1 \leq i \leq n$
- $D = < (d_1, l_1), (d_2, l_2), \ldots, (d_n, l_n) >$

1. Initialization

$$\text{Calculate } E\left[b_i^l\right] = \frac{n_d^l}{\sum_l n_d^l} \tag{3}$$

2. M step: Select the value for (4) and (5)

$$p\left(l'|l\right) = \frac{\sum_i b_i^l \cdot n_d^l}{\sum_1^C \sum_i b_i^l \cdot n_d^l} \tag{4}$$

$$\text{and } p(l) = \frac{1}{N}\sum b_i^l \tag{5}$$

to maximize the likelihood.

3. E step: Estimate the reliability of each non-expert as

$$E[b_i^l|\mathrm{D}] = p(b_i^l = 1|D) = \prod\nolimits_1^C p\left(l'|l\right)^{n_d^l} \cdot p(l) \tag{6}$$

Repeat Steps (2) and (3) until i reaches to Maximum Iteration N or differnces between iterations<0.001.

The EM system enables us to consolidate the estimation of each non-expert that display the dependability of every labeler and the learning of the model utilizing the labels assigned by these non-experts. The initial value is calculated using Eq. (1) as the ratio by counting the frequency of a specific label to the total number of labels assigned. In our technique, the Expectation step calculates the expectation for each non-expert to estimate the reliability using Eq. (6), by integerating the probability of a label with the probability of true label when observed label is given for each category and thus calculate the estimated posterior probabilities.The Maximization step, the model parameters are re-evaluated with the end goal that the probability is augmented given the information and the loads assessed in the Maximization step using Eqs. (4) and (5) which are the maximum likelihood estimators of $p(l'|l)$ and $p(l)$. $p(l'|l)$ is the likelihood of true label when observed label is given and $p(l)$ is the probability of observed label. Iteratively, the steps 2 and 3 are rehashed until the likelihood converges to a local maxima or the maximum number of iterations is reached.

3.3 Classification Model

The model uses the summary and description field to predict the impact category of a defects reported in the two datasets of Compendium and Mozilla. The Naïve Bayes (NB), Support Vector Machine (SVM) and k-Nearest Neighbor (k-NN) classification algorithms are used to categories the defects. Naive Bayes classifier [26] is one of the most effective classifier because of its performance with other competitive classifiers. It learns by computing the probability of an attribute x_i given the class y_i where $(x_i, y_i) \in D$ i.e. training data. Naïve Bayes classifier makes strong assumption that all the attributes x_i are independent.

Support Vector Machine (SVM) [27] is a supervised learning technique which is initially used for dividing hyperplane. Its capability to generalize and better performance for multiclass problem makes it suitable for categorizing software defects. k Nearest Neighbor (k-NN) is a very simple supervised technique. It is suitable for large datasets and assigns a category to new object by finding its nearest neighbor.

4 Methodology

As specified by [12] a group of five non-experts are asked to provide category to each example of the defect reports of compendium and mozilla. The categories assigned by each non-experts were processed along with the summary and description fields as in Fig. 4.

The graphical representation of the whole learning process is shown in Fig. 5.

The text provided in summary and description field are combined as summary and pre-processed using Natural Language Processing Tool Kit (NLTK) implemented in python. Figure 6 depicts all the steps of data processing. The relevant and important words from the summary field were extracted so that it can be easily used by machine learning techniques [28, 29]. Stop words were removed by downloading the stop words from nltk and by importing ***nltk.corpus.reader*** package written in python. The Text in summary field was converted to lowercase using the in-built lower () function. Porter stemmer [28] was used for stemming the words and also the tokens were formed. A bag of words was created by extracting features and ***countVectorizer*** is improted to count the frequency of a word. The value for max-features parameter of

1	summary	description	L1	L2	L3	L4	L5
2	Browser hangs while trying to access the mozilla bug re	From Bugzilla Helper:	Reliability	Reliability	Documen	Capability	Documentati
3	mozilla installer.exe fails with "-229 script error"	Whenever I try to install the latest "mozilla-win32-	Installabil	Installabil	Installabil	Installabil	Installability
4	Mozilla 0.7 fails to open: crash out with usual "error of t	The initialisation list proceeds as expected, but the	Reliability	Installabil	Performa	Installabil	Reliability
5	fatal error building rdf\chrome\tools\chromereg\regch	when trying to build_all with a fresh branch.	Installabil	Capability	Reliability	Reliability	Reliability
6	Cannot download over modem using new download ag	When I attempt to use the new download agent with	Capability	Installabil	Maintena	Installabil	Installability
7	Opening New or Reply Msg generates error: EditorShar	With newly installed M18 build (in addition to Mozilla	Capability	Reliability	Maintena	Reliability	Reliability
8	Voting for Bugs doesn't works in a local version of bugzi	I have bugzilla installed in my server, in the url above, i	Capability	Reliability	Maintena	Reliability	Capability
9	messages with attachments don't list attachments and	Build is 2001 02 07 11. If you view an message with an	Capability	Reliability	Reliability	Reliability	Capability
10	install fails with error -214	dveditz seems to know what the problem is, need to	Installabil	Installabil	Installabil	Installabil	Installability

Fig. 4 Defect report of Compendium and labeled assigned by five non-experts

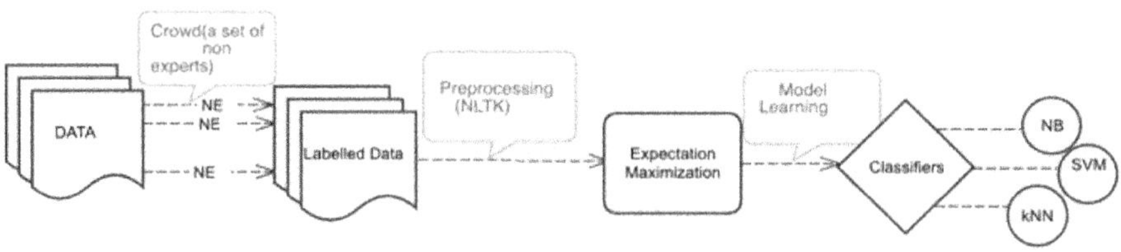

Fig. 5 Graphical representation of the process

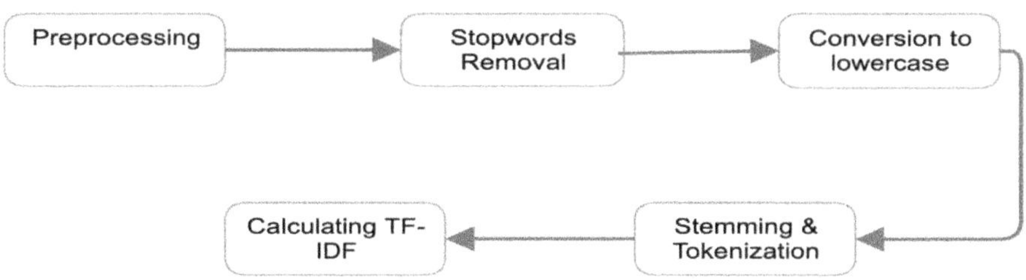

Fig. 6 Steps involved in Data Processing

countVectorizer function was set to 900. Term frequency inverse document frequency (TF-IDF) was calculated for every word. The value for parameter in EM based technique was set to 0.001. The number of iteration was set to 400. This helps to learn a classifier more accurately by using the crowd learning approaches. We have used tenfold cross validation to split the datasets. All the classifiers learn from the same dataset.

5 Experimental Framework

The various experiments on two datasets have been performed. The capabilities of EM based techniques have been explored. The different metrics are used to check the

Table 2 Different Metrics used to measure the performance of classifiers

$accuracy = \frac{\#instances\ correctly\ classified}{total\#instances}$	$Precision = \frac{\#instances\ correctly\ classified\ as\ class\ A}{total\#instances\ as\ class\ A}$
$recall = \frac{\#instances\ correctly\ classified\ as\ Class\ A}{total\#instances\ labelled\ as\ class\ A}$	$f\ measure = \frac{2*precision*recall}{precision+recall}$
$Max\ recall = max(recall)$	$min\ recall = min(recall)$

Table 3 Results of three classifiers learnt from compendium dataset and different metric based on EM

Classifier	Accuracy	Precision	Recall	F measure	Max recall	Min recall	Majority voting
NB	**0.6212**	0.5412	**0.5**310	0.4880	0.5310	0.3201	0.4016
SVM	0.5819	0.5762	0.3901	0.4645			0.4932
kNN	0.3941	0.3209	0.3821	0.4248			0.4417

Table 4 Results of three classifiers learnt from mozilla dataset and different metric based on EM

Classifier	Accuracy	Precision	Recall	F measure	Max recall	Min recall	Majority voting
NB	**0.6541**	0.6216	**0.6**152	0.4914	0.6152	0.3170	0.3754
SVM	0.5991	0.5804	0.5770	0.4032			0.3762
kNN	0.5946	0.5709	0.5610	0.4566			0.3912

capabilities of our proposed approach. The metrics used to measure the performance of classifier are shown in Table 2.

The performance of classifiers using the different metrics described in Table 2 is shown in Tables 3 and 4. Table 3 shows the accuracy, precision, recall and F measure for Naïve Bayes, SVM and kNN classifiers on the Compendium dataset. whereas the performance of these classifiers are shown in Table 4 on the Mozilla dataset.

6 Results and Discussion

The labels assigned by different non-experts are compared for both compendium and mozilaa dataset. We have used the same datasets as of [12]. For compendium datasets, we considered installability, Requirement, Usability and other. Other is a new label which is assigned to rest of the labels. The classfiers Naïve Bayes, SVM and kNN are learnt using these four categories. For Mozilla dataset, installability, maintenance, reliability and other (new label) are used to train a classifier.

So as to give a total overview of the performance of classifiers, namely Naïve Bayes, SVM and kNN, the results are presented in Tables 3 and 4. The performance of the classifiers learnt using Compendium dataset are shown in Table 3. Table 4

Table 5 Comparison of proposed approach to the Hernanedez Gonzalez [12] in terms of accuracy based on EM algorithm

Dataset	Approch	Accuracy (%)
Compendium	Reference [12]	**54**
	Proposed	62
Mozilla	Reference [12]	**54**
	Proposed	65

presents the result of classifiers learnt using Mozilla dataset. The results provide in Tables 3 and 4 measure the performance of classifiers using the same metric for both the dataset. The metrics accuracy, precision, recall, F measure and maximum and minimum recall are used in this chapter. The definition of metrics to evaluate the performance of classifiers is provided in Table 2.

Columns of Tables 3 and 4 show the majority voting, EM based method and different metrics accuracy, precision, recall, F measure, maximum and minimum recall. Whereas row represents the experiment values for each classifier. The best value for each classifier is represented in bold. The differences between minimum and maximum recall values are related to the accuracy and f measure. The high difference indicates the large values of accuracy while low difference contributes to high f measure values. Hence the performance of the classifiers can be assessed from these values across all labels.

We have also compared our results on the same dataset compendium and mozilla used by Hernández- Gonzalez's et al. [12]. They have classified their dataset by using naïve bayes, 2DB (Dependence Bayesian) and TAN (Tree Augmented Naïve Bayes). By analyzing the results, we can observe that maximum 62% accuracy is achieved in case of compendium dataset when naïve bayes classifier is learnt. An accuracy of 65% is achieved when navie bayes classifer is learnt using Mozilla dataset as shown in Table 5.

Figures 7 and 8 shows the comparision of accuracy for Compendium and Mozilla respectively.

The results comparision with the previous approaches are shown in Table 6.

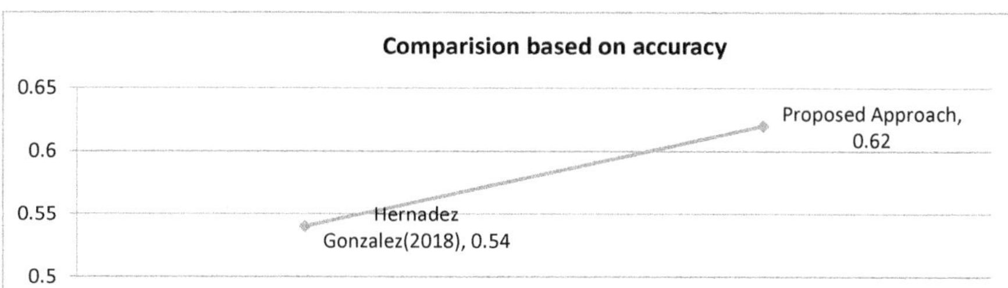

Fig. 7 Comparision based on accuracy between two approaches for compendium dataset

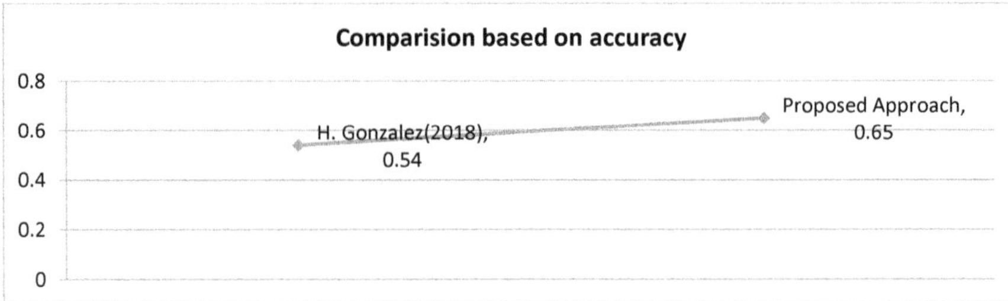

Fig. 8 Comparision based on accuracy between two approaches for mozilla dataset

Table 6 Comparision with other approaches

Approach	Accuracy (%)	Precision (%)	Categories	# Defect reports
Thung et al. [2]	77.8	69	Control and data flow, structural and non-functional	500
Thung et al. [31]	-	65.1	Control and data flow, structural and non-functional	500
Liu et al. [32]	79	75	Data, computational, interface, control/logic	1174
Hunag et al. [33]	80.7–82.9	–	ODC impact attributes	1653
Gonzalez et al. [12]	62–64	–	ODC impact attributes	1444

7 Threats to Validity

In this section we have discussed various threats to validity to our study.

7.1 Threats to Construct Validity

Threats to construct validity refer to the selection of measures and measurement tools. We have used four measures to evaluate the performance of our proposed model. These four measures are accuracy, precision, recall and f-measure. These measures are commonly used. So we can believe that there is minimal threat to construct validity.

7.2 Threats to Internal Validity

Threats to internal validity refer to the biasness of the experimenter. These defect reports are labeled manually by five people having no expertise. The distribution of classes/labels for both the datasets is not uniform. The performance of classifiers are different for both the datasets. It can due to the text describing the defects and preprocessing the textual description. As we are only using the unstructured textual defect reports, it can influence the result of the categorization.

7.3 Threats to External Validity

We have used 1444 defect reports from two projects. The number of defect reports may not be enough to generalize the results. Manual labeling of defect reports according to one of the ODC attributes is a difficult and lengthy task and the limitation to obtain a large dataset. Generalizability of the result is one of the threats to external validity.

8 Conclusion and Future Work

The chapter proposed a defect categorization approach based on EM algorithm through crowd labeled data. Two datasets from compendium and mozilla have been used to test the proposed methodology. EM method applied to learn three classifiers naive Bayes, support vector machine and k-NN. The experiment results show the performance of these classifiers. The EM-based method calculates the reliability of each non-expert. It models the problem of multiclass using multinomial distribution and maximum likelihood. Thus classifiers are learnt from the best possible configuration. The proposed approach shows the better performance as compare to exiting approach by 8 and 11% accuracy. There are various issues which can be fixed in future. To combine the knowledge of each non-experts, retrieving ground truth from crowd labeled data are such issues which must be addressed.

References

1. Boehm B, Basili VR (2005) Software defect reduction top 10 list. In: Boehm B, Rombach HD, Zelkowitz MV (eds) Foundations of empirical software engineering: the legacy of Victor R. Springer, Basili, pp 426–431
2. Thung F, Lo D, Jiang L (2012) Automatic defect categorization. In: Proceedings of 19th working conference on reverse engineering, pp 205–214
3. Hernández-González J, Inza I, Lozano JA (2015) Multidimensional learning fromcrowds: usefulness and application of expertise detection. Int J Intell Syst 30(3):326–354

4. Chillarege R, Bhandari I, Chaar J, Halliday M, Moebus D, Ray B, Wong M-Y (1992) Orthogonal defect classification—a concept for in-process measurements. IEEE Trans Softw Eng 18(11):943–956
5. Snow R, O'Connor B, Jurafsky D, Ng AY (2008) Cheap and fast—but is it good? Evaluating non-expert annotations for natural language tasks. In: Proceedings of conference on empirical methods in NLP; Honolulu, Hawaii, USA, pp 254–263
6. Sheng VS, Provost FJ, Ipeirotis PG (2008) Get another label? Improving data quality and data mining using multiple, noisy labelers. In: Proceedings of 14th international conference on knowledge discovery and data mining (ACM SIGKDD), Las Vegas, Nevada, USA, pp 614–622
7. Beñaran-Muñoz I, Hernández-González J, Pérez A (2018) Weak labeling for crowd learning
8. Whitehill J, Ruvolo P, Wu T, Bergsma J, Movellan JR (2009) Whose vote should count more: optimal integration of labels from labelers of unknown expertise. In: Proceedings of advances neural information processing systems 22 (NIPS), Vancouver, Canada, pp 2035–2043
9. Donmez P, Carbonell JG, Schneider J (2009) Efficiently learning the accuracy of labeling sources for selective sampling. In: Proceedings of the 15th international conference on knowledge discovery and data mining (KDD), pp 259–268
10. Welinder P, Branson S, Belongie S, Perona P (2010) The multidimensional wisdom of crowds. In: Proceedings of advances neural information processing systems 23 (NIPS), Vancouver, Canada, pp 2424–2432
11. Yan T, Kumar V, Ganesan D (2010a) Crowdsearch: exploiting crowds for accurate real-time image search on mobile phones. In: Proceedings of the 8th international conference on mobile systems, applications, and services. ACM, pp 77–90
12. Hernández-González J, Rodriguez D, Inza I, Harrison R, Lozano JA (2018) Learning to classify software defects from crowds: a novel approach. Appl Soft Comput 62:579–591
13. Dermatini G (2012) ZenCrowd: leveraging probabilistic reasoning and crowdsourcing techniques for large-scale entity linking
14. Chaturvedi KK, Singh VB (2012) Determining bug severity using machine learning techniques. In: 2012 CSI sixth international conference on software engineering (CONSEG) 2012 Sep 5, pp 1–6. IEEE
15. Liu W, Wang S, Chen X, Jiang H (2018) Predicting the severity of bug reports based on feature selection. Int J Softw Eng Knowl Eng 28(04):537–558
16. Otoom AF, Al-Shdaifat D, Hammad M, Abdallah EE, Aljammal A (2019) Automated labelling and severity prediction of software bug reports. Int J Comput Sci Eng 19(3):334–342
17. Yang G, Zhang T, Lee B (2018) An emotion similarity based severity prediction of software bugs: a case study of open source projects. IEICE Trans Inf Syst 101(8):2015–2026
18. Catolino G et al (2019) Not all bugs are the same: understanding, characterizing, and classifying bug types. J Syst Softw
19. Howe J (2006) The rise of crowdsourcing. Wired Mag 15(6):1–4
20. Mao K et al (2017) A survey of the use of crowdsourcing in software engineering. J Syst Softw 126:57–84
21. Sarı A, Tosun A, Alptekin GI (2019) A systematic literature review on crowdsourcing in software engineering. J Syst Softw 153(2019):200–221
22. Rodrigo EG, Aledo JA, Gámez JA (2019) Machine learning from crowds: a systematic review of its applications. Wiley Interdisc Rev: Data Min Knowl Discov 9(2):e1288
23. Raykar VC, Yu S, Zhao LH, Valadez GH, Florin C, Bogoni L, Moy L (2010) Learning from crowds. J Mach Learn Res 11:1297–1322
24. Dawid AP, Skene AM (1979) Maximum likelihood estimation of observer error-rates using the EM algorithm J. R Stat Soc Ser C: Appl Stat 28(1):20–28
25. Smyth P, Fayyad U, Burl M, Perona P, Baldi P (1994) Inferring ground truth from subjective labelling of venus images. In: Proceedings of advances in neural information processing systems (NIPS); Denver, Colorado, USA, pp 1085–1092
26. Friedmen N, Gieger D, Goldszmidt M (1997) Bayesian network classifier. Mach Learn 29:131–163

27. Hsu C-W, Lin C-J (2002) A comparison of methods for multiclass support vector machines. IEEE Trans Neural Netw 13(2):415–425
28. Vedula RMS, Bhadoria RS, Dixit M (2021) Integrating blockchain with AI. In: Multidisciplinary functions of blockchain technology in AI and IoT applications, pp 1–25. IGI Global
29. Samanta S, Pal M, Mahapatra R, Das K, Bhadoria RS (2021) A study on semi-directed graphs for social media networks. Int J Comput Intell Syst 14(1):1034–1041
30. van Rijsbergen CJ, Robertson SE, Porter MF (1980) New models in probabilistic information retrieval. British Library, London
31. Thung F, Le XBD, Lo D (2015) Active semi-supervised defect categorization. In: 2015 IEEE 23rd international conference on program comprehension. IEEE
32. Liu et al (2015) An ast-based approach to classifying defects. In: 2015 IEEE international conference on software quality, reliability and security-companion. IEEE
33. Huang et al (2015) AutoODC: automated generation of orthogonal defect classifications. Autom Softw Eng 22(1):3–46

Development of an Algorithm Using the Vikor Method to Increase Software Reliability

Shafagat Mahmudova

Abstract Software efficiency indicators play a key role in its optimization. Various ways are available to ensure software optimization. One of the key indicators of software is its reliability. Software reliability refers to the program features to perform certain functions and they are kept within certain limits under specified conditions. Software reliability is determined by its non-denial and recoverability. Software reliability is considered an important quality factor. The article uses the VIKOR (VIsekriterijumska optimizacija i KOmpromisno Resenje) method for the development of an algorithm to increase software reliability. The VIKOR method is used for different areas. Some sources provide information on the application of the VIKOR method. It refers to a multi-criteria decision method or multi-criteria decision analysis method. The alternatives here are ranked and the one closest to the ideal so-called compromise is determined. As a result of the author's research, six important criteria for software reliability are identified and alternatives are used. The fuzzy VIKOR method is used for multi-criteria evaluation of software. The work done is considered to be novel, and the advantage is that the selected criteria have not yet been used for this type of task, this positively changes its efficiency. The experiments perform positive results.

Keywords Software · Efficiency characteristics · Optimization · VIKOR · Multi-criteria method

1 Introduction

To develop high quality software systems, various technologies and methods are used. Optimal software is created through different possible ways. In previous articles [1] offered an algorithm to select the best software using the TOPSIS (Technique for Order of Preference by Similarity to Ideal Solution) method. AHP (Analytic Hierarchy Process) method Mahmudova and Jabrailova [2] offered optimizing the

S. Mahmudova (✉)
Institute of Information Technology of ANAS, Baku, Azerbaijan
e-mail: shafagat_57@mail.ru

V. Kumar and H. Pham (eds.), *Predictive Analytics in System Reliability*,
Springer Series in Reliability Engineering,
https://doi.org/10.1007/978-3-031-05347-4_15

software, and good outcomes are obtained as a result of experiments. The TOPSIS method is applied to rise the efficiency of software through its efficiency characteristics and making critical decisions in problem solutions. It refers to a multi-criteria decision-making analysis method offered by Hwang and Yoon [3], and includes superior features compared to others.

The best alternative based on compromise solution is identified through TOPSIS. Its chief concept is that the chosen alternative should be at the shortest Euclidean distance from the positive ideal solution and at the farthest Euclidean distance from the negative ideal solution. The negative ideal solution maximizes the loss criterion and minimizes the profitability criterion. And the positive ideal solution maximizes the profitability criterion and minimizes the loss criterion. The method defines an index close to the positive ideal solution and far from the negative ideal solution. In conclusion, the closest alternative to the positive ideal solution is selected. The compromise solution can be considered the selection of a solution at the farthest Euclidean distance from the negative ideal solution and at the shortest Euclidean distance from the positive ideal solution.

Software efficiency (SE) (ISO/IEC standard 25010: 2011 (state standard R ISO/MEK 25010–2015) determines the product quality model, since it has eight top-level characteristics.

The efficiency characteristics of software are:

1. Functionality;
2. Productivity;
3. Compliance;
4. Ease of use;
5. Reliability;
6. Security;
7. Accompanying;
8. Mobility.

As noted, reliability is one of the key performance characteristics of software.

Software reliability refers to the features of a program to perform certain functions, and they are kept within certain limits under specified conditions. In other words, the reliability of a program is the probability that software will work without any failure for a certain period of time. Opricovic and Tzeng [4] proposed the software reliability determined by its non-denial and recoverability. There may be important factors that affect the reliability of software. The reliability of software is its ability to maintain its functioning in the course of data processing on computer. The reliability of software can be assessed by the probability that it will operate without malfunction under certain environmental conditions during the observation period.

Different models have been developed based on different sets of assumptions. Several models have been developed under practical conditions considering testing effort, test coverage, time delay error correction, and error reduction factor [5].

Software accuracy refers to its compliance with the specifications. One of the important features of software reliability is that it can be restored due to errors and consequences in the program. Recovery after a software failure is the ability to correct the program text, correct the data, and make changes to the organization of the computation process. The recovery capability of software can be assessed by the average time it takes to troubleshoot a program and restore it to working condition. Software recovery depends on several factors: the complexity of the software structure, the algorithmic language in which software is developed, the style of programming, the quality of software documents, and so on. Causes of software failure and the main causes of direct software failure lie in tie followings.

Software reliability is considered an important quality factor. "Software reliability refers to the failure-free operation of software over a specified period of time and in a specified environment" [6].

One of the important features of software reliability is that it can be restored due to errors and consequences in the program.

1. errors hidden in software itself;
2. falsification of used input data;
3. user error;
4. device failure on which the computing process is performed.

The American National Standards Institute (ANSI) defines the reliability of software as: the probability that a program will run flawlessly over a period of time in a given environment. It is difficult to get the reliability of the program, because the high complexity of the program does not allow it.

The following information should be considered to improve the reliability of software:

- Computer's configuration;
- Performance and reliability, for example, how software responses when a button is pressed, how many problems it encounters while software is running, how fast the data is sent over the network;
- Most commonly used tools in software program.

The development of an algorithm for software reliability is one of the foremost issues.

An algorithm is a sequence of operations to be performed to solve a task. There are three main types of algorithms used to solve different types of problems on a computer:

- Linear algorithms;
- Branching algorithms;
- Periodic algorithms.

Linear algorithms consist of a series of operations that represent a simple computational process, and they are performed in the sequence in which they are written.

Branching algorithms contain one or more logic steps. At this stage, it is checked whether certain quantities meet any conditions, and the direction of the next step is selected accordingly. That is, if intended condition is met, it moves in one direction, if not, it moves on another direction. Thus, branching occurs in the algorithm.

Periodic algorithm. Programming often requires a large number of repetitions of the same group of operations. In this case, the cyclic algorithm is used. Cycles may be simple and complex. A simple cyclic algorithm includes one cycle. If any algorithm involves several internal cycles, then such cycles are called complex.

Proper construction of algorithm is one of the main conditions for solving any problem.

One of the important issues is to get software backup.

Handy Backup ™ is used to back up data and programs. Advantages of Handy Backup ™ includes:

- Backup in original format;
- Simple and convenient interface:
- Availability of different backup methods;
- Cloud support;
- Work in advanced mode.

2 Software Reliability Models

Software reliability models show the form of a random process, as it periodically determines the behavior of software failures. Models of software reliability appeared when people tried to understand its features, such as why the software is faulty and so on. Neufelder [7] proposed that people have tried to quantify the reliability of software.

More than 200 software reliability models have been developed since the early 1970s, but the question of how to assess the reliability of software remains unresolved.

A list of software reliability models is shown in Table 1.

Reliability determines the end result of software. During fierce competition, any software should not only provide the necessary functionality, but also provide some additional benefits to end users. Developing software is a tedious and time-consuming process, like an experiment. Thus, ensuring the reliability of software should be the primary goal of the appropriate model specified, adopted, and selected by the organization listed above.

3 About the VIKOR Method

Different methods are used to determine the reliability of software. One of them is the VIKOR method. Brief information about this method is given below.

The VIKOR method refers to a multi-criteria decision (MCD) or multi-criteria decision analysis method. The VIKOR method was developed for multi-criteria optimization of complex systems. It determines compromise ranking list and the compromise solution obtained with the initial (given) weights. This method focuses on ranking and selecting from a set of alternatives in the presence of conflicting

Table 1 Software reliability models

Model	Number of inputs	Industry supported	Effort required to use the model	Relative accuracy	Year developed/last updated
Industry tables	1	Several	Quick	Varies	1992, 2015
CMMI® tables	1	Any	Quick	Low at low CMMi®	1997, 2012
Shortcut model	23	Any	Moderate	Medium	1993, 2012
Full-scale model	94–299	Any	Detailed	Medium–High	1993, 2012
Metric based models	Varies	Any	Varies	Varies	NA
Historical data	A minimum of 2	Any	Detailed	High	NA
Rayleigh model	3	Any	Moderate	Medium	NA
RADC TR-92-52	43–222	Aircraft	Detailed	Obsolete	1978, 1992
Neufelder model	156	Any	Detailed	Medium to high	2015

criteria. Chang et al. [8] proposed multiple criteria decision-making (MCDM) is a subdiscipline of operations research that explicitly considers multiple criteria in decision-making environments. VIKOR ranks the alternatives and determines the one closest to the so-called compromise ideal. Yen-Chu Chen and Po-Lung Yu first offered the idea of a compromise solution in 2012 [9]. MCDM (Multiple criteria decision-making) is a subdiscipline of operations research that explicitly considers multiple criteria in decision-making environments.

It was stated that a compromise was acceptable, originally developed by Seraphim Oprikovic to resolve conflict resolution problems and diverse (different sections) criteria, that the decision-maker wanted the solution closest to the ideal and evaluated all alternatives based on established criteria. Opricovic and Gwo-Hshiung [10] propose VIKOR method to evaluate alternatives and identifies a solution called compromise, which means the closest to the ideal.

Cochrane JL. and Milan Zeleny first presented the idea of a compromise solution in the MCD in 1973 [11].

Lucien and Opricovic [12] developed the main ideas about VIKOR in his dissertation in 1979, and information on its application was published in 1980. The name VIKOR originated from the Serbian language in 1990: Multi-value and Optimization of Compromise solution (VIseKriterijumska Optimizacija I Kompromisno Resenje, VIKOR). In 1998, real expressions were introduced. Sayadi et al. [13] proposed the document adopted in 2004 contributed to the international recognition of the VIKOR method [14].

Vahdani and Mousavi [15] is proposed methodology as a compromised method to solve the Multi-Objective Large-Scale Nonlinear Programming problems with block angular structure involving fuzzy coefficients.

4 Literature Review

Some tasks in which the VIKOR method is applied are reviewed below.

1. Hajiagha et al. [16] proposed VIKOR method used in linear programming task. Real decision-making problems often involve the consideration of many opposing goals. MCD is an experimental basis in relevant fields. The problem of fuzzy MCD, in which all parameters are fuzzy, is examined, and a solution using the multi-criterion VIKOR method is offered. The proposed method seeks to find a fuzzy effective solution to the problem by minimizing the distance from ideal and anti-ideal solutions. Applying this method can reveal the effective boundary of the problem. The applicability of the proposed method is shown in the example and the application is generalized to the investment problem. Both examples show the usefulness of the proposed method.
2. Mary et al. [17] proposed a method based on the VIKOR method as a compromise method for solving large-scale nonlinear programming tasks. The proposed method was first introduced to solve large-scale nonlinear programming in a fuzzy environment. The problem involves fuzzy ratios in both objective functions and constraints. In this method, the aggregate function based on the LP metric approaches the "ideal" solution based on a special "proximity" dimension. The solution process consists of two stages. The first uses the decomposition algorithm to reduce the q-dimensional space to a one-dimensional space. Then, to solve the problem, multi-purpose identical nonlinear programming is obtained from each fuzzy nonlinear model. The second one solves the problem of large-scale single-purpose nonlinear programming to find the final solution. An illustrative example is provided to substantiate the proposed method.
3. Heydari et al. [18] proposes the VIKOR method as a Multi-Attribute Decision Making (MADM) method to solve decision-making problems with separate and conflicting criteria. This method seeks to list and select a number of alternatives based on a certain "proximity" metric to an "ideal" solution. A multi-criteria method for compromise sorting is developed based on the l-p metric used in the compromise programming method as an aggregate function. In this paper, the VIKOR method is extended to solve large-scale non-linear programming tasks with block-angle structure. The proposed approach applies the Dantzig-Wolfe fragmentation algorithm along with the Y-dimensional target area reduced to a one-dimensional area by expanding the concepts of the VIKOR method to make decisions in a sustainable environment. Finally, the paper presents an example to illustrate and clarify the main results obtained in this study.
4. Opricovic [19] proposed the VIKOR and TOPSIS multi-criteria decision methods based on a set of aggregate functions that represent the "ideal proximity" arising from compromise programming. The VIKOR uses linear normalization, while the TOPSIS uses vector normalization in order to exclude criteria function units. VIKOR's compromise ranking method determines the maximum "group benefit" for the "majority" and a compromise solution for

the "competitor". The TOPSIS method determines the solution at the shortest distance to the ideal solution and the longest distance to the negative ideal solution, but does not take into account the relative importance of these distances. A comparative analysis of these two methods is illustrated by an example showing similarities and some differences.

5. The issue of emissions has forced energy systems to use cleaner energy sources such as renewable and hydroelectric technologies. However, in recent decades, the optimal use of the reservoir has been highlighted due to water insufficiency in many areas. In this regard, Simab et al. [20] proposed a multi-purpose model for the short-term hydrothermal planning problem when pumped storage technology is available. It uses VIKOR method to solve the task. The effectiveness of the proposed model is tested by comparing the results obtained with four sample studies using different methods.
6. The linguistic ambiguity of a particular fuzzy set derived from linguistic terms may represent the qualitative preferences of decision-makers, as well as their uncertainties and hesitations. In this study, a new VIKOR method is used to solve multi-criterion decision tasks. Dong et al. [21] propose an evaluation sample of a smart transport system to demonstrate the effectiveness and expediency of the proposed method.
7. Digital control machines are used for high-precision, repetitive, complex and dangerous production operations. However, there are several decision-making criteria to be considered when choosing the right one. In this study, a multi-criteria group decision-making method based on the fuzzy VIKOR is developed to solve the problem. Language variables represented by triangular fuzzy numbers are used to replicate decision-makers' preferences related to the weights of criterion significance and the evaluation of their effectiveness. This study develops two algorithms based on a fuzzy linguistic approach. Wu et al. [22] proposed a common method based on these two algorithms and the VIKOR method.
8. Alguliyev et al. [23] proposed a modified fuzzy VIKOR method for multi-criteria assessment of information culture of individuals. The VIKOR method is considered to be more appropriate for solving the individual selection problem. The paper proposes a modified fuzzy VIKOR method to rank the alternatives. It presents comparative analysis of the results of fuzzy and modified fuzzy VIKOR methods. Experience shows that the proposed modified fuzzy VIKOR method has a number of advantages over the conventional fuzzy VIKOR method. The presented model is efficient in terms of computational complexity.

5 Application of the Vikor Method

Opricovic [19] proposed the VIKOR procedure includes the following steps:

Step 1. For all criterion functions, i $= 1, 2...,$ n; the best value f_i^* and the worst value $f_i^\wedge$ are set,

$$f_i^* = \max\left(f_{ij}, j = 1, \ldots, J\right), f_i^\wedge = \min\left(f_{ij}, j = 1, \ldots, J\right),$$

if the i-th functions is benefit;

$$f_i^* = \min\left(f_{ij}, j = 1, \ldots, J\right), f_i^\wedge = \max\left(f_{ij}, j = 1, \ldots, J\right),$$

if the i-th functions is cost.

Step 2. The values of S_j and R_j, $j = 1,2 \ldots, J$ are calculated according to their relationship:

$S_j = \text{sum}\,[w_i(f_i^* - f_{ij})/(f_i^* - f_i^\wedge), i = 1\ldots n]$, weighted and normalized Manhattan distance;

$R_j = \max\,[w_i(f_i^* - f_{ij})/(f_i^* - f_i^\wedge), i = 1\ldots n]$, weighted and normalized Chebyshev distance;

where w_i are the weights of criteria, expressing the choice of DM as the relative importance of the criteria.

Step 3. The values of Q_j, $j = 1,2, \ldots, J$ are calculated in proportion

$$Q_j = v(S_j - S^*)/(S^\wedge - S^*) + (1 - v)(R_j - R^*)/(R^\wedge - R^*),$$

where

$$S^* = \min\,(S_j, j = 1\ldots, J),\ S^\wedge = \max(S_j, j = 1, \ldots, J),$$
$$R^* = \min\,(R_j, j = 1\ldots, J),\ R^\wedge = \max\,(R_j, j = 1, \ldots, J);$$

and is presented as a weight for the maximum beneficial strategy of the group, while $1 - v$ is the weight of the individual strategy. These strategies can be compromised if $v = 0.5$,

where

$$v = (n + 1)/2xnx\,(v + 0.5(n - 1)/n)$$

is varied, because the criterion (from 1 of n) is related to R and is included to S.

Step 4. Alternatives shall be ranked from the minimum value of S, R and Q to the maximum value. The result presents three rating lists.

Step 5. An alternative A (1) with the size Q (minimum) is offered as a compromise solution if the following two conditions are met.

As a compromise solution, the best-rated alternative A (1) with the size Q (minimum) is offered if the following two conditions are met:

"Acceptable advantage": $Q\,(A\,(2) - Q\,(A\,(1)) >= DQ$

where: A (2) is the alternative with second position in the ranking list by Q;

$DQ = 1/(J - 1)$ C2 "Acceptable stability when making decisions": The alternative must have the best rating by A (1), S and/or R. This compromise solution is stable throughout the decision-making process, and it can be a strategy of maximum group utility (v > 0.5 if necessary) or (v is approximately 0.5 "by consensus" or v < 0.5 "by veto"). If one of the conditions is not satisfied, a number of compromise solutions are proposed, which are: – Alternatives A (1) and A (2) only if conditions C2 is not satisfied; – Alternatives A (1), A (2), …, A (M), if condition C1 is not satisfied; A (M) is determined by the relation Q (A (M)) – Q (A (1)) < DQ for maximum M.

Alternative A (1) shall be rated best by S or and R. This compromise solution is stable within the decision-making process, it can be the group's maximum beneficial strategy (v > 0.5 if necessary) or v is approximately 0, 5 "by consensus" or v < 0.5 "by veto".

As a result, a compromise solution can be provided by the decision makers, thus ensuring the maximum benefit of the majority (denoted by min S) and the minimum failure of the opponent (denoted by min R). The metrics S and R are integrated into Q for a compromise solution based on an agreement with mutual concessions.

6 Problem Statement and Experiments

A list of alternatives is presented below:

1. Very weak;
2. Weak;
3. Less weak;
4. Unsatisfactory;
5. Not good;
6. Good;
7. Excellent.

In this case, six criteria for software reliability and seven alternatives are used with the VIKOR method to determine the indicator that is closest to the ideal compromise. Here, the reliability criteria of three software are used.

The problem is expressed as follows: Determine the best (compromise) solution as multicriteria A1, A2, …, Am out of a set of possible alternatives A evaluated according to the function of criteria N. Input data is the elements of the solution matrix f_{ij}, in which f_{ij} is the value of the i-th criterion function for alternative A_i.

Step 1. For all criterion functions, i = 1, 2 …, 6; the best is set to fi* and the worst is set to fi^. Here, the criteria of reliability include the criteria functions. j = 1 … 6.

$$F = (f_{ij})_{mxn}$$

Here, m shows the number of alternatives, and n denotes the number of criteria.

Table 2 Values of alternatives and criteria

Alternatives	f_{i1}	f_{i2}	f_{i3}	f_{i4}	f_{i5}	f_{i6}
A_1 (very weak)	0.15	0.22	0.13	0.14	0.15	0.16
A_2 (weak)	0.21	0.12	0.23	0.24	0.25	0.26
A_3 (less weak)	0.31	0.32	0.33	0.34	0.35	0.36
A_4 (unsatisfactory)	0.11	0.12	0.11	0.11	0.11	0.13
A_5 (not good)	0.41	0.42	0.43	0.44	0.45	0.46
A_6 (good)	1.41	1.42	1.43	1.44	1.45	1.46
A_7 (excellent)	2.41	2.42	2.43	2.44	2.45	2.46
Max	2.41	2.42	2.43	2.44	2.45	2.46
Min	0.11	0.12	0.11	0.11	0.11	0.13

$$F = \begin{bmatrix} f_{11} & f_{21}\ldots & f_{1n} \\ f_{21} & f_{22}\ldots & f_{21} \\ & & \\ f_{m1} & f_{m2}\cdots & f_{mn} \end{bmatrix} \quad (1)$$

A positive ideal candidate and a negative ideal solution are anti-ideal candidates: $f_i^* = \max\left(f_{ij}, j = 1, \ldots, 7\right)$, $f_i^\wedge = \min\left(f_{ij}, j = 1, \ldots, 7\right)$, if the i-th functions is benefit;

$f_i^* = \min\left(f_{ij}, j = 1, \ldots, 7\right)$, $f_i^\wedge = \max\left(f_{ij}, j = 1, \ldots, 7\right)$, if the i-th functions is cost.

The values of alternatives and criteria are shown in Table 2.

$$\max(f_i^*) = 2.46;\ \min(f_i^\wedge) = 0.11$$

$$\min(f_i^*) = 0.11; \max(f_i^\wedge) = 2.46$$

Step 2. S_j j = 1,2 ..., 7 values are calculated according to the ratio of ions:

$S_j = \text{sum}\,[w_i(f_i^* - f_{ij})/(f_i^* - f_i^\wedge)$, i = 1... n is the weighted and normalized Manhattan distance.

Here, w_i are the relative weights (Table 3), determined by the decision maker, the sum equals to 1.

Alternatives and normalized Manhattan distances are shown in Table 4.

Alternatives and normalized Manhattan distance values are shown in Table 5.

$$R_j = \max\,[w_i(f_i^* - f_{ij})/(f_i^* - f_i^\wedge),\ i = 1\ldots n], j = 1, \ldots, m \quad (2)$$

weighted and normalized Chebyshev distance;

Table 3 Values of relatively important weights

No	w_i
1	0.1
2	0.2
3	0.1
4	0.2
5	0.2
6	0.1
7	0.1
Total =	1

where, w_i is the weight of the criterion, expressing the choice of DM as the relative importance of the criteria.

Table 6 shows the alternatives and the values of the normalized Chebyshev distance calculated according to formula (2).

Step 3. The values of Q_j, $j = 1,2, ..., J$ calculate according to the following relationship:

$$Q_j = 0.5(S_j - S^*)/(S^\wedge - S^*) + (1 - 0.5)(R_j - R^*)/(R^\wedge - R^* \quad (3)$$

Here,

$$S^* = \min(S_j,\ j = 1..., 7),\ S^\wedge = \max(S_j,\ j = 1, ..., 7),$$
$$R^* = \min(R_j,\ j = 1..., 7),\ R^\wedge = \max(R_j\ j = 1, ..., 7); \quad (4)$$

and is presented as a weight for the maximum beneficial strategy of the group, and 1 – v is the weight of the individual strategy. These strategies can be compromised when v = 0.5, where v = (n + 1) / 2n (v + 0.5 (n – 1) / n = 1) is varied, because the criterion (from 1 to n) is related to R and included to S.

In Table 7, the values of S*, S^, R*, R^ are calculated according to formula (4).

$$v = (1 + 1)/2x1x(0.5 + 0.5(1 - 1)/1)$$

Table 8 shows the values of Q_j, $j = 1, 2, ..., 6$ calculated by formula (3).

Table 9 shows the ranked values of S_j, Table 10 shows the ranked values of R_j, and Table 11 and Fig. 1 shows the ranked values of Q_j.

Alternatives are denoted as A (J) j = 1, …, 7.

Here: A (2) is an alternative ranked second in the ranking list with Q;

$$DQ = 1/(J - 1).\ J = 7$$

$$DQ = 0.16667$$

Table 4 Alternatives and normalized Manhattan distance

Alternatives	f_{i1}	f_{i2}	f_{i3}	f_{i4}	f_{i5}	f_{i6}
A_1 (very weak)	0.01*(5.46–0.11)/(5.46–0.01)	0.01*(5.46–0.12)/(5.46–0.01)	0.01*(5.46–0.13)/(5.46–0.01)	0.01*(5.46–0.14)/(5.46–0.01)	0.01*(5.46–0.15)/(5.46–0.01)	0.01*(5.46–0.16)/(5.46–0.01)
A_2 (weak)	0.02*(5.46–0.21)/(5.46–0.01)	001*(5.46–0.22)/(5.46–0.01)	0.01*(5.46–0.23)/(5.46–0.01)	0.01*(5.46–0.24)/(5.46–0.01)	0.01*(5.46–0.25)/(5.46–0.01)	0.01*(5.46–0.26)/(5.46–0.01)
A_3 (less weak)	0.01*(5.46–0.31)/(5.46–0.01)	0.01*(5.46–0.32)/(5.46–0.01)	0.01*(5.46–0.33)/(5.46–0.01)	0.01*(5.46–0.34)/(5.46–0.01)	0.01*(5.46–0.35)/(5.46–0.01)	0.01*(5.46–0.36)/(5.46–0.01)
A_4 (unsatisfactory)	0.02*(5.46–0.01)/(5.46–0.01)	0.02*(5.46–0.02)/(5.46–0.01)	0.02*(5.46–0.03)/(5.46–0.01)	0.02*(5.46–0.04)/(5.46–0.01)	0.02*(5.46–0.05)/(5.46–0.01)	0.02*(5.46–0.06)/(5.46–0.01)
A_5 (not good)	0.02*(5.46–0.41)/(5.46–0.01)	0.02*(5.46–0.42)/(5.46–0.01)	0.02*(5.46–0.43)/(5.46–0.01)	0.02*(5.46–0.44)/(5.46–0.01)	0.02*(5.46–0.45)/(5.46–0.01)	0.02*(5.46–0.46)/(5.46–0.01)
A_6 (good)	0.02*(5.46–3.41)/(5.46–0.01)	0.02*(5.46–3.42)/(5.46–0.01)	0.02*(5.46–3.43)/(5.46–0.01)	0.02*(5.46–3.44)/(5.46–0.01)	0.02*(5.46–3.45)/(5.46–0.01)	0.02*(5.46–3.46)/(5.46–0.01)
A_7 (excellent)	0.02*(5.46–5.41)/(5.46–0.01)	0.02*(5.46–5.42)/(5.46–0.01)	0.02*(5.46–5.43)/(5.46–0.01)	0.02*(5.46–5.44)/(5.46–0.01)	0.02*(5.46–5.45)/(5.46–0.01)	0.02*(5.46–5.46)/(5.46–0.01)

Table 5 Alternatives and normalized Manhattan distance values

Alternatives	f_{i1}	f_{i2}	f_{i3}	f_{i4}	f_{i5}	f_{i6}	S_j	S^*	$\hat{S}$
A_1 (very weak)	0.09830	0.23664	0.09915	0.09872	0.08290	0.09787	0.09830	0.08290	0.23664
A_2 (weak)	0.19149	0.19915	0.18979	0.18894	0.19660	0.18723	0,19,149	0.18723	0.19915
A_3 (less weak)	0.09149	0.09106	0.09064	0.09021	0.08979	0.08936	0.09149	0.08936	0.09149
A_4 (unsatisfactory)	0.20000	0.19915	0.20000	0.20000	0.20000	0.19830	0.20000	0.19830	0.20000
A_5 (not good)	0.17447	0.17362	0.17277	0.17191	0.17106	0.17021	0.17447	0.17021	0.17447
A_6 (good)	0.08936	0.08851	0.08766	0.08681	0.08596	0.08511	0.08936	0.08511	0.08936
A_7 (excellent)	0.00426	0.00340	0.00255	0.00170	0.00085	0.00000	0.00426	0.00000	0.00426
							$Sum(S_j) = 5.17699$	$Min(S^*) = 0.00000$	$Max(\hat{S}) = 0.23664$

Table 6 Alternatives and the values of the normalized Chebyshev distance

Alternatives	f_{i1}	f_{i2}	f_{i3}	f_{i4}	f_{i5}	f_{i6}	R_j	R^*	$R^{\wedge}$
A_1 (very weak)	0.09830	0.23664	0.09915	0.09872	0.08290	0.09725	0.09830	0.08290	0.23664
A_2 (weak)	0.19149	0.19915	0.18979	0.18894	0.19660	0.18723	0.19149	0.18723	0.19915
A_3 (less weak)	0.09149	0.09106	0.09064	0.09021	0.08979	0.08936	0.09149	0.08936	0.09149
A_4 (unsatisfactory)	0.20000	0.19915	0.20000	0.20000	0.20000	0.19830	0.20000	0.19830	0.20000
A_5 (not good)	0.17447	0.17362	0.17277	0.17191	0.17106	0.17021	0.17447	0.17021	0.17447
A_6 (good)	0.08936	0.08851	0.08766	0.08681	0.08596	0.08511	0.08936	0.08511	0.08936
A_7 (excellent)	0.00426	0.00340	0.00255	0.00170	0.00085	0.00000	0.00426	0.00000	0.00426
							$Max(R_j) = 0.23664$	$Min(R^*) = 0.00000$	$Max(R^{\wedge}) = 0.23664$

Table 7 Values of S*, S^, R*, R^

S*	S^	R*	R^
0.00000	0.23664	0.00000	0.23664

Table 8 Values of Q j

Alternatives	S_j	R_j	Q_j
A_1 (very weak)	0.71358	0.08290	1.68291
A_2 (weak)	1.15319	0.18723	2.83220
A_3 (less weak)	0.54255	0.08936	1.33518
A_4 (unsatisfactory)	1.19745	0.19830	2.94909
A_5 (not good)	1.03404	0.17021	2.54449
A_6 (good)	0.52340	0.08511	1.28573
A_7 (excellent)	0.01277	0.00000	0.02697

Table 9 Ranked values of S_j

Alternatives	S_j
A(1)	0.01277
A(2)	0.52340
A(3)	0.54255
A(4)	0.71358
A(5)	1.03404
A(6)	1.15319
A(7)	1.19745

Table 10 Ranked values of R_j

Alternatives	R_j
A(1)	0.00000
A(2)	0.08290
A(3)	0.08511
A(4)	0.08936
A(5)	0.17021
A(6)	0.18723
A(7)	0.19830

Table 11 Ranked values of Q_j

Alternatives	Q_j
A(1)	0.02697
A(2)	1.28573
A(3)	1.33518
A(4)	1.68291
A(5)	2.54449
A(6)	2.83220
A(7)	2.94909

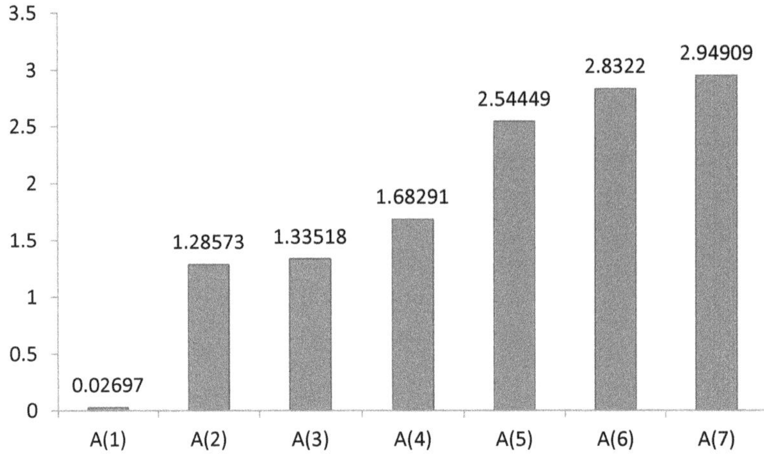

Fig. 1 Ranked values of Q_j

$$(Q\ (A(2) - Q\ (A1)) = 1.25876$$

If the values A (2) and A (1) are placed in the formula (Q (A(2) − Q (A(1)) >= DQ, then 1.25876 >= 0.16667 for C (1). If one of the conditions is not met, a number of compromise solutions are proposed, which are: – Alternatives A (1) and A (2) if only conditions C2 are not met, or alternatives A (1), A (2), …, A (M), if condition C1 is not met; A (M), determined by the relation Q (A(M)) − Q(A(1)) < DQ for maximum M.

Q (A (M)) – Q (A (1)) = 2.92212.

2.92212 >= 0.16667 meets the condition.

As a result, a compromise solution can be provided by those who make decisions, as it provides the maximum utility of the majority (represented by min S) and the minimum failure of the individual competitor (represented by min R). Measures S and R are integrated into Q for a compromise solution, which is the basis of an agreement established by mutual concessions.

7 Conclusion

The fuzzy VIKOR method is designed to solve the problem in a fuzzy environment, where both criteria and weights may be fuzzy. Triangular fuzzy numbers are used to control uncertain numerical quantities. Fuzzy VIKOR is based on fuzzy work that represents and combines the distance of an ideal solution alternative. Fuzzy operations and fuzzy ranking procedures play a key role in the development of fuzzy

VIKOR algorithm. Applying this method, the reliability of software can be achieved. The given algorithm can be applied to any software.

References

1. Mahmudova Sh (2020) Application of the TOPSIS method to improve software efficiency and to optimize its management. Soft Comput (SI 24(1)):697–708. https://doi.org/10.1007/s00500-019-04549-4
2. Mahmudova Sh, Jabrailova Z (2020) Development of an algorithm using the AHP method for selecting software according to its functionality. Soft Comput (SI 24(11)):8495–8502. https://doi.org/10.1007/s00500-020-04902-y
3. Hwang CL, Yoon K (1981) Multiple attribute decision making. Springer, New York, pp 58–191. https://doi.org/10.1007/978-3-642-48318-9
4. Opricovic S, Tzeng GH (2004) Compromise solution by MCDM methods: a comparative analysis of VIKOR and TOPSIS. Eur J Oper Res (SI 156(2)):445–455. https://doi.org/10.1016/S0377-2217(03)00020-1
5. Kumar V, Saxena P, Garg H (2021) Selection of optimal software reliability growth models using an integrated entropy–technique for order preference by similarity to an ideal solution (TOPSIS) approach. Math Methods Appl Sci. https://doi.org/10.1002/mma.7445
6. Saxena P, Kumar V, Ram M (2021) Ranking of software reliability growth models: a entropy-ELECTRE hybrid approach. Reliab: Theory Appl (SI 2(64)):95–113
7. Neufelder A (2017) Cold hard truth about reliable software—version 6g, 77
8. Chang CT, Tan KH, Lu HC (2014) Multiple criteria decision making theory, methods, and applications in engineering. Math Problems Eng (SI 2014):1–3. https://doi.org/10.1155/2014/431037
9. Chen Y-C, Huang H-S, Yu PL (2012) Empower MCDM by habitual domains to solve challenging problems in changeable spaces. Int J Inf Technol Decis Making (SI 11(02)):457–490. https://doi.org/10.1142/S0219622012400111
10. Opricovic S, Gwo-Hshiung T (2007) Extended VIKOR method in comparison with outranking methods. Eur J Oper Res (SI 178(2)):514–529. https://doi.org/10.1016/j.ejor.2006.01.020
11. Cochrane JL, Zeleny M (1973) Multiple criteria decision making. University of South Carolina Press, Columbia
12. Lucien D, Opricovic S (1980) Multiobjective optimization in river basin development. Water Resour Res (SI 16(1)):14–20
13. Sayadi MK, Heydari M, Shahanaghi K (2009) Extension of VIKOR method for decision making problem with interval numbers. Appl Math Model (SI 33(5)):2257–2262. https://doi.org/10.1016/j.apm.2008.06.002
14. Opricovic S, Gwo-Hshiung T (2004) The compromise solution by MCDM methods: a comparative analysis of VIKOR and TOPSIS. Eur J Oper Res (SI 156(2)):445–455
15. Vahdani B, Salimi M, Mousavi SM (2015) A compromise decision-making model based on VIKOR or multi-objective large-scale nonlinear programming problems with a block angular structure under uncertainty. Trans E: Ind Eng (SI 22(6)):2571–2584
16. Hajiagha SHR, Mahdiraji HA, Zavadskas EK, Hashemi SS (2014) Fuzzy multi-objective linear programming based on compromise VIKOR method. Int J Inf Technol Decis Making (SI 13):679–698. https://doi.org/10.1142/S0219622014500667
17. Mary V, Michele H, Neel S, JoAnn J (2016) Quality improvement initiatives lead to reduction in nulliparous term singleton vertex cesarean delivery rate. Joint Comm J Qual Patient Saf/Joint Comm Resour (SI 43(2)). https://doi.org/10.1016/j.jcjq.2016.11.008
18. Heydari M, Sayadi MK, Shahanaghi K (2010) Extended VIKOR as a new method for solving multiple objective large-scale nonlinear programming problems. Rairo-Oper Res (SI 44(2)):139–152. https://doi.org/10.1051/ro/2010011

19. Opricovic S (2011) Fuzzy VIKOR with an application to water resources planning. Expert Syst Appl (SI 38):12983–12990. https://doi.org/10.1016/j.eswa.2011.04.097
20. Simab M, Javadi MS, Nezhad AE (2018) Multi-objective programming of pumped-hydro-thermal scheduling problem using normal boundary intersection and VIKOR. Energy (SI 143):854–866. https://doi.org/10.1016/j.energy.2017.09.144
21. Dong JY, Yuan FF, Wan SP (2017) Extended VIKOR method for multiple criteria decision-making with linguistic hesitant fuzzy information. Comput Ind Eng (SI 112):305–319. https://doi.org/10.1016/j.cie.2017.07.025
22. Wu ZB, Ahmad J, Xu JP (2016) A group decision making framework based on fuzzy VIKOR approach for machine tool selection with linguistic information. Appl Soft Comput (SI 42):314–324. https://doi.org/10.1016/j.asoc.2016.02.007
23. Alguliyev RM, Aliguliyev RM, Mahmudova RS (2015) Multicriteria personnel selection by the modified fuzzy VIKOR method. Sci World J (SI 2015) 1–17. https://doi.org/10.1155/2015/612767

Mathematical Modeling for Evaluation Reliability of a Bleaching System

Subhi Tyagi, Akshay Kumar, Nupur Goyal, and Mangey Ram

Abstract The current research deals with the various reliability measures analysis for a complex bleaching system. The system has a complex structure with three subsystems A, B and C associated to each other in series arrangement. The subsystem A has only one unit and subsystem B and C have two identical units which are connected in parallel configuration with each other. In both the cases (B and C), second unit is in standby. The assumed bleaching framework has three type of states i.e., working, partially working and failed. The framework is assumed to be repaired from the degraded and failed states. Mathematical model of the designed framework is solved by means of supplementary variable technique and Markov process. Laplace transform of numerous differential equations is obtained. Various reliability measures such as reliability, availability, mean time to failure and expected cost are evaluated, graphical depiction of the reliability characteristics are also illustrated for the considered system.

Keywords Availability · Reliability · Mean time to failure · Cost analysis · Complex bleaching system · Markov process

---

S. Tyagi · N. Goyal
Department of Mathematics, Graphic Era Deemed to be University, Dehradun, India

A. Kumar (✉)
Department of Mathematics, Graphic Era Hill University, Dehradun, India
e-mail: akshaykr1001@gmail.com

M. Ram
Department of Mathematics, Computer Science and Engineering, Graphic Era Deemed to be University, Dehradun, India

Institute of Advanced Manufacturing Technologies, Peter the Great St. Petersburg Polytechnic University, 195251 Saint Petersburg, Russia

V. Kumar and H. Pham (eds.), *Predictive Analytics in System Reliability*,
Springer Series in Reliability Engineering,
https://doi.org/10.1007/978-3-031-05347-4_16

1 Introduction

Reliability has always been a highly concerned topic in every field. Whether in the field of army for weapons or in the field of engineering systems or in the field of medical, everywhere reliability is the most important aspect of today's world. In case of any machinery or an engineering system, researchers always focussed about the system's higher reliability with low maintenance cost. Now a days, systems are getting compact and complex which leads to the upsurge in the equipment cost. Due to this, the importance of the system's maintenance and its desirable work under fixed time has also increased.

Gupta and Tyagi [1] had investigated a standby complex redundant system. The system was considered to have the human failure in two states i.e., working and failed. Availability and MTTF of the presented complex framework were evaluated by using supplementary variable technique, also for many states Laplace transformations were obtained. Authors had also used Abel's theorem to calculate the various time independent probabilities. A model was presented by Pham et al. [2] where the components did not fail fully but degraded to several stages and failed due to catastrophic failures. The presented model was a *k*-out-of-*n*:G type whose reliability and mean time to failure (MTTF) were determined. Dhillon [3] presented various aspects of human reliability and errors in the medical systems. Mathematical concepts for analysing human reliability were discussed. Human errors in medication and anaesthesia were also considered and some topics for example medication facts, types and causes of medication errors, medication errors in hospitals, medical error reduction etc. were also studied. Oliveira et al. [4] had established a technique to evaluate the reliability of the system whose component's failure rate was considered to be time dependent. Dhillon [5] had covered the reliability, maintainability, and safety issues for the mining equipment. Author had discussed about that mining equipment because of the growing complexity and sophistication in the equipment designing. Liang et al. [6] studied the consecutive *k*-out-of-*n* repairable frameworks. Authors gave some more general results and formula for various measures such as reliability, rate of existence of failure in the system etc. Lisnianski [7] had discussed about the limitation of universal generating function (UGF) technique in the evaluation of reliability. So, author proposed a new discrete state continuous time Markov process to estimate the reliability of dynamic multi state system (MSS). The proposed technique is called L_Z transformation. Garg et al. [8] analysed the reliability with the help of vague lambda-tau methodology for industrial system in which the collected information about the components of the system was uncertain and the nature of the information was also inaccurate. Also, rather than fuzzy set theory author had used intuitionistic fuzzy set theory to control the uncertainty in the data. Ram and Kumar [9] had applied the probabilistic approach on a coal handling component of a thermal power plant for the analyzation of the reliability and sensitivity. The coal handling system was considered which had two subsystems allied in series configuration and each subsystem had two units connected in parallel. Authors had considered the failure and repair rates of the coal handling unit constant and evaluated various reliability measures. Ram et al. [10] considered a standby framework and estimated the reliability and other measures using Markov process. The considered standby system contains waiting time repair. Reliability was obtained with the assistance of Laplace transformation and supplementary variable technique. Ram and Nagia [11] had examined about the various reliability measures of the satellite communication framework. The system comprised of satellite, earth station and terrestrial system and failure and repair rates were assumed to be constant. Cases and graphical representations were also presented. Singh et al. [12] had assumed a framework having two subsystems 1 and 2, connected in series configuration and one controller was connected with each subsystem for better functioning. Subsystem 1 was a k-out-of-n:G type and subsystem 2 had two units joined in parallel. Transitional state probabilities, asymptotic behaviour and few reliability characteristics were evaluated with the assistance of supplementary variable technique, Laplace transformation and copula method. Kumar and Ram [13] had considered a system consisting standby and k-out-of-n redundancies. System had two subsystems A and B where A contained a standby redundant unit and B was a 2-out-of-3:F type. Both A and B were connected in series configuration and many reliability measures such as reliability, availability, MTTF were calculated. Li [14] had discussed two simple yet main redundancies i.e., active redundancy and standby redundancy for a considered system. Author had also discussed the pros and cons of the two redundancies. Markov model technique was applied to calculate the mean time between failure of the proposed framework and compared the redundancies from the reliability viewpoint. Dhillon [15] had discussed about the important topic of transportation safety and its system's reliability. The main motive of the author was to eliminate the need for consultation for many sources for getting the desired information on the topic. Some transportation history along with system's reliability and safety measures were discussed. Author had used Boolean algebra laws, probability distribution, Markov process, fault tree analysis etc. for reliability, maintainability and safety models. Li [16] had introduced calculation of the redundancy of a dormant k-out-of-n framework. Due to the character of the failure, dormant failure can't be detected. So, authors assumed that failure as a blind point while designing for reliability and maintainability. Also, some case studies were given in the mass transit train reliability and safety design to apply the designed methodology. Amrutkat and Kamalja [17] had discussed about various reliability measures and discussed about their importance in a system. Authors also overviewed some extended importance reliability measures for few popular systems. Shekhar et al. [18] considered a redundant machining framework which was comprised of various functioning machines and studied the performance and reliability characteristics of the system. Authors had also included the conception of switching failure and geometric reneging. Also, a numerical example based on the theoretical model was illustrated for the practicability purpose of the theoretical system. Nakagawa et al. [19] and Zhao et al. [20] determined the replacement policies of minimal repairable elements and also discussed the Barlow Proschan of generalization models. Jain et al. [21] deals with a module-based software reliability development model. The considered model contained imperfect debugging and fault reduction factor together. Authors had considered three stage process i.e., isolation, observation, and removal process for each module. Li et al. [22] presented a system responsibility growth analysis mistreatment actual field failure knowledge, and first objective of the system responsibility growth was to enhance the accomplishment of system responsibility performance throughout system responsibility demonstration to realize the expected responsibility commitment of the framework. Gaonkar et al. [23] had computed the travel time reliability for any kind of transportation vehicle under fuzzy type of data and advocates its probabilistic approach. Dhillon and Misra [24] had taken a redundant system with two units in parallel. Then authors had presented four mathematical models with critical human fault and estimated the reliability state probabilities and MTTF for those models. Authors had also shown the graphs of the evaluated reliability measures.

In this present research, a bleaching system's reliability characteristics are analysed by means of supplementary variable technique, Laplace transformation and Markov process. In this system the repair ability is assumed for the degraded and failed states. The manuscript is arranged as follows: the mathematical model details are given in Sect. 2. In Sect. 2.1, the description of the bleaching system is explained followed by the transition diagram of the system in Sect. 2.2. In Sect. 2.3, assumptions and notations are given. In Sect. 3, the mathematical modelling of the considered bleaching system is discussed. Numerical calculations of the various reliability measures are evaluated in Sect. 4. In Sect. 4.1, the availability of the system is analysed briefly, in Sect. 4.2, reliability is analysed followed by its graphical representation. In Sect. 4.3, the analysation of MTTF is done in the tabular form as well as graphical form. In Sect. 4.4, expected profit is evaluated. In Sect. 5, the results related to the reliability measures of the bleaching framework are discussed and explained. Lastly, in Sect. 6, the conclusion of the presented research is given.

2 Mathematical Model Details of the Designed System

2.1 System Description

A bleaching system is considered in this research which is a complex series–parallel structure. The system contains three units A, B and C connected to each other in series arrangement as given in Fig. 1. Subsystem A has only one unit where subsystem B have two identical units connected in parallel i.e., second unit of the subsystem B is of standby manner. When the first unit of subsystem B fails, the second standby unit will start working in place of the first one. At last subsystem C also have two identical units connected in parallel arrangement where the second unit of the Subsystem C is in standby. When the first unit of the subsystem C stops working, the second standby unit will begin to work in place of the first unit. The considered bleaching system has three types of states i.e., working, partially working and completely failed. In good state, the system will work properly. In degraded state, the system is assumed

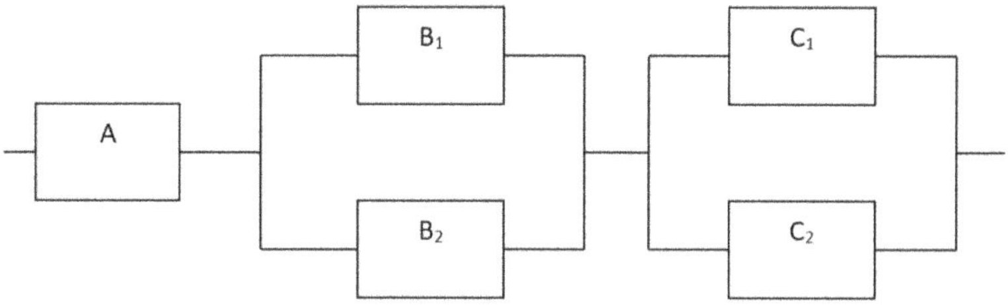

Fig. 1 Block diagram of bleaching system

to work partially. That means in either of the subsystem B or C where two units are connected in parallel, if one of them fails then the system is in degraded sate. In failed state, the system will stops working. If both the units of the subsystems B or C fail, the system will go to the failed state. The subsystem A is assumed to fail with the rate of λ_1 and reached to the state P_3. Since, P_3 is a failed state, the repair rate from this state is considered to be $\emptyset(x)$, the repair rate from all the failed states are considered as same. The subsystem B have two units connected in parallel, so, on the failure of first unit with the rate λ_2, the system will go in degraded state P_1 and when the second unit fails the system will fail and goes to the state P_4. From the degraded state P_1, the system will go under repair with the rate μ and go back to the good state P_0 and also from the failed state P_4, the system will go under repair with the rate $\emptyset(x)$ and go back to the good state P_0. Subsystem C also has the same condition of failure and degradation as B, on failure of the first unit with rate λ_3, it goes to the degraded state P_2 and the repair rate from this state to the state P_0 is also μ. Then after the failure of the second unit of C, the system will fail and goes to the state P_5 and the repair rate from this state to the good state P_0 is $\emptyset(x)$.

2.2 State Transition Diagram of the Bleaching System

State transition diagram of bleaching framework is designed on the basis of its working and shown in Fig. 2.

2.3 Assumptions and Notations

The considered bleaching framework structured by making the following assumptions:

1. The framework is assumed to be in good working condition in initial state.
2. Repair facility is available for both degraded and failed states.
3. After repair, the unit is assumed as good as new.
4. All failure and repair rates are assumed constant.

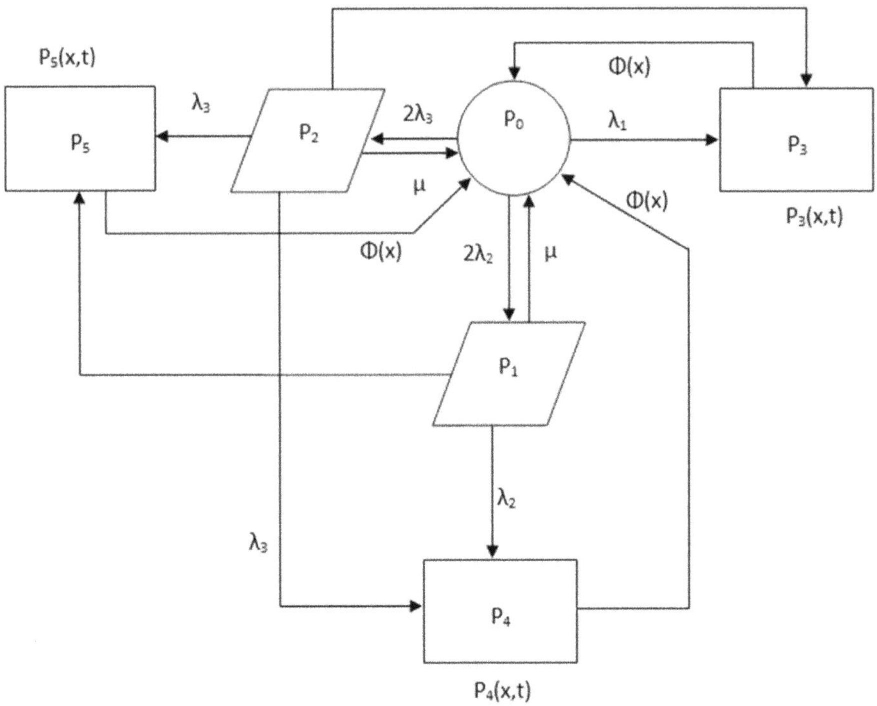

Fig. 2 State transition diagram of the bleaching system

These notations (shown in Table 1) have been used for the considered bleaching system.

3 Mathematical Modelling

3.1 Formulation of the Model

The following differential equations have been drawn from the above state transition diagram of the bleaching system.

$$\left[\frac{d}{dt} + 2\lambda_3 + 2\lambda_2 + \lambda_1\right] P_0(t) = \int_0^\infty \phi(x)[P_3(x,t) + P_4(x,t) + P_5(xt)]dx + \mu[P_1(t) + P_2(t)] \quad (1)$$

$$\left[\frac{d}{dt} + \lambda_2 + \lambda_3 + \mu\right] P_1(t) = 2\lambda_2 P_0(t) \quad (2)$$

Table 1 Notations

t	Time scale
s	Laplace transform variable
x	Supplementary variable
$\lambda_1/\lambda_2/\lambda_3$	Failure rate of unit A/B/C
μ	Repair rates of the system from degraded state
$P_0(t)$	State probability when the system is in good working condition
$P_1(t)$	State probability when the system is functioning with one failed unit of B
$P_2(t)$	State probability when the system is functioning with one failed unit of C
$P_3(t)$	Probability of the completely failed state because of the failure of subsystem A
$P_4(t)$	Probability of the completely failed state because of the failure of the second unit of subsystem B after the failure of its first unit
$P_5(t)$	Probability of the completely failed state because of the failure of the second unit of subsystem C after the failure of its first unit
$\phi(x)$	Repair rate from the failed states
K_1, K_2	Revenue, service cost per unit time respectively

$$\left[\frac{d}{dt} + \lambda_1 + \lambda_2 + \lambda_3 + \mu\right] P_2(t) = 2\lambda_3 P_0(t) \tag{3}$$

$$\left[\frac{\partial}{\partial x} + \frac{\partial}{\partial t} + \phi(x)\right] P_3(x, t) = 0 \tag{4}$$

$$\left[\frac{\partial}{\partial x} + \frac{\partial}{\partial t} + \phi(x)\right] P_4(x, t) = 0 \tag{5}$$

$$\left[\frac{\partial}{\partial x} + \frac{\partial}{\partial t} + \phi(x)\right] P_5(x, t) = 0 \tag{6}$$

Boundary conditions

$$P_3(0, t) = \lambda_1[P_0(t) + P_2(t)] \tag{7}$$

$$P_4(0, t) = \lambda_2[P_1(t) + P_2(t)] \tag{8}$$

$$P_5(0, t) = \lambda_3[P_1(t) + P_2(t)] \tag{9}$$

Initial Conditions

$$P_0(0) = 1 \tag{10}$$

and all other state probabilities are zero at $t = 0$.

3.2 Solution of the Model

Solution of the model is given by taking Laplace transformation (which converts the variable t into s) from Eqs. (1) to (9) using Eq. (10).

$$[s + 2\lambda_3 + 2\lambda_2 + \lambda_1]\overline{P_0}(s) = 1 + \mu\left[\overline{P_1}(s) + \overline{P_2}(s)\right] + \int_0^\infty \phi(x)[\overline{P_3}(x, s) + \overline{P_4}(x, s) + \overline{P_5}(x, s)]dx \tag{11}$$

$$[s + \lambda_2 + \lambda_3 + \mu]\overline{P_1}(s) = 2\lambda_2\overline{P_0}(s) \tag{12}$$

$$[s + \lambda_1 + \lambda_2 + \lambda_3 + \mu]\overline{P_2}(s) = 2\lambda_3\overline{P_0}(s) \tag{13}$$

$$\left[\frac{\partial}{\partial x} + s + \phi(x)\right]\overline{P_3}(x, s) = 0 \tag{14}$$

$$\left[\frac{\partial}{\partial x} + s + \phi(x)\right]\overline{P_4}(x, s) = 0 \tag{15}$$

$$\left[\frac{\partial}{\partial x} + s + \phi(x)\right]\overline{P_5}(x, s) = 0 \tag{16}$$

Rewriting (14), (15) and (16) as

$$\left[\frac{\partial}{\partial x} + s + \phi(x)\right]\overline{P_i}(x, s) = 0 \tag{17}$$

For $i = 3, 4, 5$.
Boundary condition

$$\overline{P_3}(0, s) = \lambda_1\left[\overline{P_0}(s) + \overline{P_2}(s)\right] \tag{18}$$

$$\overline{P_4}(0, s) = \lambda_2\left[\overline{P_1}(s) + \overline{P_2}(s)\right] \tag{19}$$

$$\overline{P_5}(0, s) = \lambda_3\left[\overline{P_1}(s) + \overline{P_2}(s)\right] \tag{20}$$

$$\overline{P_0}(s) = \frac{1}{D(s)} \tag{21}$$

where

$$D(s) = \lambda_1 - \left(\frac{2\lambda_2}{s+\lambda_2+\lambda_3+\mu} + \frac{2\lambda_3}{s+\lambda_1+\lambda_2+\lambda_3+\mu}\right)\left(\mu + \overline{s_\phi}(s)(\lambda_2+\lambda_3)\right)$$
$$+ s + 2\lambda_3 + 2\lambda_2 + -\lambda_1\overline{s_\phi}(s)\left(1 + \frac{2\lambda_3}{s+\lambda_1+\lambda_2+\lambda_3+\mu}\right) \tag{22}$$

From (12)

$$\overline{P_1}(s) = \frac{2\lambda_2}{s+\lambda_2+\lambda_3+\mu}\overline{P_0}(s) \tag{23}$$

From (13)

$$\overline{P_2}(s) = \frac{2\lambda_3}{s+\lambda_1+\lambda_2+\lambda_3+\mu}\overline{P_0}(s) \tag{24}$$

From (18) and (23)

$$\overline{P_3}(s) = \left(\frac{1-\overline{s_\phi}(s)}{s}\right)\lambda_1\left(1 + \frac{2\lambda_3}{s+\lambda_1+\lambda_2+\lambda_3+\mu}\right)\overline{P_0}(s) \tag{25}$$

$$\overline{P_4}(s) = \left(\frac{1-\overline{s_\phi}(s)}{s}\right)\lambda_2\left(\frac{2\lambda_2}{s+\lambda_2+\lambda_3+\mu} + \frac{2\lambda_3}{s+\lambda_1+\lambda_2+\lambda_3+\mu}\right)\overline{P_0}(s) \tag{26}$$

$$\overline{P_5}(s) = \left(\frac{1-\overline{s_\phi}(s)}{s}\right)\lambda_3\left(\frac{2\lambda_2}{s+\lambda_2+\lambda_3+\mu} + \frac{2\lambda_3}{s+\lambda_1+\lambda_2+\lambda_3+\mu}\right)\overline{P_0}(s) \tag{27}$$

$$\overline{P_{up}}(s) = \overline{P_0}(s) + \overline{P_1}(s) + \overline{P_2}(s) \tag{28}$$

$$\overline{P_{down}}(s) = \overline{P_3}(s) + \overline{P_4}(s) + \overline{P_5}(s) \tag{29}$$

It is noticed that

$$\overline{P_{up}}(s) + \overline{P_{down}}(s) = \frac{1}{s}. \tag{30}$$

Table 2 Availability of the system

Time (t)	Availability
0	1.00000
1	0.92589
2	0.91192
3	0.90974
4	0.90965
5	0.90979
6	0.90989
7	0.90993
8	0.90995
9	0.90996
10	0.90996

4 Numerical Calculations

In this section, particular cases related to the bleaching system are taken and several reliability characteristics are analysed with respect to time and other measures as follows.

4.1 Availability Analysis

Availability function of the framework is generally obtained when the framework is not in a completely failed state. The probability of the system performing the necessary function at any instant or during a particular time interval when the system is operated or installed according to a defined standard can be used to characterize as its availability. Availability of the bleaching system is

$$A = 0.10389e^{(-1.40000*t)} - 0.031579e^{(-0.95000*t)} + 0.17716e^{(-1.95000)} + 0.90997 \tag{31}$$

The availability of the assumed bleaching structure is obtained by putting $t = 0$ to 10 in Eq. (31) and shown in Table 2 and Fig. 3.

4.2 Reliability Analysis

The reliability function of the framework is generally concerned with the amount of time in which framework will work without failure after it starts working. Reliability of the system is

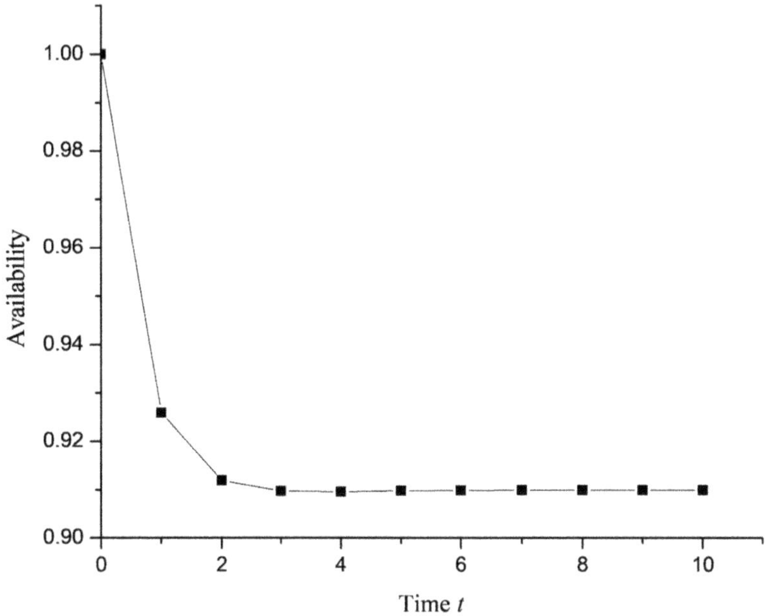

Fig. 3 Availability versus time

$$R = 0.36364e^{(-0.4000*t)} + (0.60000t + 0.636364)e^{(-0.95000*t)} \quad (32)$$

Now, putting $t = 0$ to 10 in Eq. (32) and getting the reliability of the framework and shown in Table 3 and Fig. 4.

Table 3 Reliability of the system

Time (t)	Reliability
0	1.00000
1	0.72191
2	0.43805
3	0.25045
4	0.14134
5	0.08067
6	0.04716
7	0.02837
8	0.01754
9	0.01110
10	0.00716

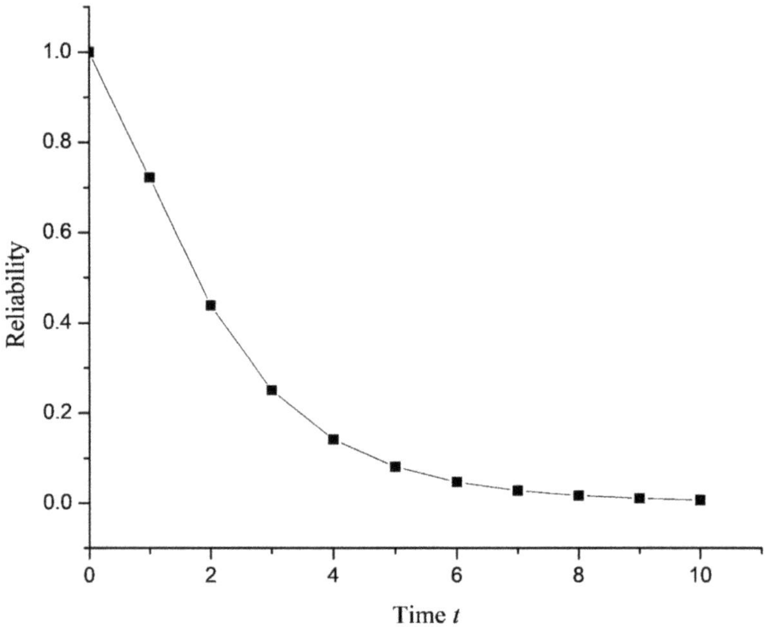

Fig. 4 Reliability versus time

4.3 Analysis of Mean Time to Failure (MTTF)

MTTF is the mean time expected until the first failure of the system occurs. Specifically, MTTF is used for the non-repairable systems. So, by taking all repair rates equals to zero, one can evaluate mean time to failure as the function of failure rates.

$$MTTF = \lim_{s \to 0} \overline{P_{up}}(s) = \frac{1 + \frac{2\lambda_2}{\lambda_2 + \lambda_3} + \frac{2\lambda_3}{2\lambda_3 + 2\lambda_2 + \lambda_1}}{2\lambda_3 + 2\lambda_2 + \lambda_1}$$

After setting the values of failure rates, MTTF of the bleaching system is shown in Table 4 and Fig. 5.

4.4 Expected Profit

The standard equation of expected profit is given by

$$E_p(t) = K_1 \int_0^t P_{up}(t)dt - tK_2 \tag{33}$$

Table 4 MTTF of the system

Variation in failure rates	λ_1	λ_2	λ_3
0.1	2.40741	2.24377	4.29752
0.2	2.10000	2.01890	2.93333
0.3	1.85950	1.81069	2.24377
0.4	1.66667	1.63223	1.82231
0.5	1.50888	1.48163	1.53635
0.6	1.37755	1.35437	1.32897
0.7	1.26667	1.24608	1.17142
0.8	1.17188	1.15313	1.04755
0.9	1.08997	1.07266	0.94754

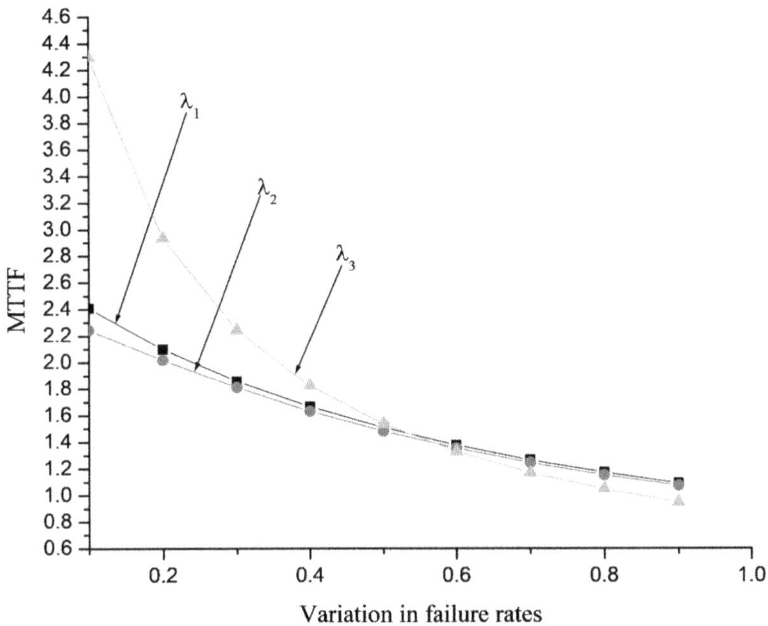

Fig. 5 MTTF as a function of failure rates

Substituting the values of inverse Laplace transform and integrating, we will get the value of the expected profit as shown in Table 5 and Fig. 6.

$$E_p(t) = K_1\left(-0.074211e^{(-1.4000*t)} + 0.03324e^{(-0.95000*t)} - 0.00908e^{(-1.95000*t)} + 0.90997t - 0.05006\right) - tK_2 \quad (34)$$

Varying the value of t from 0 to 9, setting different value of K_2, we get Table 5 and Fig. 6 which shows the expected profit of the bleaching system.

Table 5 The values of profit of the bleaching system

t	$K_2 = 0.2$	$K_2 = 0.4$	$K_2 = 0.6$	$K_2 = 0.8$
0	−0.10011	−0.10011	−0.10011	−0.10011
1	0.65317	0.45317	0.25317	0.05317
2	1.37015	0.97015	0.57015	0.17015
3	2.08063	1.48063	0.88063	0.28063
4	2.79027	1.99028	1.19028	0.39028
5	3.50000	2.50000	1.50000	0.50000
6	4.20984	3.00984	1.80984	0.60984
7	4.91975	3.51975	2.11975	0.71975
8	5.62969	4.02969	2.42969	0.82969
9	6.33966	4.53966	2.73966	0.93966

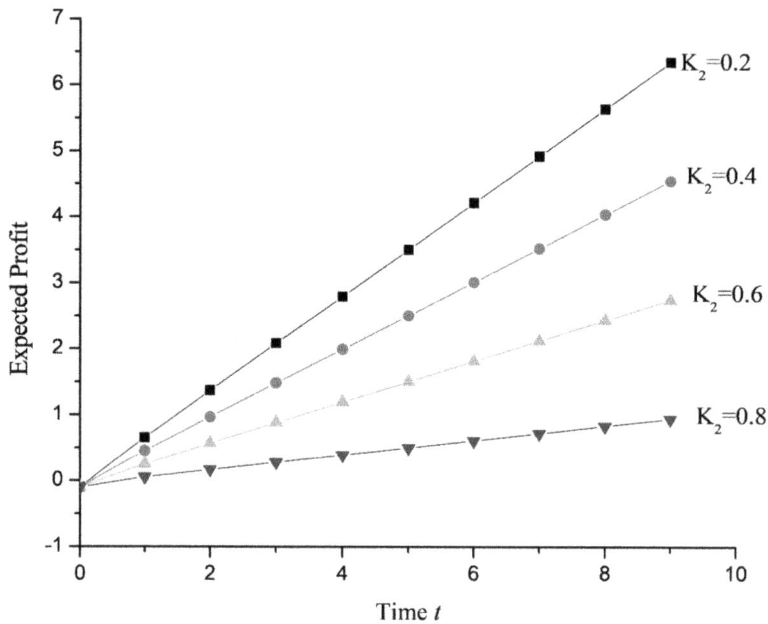

Fig. 6 Expected profit of the system

5 Result Discussion

Some reliability measures for example reliability, availability, MTTF and expected cost are evaluated for the proposed framework by using Markov process. From the above Tables 2, 3, 4 and 5 and Figs. 3, 4, 5 and 6 the following results can be concluded. Table 2 provides the values of availability of the proposed framework at different time period (from 0 to 10). Figure 3 is the graphical representation of Table 2 i.e.; it shows the variation in availability with respect to the increasing time. Figure 3 depicts that the availability of the proposed system decreases immediately with time up to 1 and after 3 it almost becomes constant with the increasing time.

Table 3 gives the values of the reliability of the model at different time period (from 0 to 10), considering some fixed values of all failure and repair rates. Figure 4 concluded the above Table 3 and shows the behaviour of the system's reliability with the increasing time period. The graph in Fig. 4 shows that the reliability of the system decreases exponentially with the passing time.

Table 4 shows the MTTF with respect to variation in failure rates and Fig. 5 is the depiction of Table 4 in graphical form. Clearly, the value of λ_3 is higher than λ_2 and λ_1 and the value of λ_1 is higher than λ_2. The curve of λ_3 is falling more steeper than λ_1 and λ_2.

Table 5 gives the data about the variation on profit with different values of service costs and graphical representation of the Table 5 is in Fig. 6. From Fig. 6, it can be concluded that on increasing service cost expected profit is decreasing. So, the expected cost is inversely proportional to service cost, the expected profit increases with increasing time for the proposed system.

6 Conclusion

A bleaching framework of a chapter mill is considered in this research and many reliability measures for example availability, reliability, MTTF, expected profit are evaluated by means of supplementary variable technique, Laplace transform and Markov process. Reliability measures are calculated for different time duration. From the availability and reliability graphs, it can be resulted that both the parameters decrease suddenly with the passage of time and after some time it almost becomes constant. It is also seen that the expected profit is inversely proportional to service cost. In future, researchers can extend this work by evaluate the sensitivity of reliability measures for bleaching system to improve the design and structure of system.

Declaration of Competing Interest There is no conflict of interests associated with this chapter.

References

1. Gupta PP, Tyagi L (1986) MTTF and availability evaluation of a two-unit, two-state, standby redundant complex system with constant human failure. Microelectron Reliab 26(4):647–650
2. Pham H, Suprasad A, Misra RB (1996) Reliability and MTTF prediction of k-out-of-n complex systems with components subjected to multiple stages of degradation. Int J Syst Sci 27(10):995–1000
3. Dhillon BS (2003) Human reliability and error in medical system, vol 2. World Scientific

4. Oliveira EA, Alvim ACM, e Melo PF (2005) Unavailability analysis of safety systems under aging by supplementary variables with imperfect repair. Ann Nucl Energy 32(2):241–252
5. Dhillon BS (2008) Mining equipment reliability. Springer, London, pp 57–70
6. Liang X, Xiong Y, Li Z (2010) Exact reliability formula for consecutive *k*-out-of-*n* repairable systems. IEEE Trans Reliab 59(2):313–318
7. Lisnianski A (2012) L z-Transform for a discrete-state continuous-time Markov process and its applications to multi-state system reliability. In: Recent advances in system reliability. Springer, London, pp 79–95
8. Garg H, Rani M, Sharma SP (2013) Reliability analysis of the engineering systems using intuitionistic fuzzy set theory. J Qual Reliab Eng 2013:10. Article ID 943972. https://doi.org/10.1155/2013/943972
9. Ram M, Kumar A (2013) Reliability measures improvement and sensitivity analysis of a coal handling unit for thermal power plant. Int J Eng 26(9):1059–1066
10. Ram M, Singh SB, Singh VV (2013) Stochastic analysis of a standby system with waiting repair strategy. IEEE Trans Syst Man Cybern Syst 43(3):698–707
11. Ram M, Nagia N (2013) Reliability characteristics of a satellite communication system including earth station and terrestrial system. Int J Perform Eng 9(6):667–676
12. Singh VV, Singh SB, Ram M, Goel CK (2013) Availability, MTTF and cost analysis of a system having two units in series configuration with controller. Int J Syst Assur Eng Manag 4(4):341–352
13. Kumar A, Ram M (2016) System reliability measures in the presence of common cause failures. Int J Ind Syst Eng 24(1):44–61
14. Li J (2016) Reliability calculation for dormant *k*-out-of-*n* systems with periodic maintenance. Int J Math Eng Manag Sci 1(2):68–76
15. Dhillon BS (2016) Transportation systems reliability and safety. CRC Press
16. Li J (2016) Reliability comparative evaluation of active redundancy vs. standby redundancy. Int J Math Eng Manag Sci 1(3):122–129
17. Amrutkar KP, Kamalja KK (2017) An overview of various importance measures of reliability system. Int J Math Eng Manag Sci 2(3):150–171
18. Shekhar C, Jain M, Raina A, Mishra R (2017) Sensitivity analysis of repairable redundant system with switching failure and geometric reneging. Decis Sci Lett 6(4):337–350
19. Nakagawa T, Chen M, Zhao X (2018) Note on history of age replacement policies. Int J Math Eng Manag Sci 3(2):151–166
20. Zhao X, Qian C, Nakamura S, Nakagawa T (2018) A summary of replacement policies with number of failures. Int J Math Eng Manag Sci 3(2):136–150
21. Jain M, Jain A, Gupta R (2018) Analysis of module-based software reliability growth model incorporating imperfect debugging and fault reduction factor. In: Quality, IT and business operations. Springer, Singapore, pp 69–80
22. Li J, Collins G, Govindarajulu R (2019) System reliability growth analysis during warranty. Int J Math Eng Manag Sci 4(1):85–94
23. Gaonkar RSP, Nigalye AV, Pai SP (2021) Possibilistic approach for travel time reliability evaluation. Int J Math Eng Manag Sci 6(1):223–243
24. Dhillon BS, Misra RB (1984) Reliability evaluation of systems with critical human error. Microelectron Reliab 24(4):743–759

An Effort Allocation Model for a Three Stage Software Reliability Growth Model

Sujit Kumar Pradhan, Anil Kumar, and Vijay Kumar

Abstract This chapter investigates the optimal efforts allocation plan to minimize the total cost during the testing phase of the software development life cycle using three stages fault detection, isolation, and removal under a dynamic environment. We have used three-stage modelling to allocate resources, which incorporates different efforts, i.e. detection effort, isolation effort, and removal effort. We have used the optimal control-theoretic approach to find the optimal policies by considering effort as a control parameter. We also discussed the variations in the future cost of the model by assuming that the cost of detection, isolation and removal follows the learning curve phenomenon. The theoretical results and optimal control theory-based optimized policy is supported by a numerical example.

1 Introduction

Computers and computer-based systems interpenetrate every feature of our daily lives. It has benefited our society and increased our productivity. The good functioning of any computer system depends upon its software components. Software plays a vital role in both real life and industrial organizations. It is becoming more challenging to develop a highly reliable software system. To assure software reliability, a lengthy testing process is usually needed before releasing software to market. Hence, faults are detected and corrected during the software development life cycle to increase software reliability.

S. K. Pradhan · A. Kumar
Department of Mathematics, BITS Pilani-KK Birla Goa Campus, Zuarinagar 403726, Goa, India
e-mail: p20180407@goa.bits-pilani.ac.in

A. Kumar
e-mail: anilpundir@goa.bits-pilani.ac.in

V. Kumar (✉)
Department of Mathematics, Amity Institute of Applied Sciences, Amity University Uttar Pradesh, Noida 201313, India
e-mail: vijay_parashar@yahoo.com

V. Kumar and H. Pham (eds.), *Predictive Analytics in System Reliability*,
Springer Series in Reliability Engineering,
https://doi.org/10.1007/978-3-031-05347-4_17

In the software industries, prediction and estimation of software reliability enable to meet complexities of software development. It is becoming more challenging for software managers to develop highly reliable software systems efficiently. To obtain fault-free software, there is a requirement of the process to track fault content and reliability. Hence, a mathematical relationship termed a software reliability growth model (SRGM) describing the process of finding and removing errors to increase software reliability is introduced.

In the last few years, many software reliability growth models (SRGMs) [1–19] have been proposed to distinguish the growth of software reliability during the software development process. SRGMs are divided into two categories: perfect debugging model and imperfect debugging model. Perfect debugging models are based on the assumption that faults detected during the testing phase are removed immediately, and no new faults are introduced into the software [1, 11, 20]. The other category is imperfect debugging, i.e. at all times, the testing team may not be able to remove the original fault with certainty, or new faults can be added into the software during the testing phase [4, 12, 21].

Further, the imperfect debugging model can be classified into two types based on error generation and imperfect fault removal. Yamada et al. [21] proposed an imperfect debugging model with exponential or linear fault content function of the testing time. Pham et al. [15] presented an imperfect debugging SRGM with time-dependent fault content function. Zhu et al. [16] proposed a two-phase SRGM that incorporates both software fault dependency and imperfect fault removal.

Software faults are classified based on kind of failures. Software fault classes are discussed. Firstly, solid faults are known as Bohrbugs, and soft faults are known as Mandelbugs. Bohrbugs are the faults that can be easily isolated, and Mandelbugs are the opposite of Bohrbugs [22, 23]. Secondly, software faults can be classified as independent and related faults. Laprie et al. [24] presented an SRGM in which they have discussed that software faults are either related or independent.

Software reliability is defined as the probability of failure-free software in a particular period. For developing reliable software, different factors affect software reliability. The two main factors, i.e. initial faults and fault detection rate, affect software reliability. Software reliability can be improved by testing process factors, i.e. testing effort and imperfect debugging. The most common method to estimate software reliability is during the testing phase. The testing effort is the total resources consumed during the testing phase. Generally, testing efforts are measured by human power, the number of CPU hours, etc. [5, 21, 25]. The software testing effort curve is described by the traditional Rayleigh, Weibull, Exponential, or S-shaped curve [26]. Huang et al. [10, 27] developed SRGMs where testing effort is incorporated with logistic function.

Ji et al. [7] developed software by applying optimal control theory. Kumar and Sahni [28] used FDP and FCP to reduce total cost during the development period of SRGM under a dynamic environment. Kapur et al. [29] have discussed a model to allocate the resources and minimize the total cost during the testing process of SRGM. Kumar and Sahni [30] presented an SRGM to estimate the testing efforts in a dynamic environment under the condition that debugging costs associated with each release follows a learning curve. Kapur et al. [31] developed a model for profit maximization of SRGM under the influence of promotional effort. Pradhan et al. [32] have discussed a resource allocation model to minimize the total cost of a two-stage SRGM incorporating testing effort and imperfect fault removal. Kumar et al. [33] proposed a model incorporating resource allocation and testing time in a two-stage process of fault detection and fault correction. Ji et al. [34] presented a model to enhance the lifetime of software systems by control theory. Saxena et al. [35] have discussed the ranking of SRGM by an entropy-ELECTRA hybrid approach. Kumar et al. [36] have discussed a selection of optimal SRGM using an integrated entropy technique for order preference by similarity to an ideal solution approach.

In the present work, we have incorporated a different fault detection effort, isolation effort and removal effort along with different fault detection rate, isolation rate and removal rate in the model discussed in Sect. 2. Yamada and Osaki [37] presented an SRGM to minimize the total cost under static conditions. A problem arises when the software development process is carried out not under static but dynamic conditions. We also study the control problem to allocate resources optimally by examining the behaviour of the model parameters.

The remaining chapter is organized as follows: In Sect. 2, we briefly discussed the model developed by Kapur et al. [38]. Section 3 deals with the model development, and we introduced an optimal control problem. In Sect. 4, the optimal policies are developed, and optimal solutions are given. In Sect. 5, numerical analysis is performed by taking some base value of parameters by varying efforts (detection effort, isolation effort and removal effort). Section 6 concludes the paper with conclusions and some possible research on this topic.

Notations

T	The complete life cycle of the software.
$m_d(t)$	The cumulative number of detected faults at time t.
$m_i(t)$	The cumulative number of isolated faults at time t.
$m_r(t)$	The cumulative number of removed faults at time t.
a	The total fault.
b_1	The fault detection rate.
b_2	The fault isolation rate.
b_3	The fault removal rate.
$w_1(t)$	The fault detection effort at time t.
$w_2(t)$	The fault isolation effort at time t.
$w_3(t)$	The fault removal effort at time t.
$x_1(t)$	The number of detected faults at any point of time t.
$x_2(t)$	The number of isolated faults at any point of time t.
$x_3(t)$	The number of removed faults at any point of time t.
$\tilde{c}_1(t)$	Per unit detection cost at time t.
$\tilde{c}_2(t)$	Per unit isolation cost at time t.
$\tilde{c}_3(t)$	Per unit removal cost at time t.

2 Software Reliability Growth Model

Kapur et al. [38] presented a software reliability growth model by assuming the following assumptions.

1. The error occurrence in SRGM follows the non-homogeneous Poisson process.
2. The SRGM is modelled as a three-stage process, i.e. fault detection, fault isolation, and fault removal.
3. The time delay between the failure observation and subsequent removal is negligible.
4. No new faults are introduced into the software during the isolation/removal process.
5. Failure rate of the software is equally affected by faults remaining in the software.
6. The fault isolation/removal rate concerning testing effort intensity is proportional to the number of observed failures whose causes are yet to be identified.

Model. Based on the above assumptions, the following SRGM is considered for the study, which incorporates testing effort, i.e. detection effort, isolation effort, and removal effort. From the practical point of view, the testing effort is an important aspect of software reliability. The total number of software faults that cause failure concerning the testing effort is proportional to the remaining faults and fault detection/isolation/removal rate. The model is presented below:

$$\frac{dm_d(t)}{dt} = w_1(t)b_1\left[a - m_d(t)\right], \qquad m_d(0) = 0,\ 0 \le t \le T, \tag{1}$$

$$\frac{dm_i(t)}{dt} = w_2(t)b_2\left[m_d(t) - m_i(t)\right],\ m_i(0) = 0,\ 0 \le t \le T, \tag{2}$$

$$\frac{dm_r(t)}{dt} = w_3(t)b_3\left[m_i(t) - m_r(t)\right],\ m_r(0) = 0,\ 0 \le t \le T. \tag{3}$$

3 Model Development

The fault detection, isolation, and removal phase aim to detect, isolate and correct faults respectively and make the software more reliable during the development process of the software. The resources spent in the three-stage process, illustrated in Fig. 1, can affect the software's reliability. So resources should be allocated optimally. We aim to develop an optimal resource allocation plan to minimize the total cost of the software during the three-stage process of a software development life cycle under dynamic conditions. The mathematical expression for the resource over the interval $[0, T]$ is written as

$$0 \le w_j(t) \le 1,\ j = 1, 2, 3. \tag{4}$$

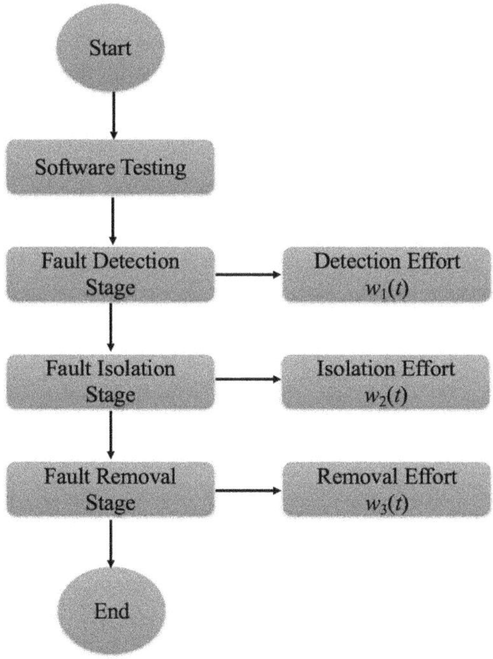

Fig. 1 Allocation of total resources during testing phase of SDLC

Cost minimization model. The main objective is to minimize the total cost, i.e. detection, isolation, and removal cost during the software development process. Detection, isolation, and removal costs are three different cost functions in the detection, isolation, and removal stages. The proposed model incorporates detection, isolation, and removal effort as a control parameter. Then the model can be represented mathematically over the interval $[0, T]$ as follows:

$$\min\left[\int_0^T \left\{\tilde{c}_1(t)x_1(t) + \tilde{c}_2(t)x_2(t) + \tilde{c}_3(t)x_3(t)\right\}dt\right] \tag{5}$$

subject to

$$x_1(t) = \frac{dm_d(t)}{dt} = w_1(t)b_1\left[a - m_d(t)\right], \qquad 0 \le t \le T, \tag{6}$$

$$x_2(t) = \frac{dm_i(t)}{dt} = w_2(t)b_2\left[m_d(t) - m_i(t)\right], \quad 0 \le t \le T, \tag{7}$$

$$x_3(t) = \frac{dm_r(t)}{dt} = w_3(t)b_3\left[m_i(t) - m_r(t)\right], \quad 0 \le t \le T, \tag{8}$$

$$\frac{m_r(T)}{a} \ge m_{rT} \Rightarrow m_r(T) \ge m_t\ (= am_{rT}), \tag{9}$$

with the conditions $m_d(0) = 0$, $m_i(0) = 0$ and $m_r(0) = 0$.

In the above control problem, we have considered the required reliability to be at least m_t when $0 < \frac{m_r(T)}{a} < 1$. This means that the software development team wants to reach at least m_t at the end of software development within the planning period.

4 Optimal Solution

The dynamic optimal control problem in Eq. (5) may be solved by Pontryagin maximum principle. To apply the Pontryagin maximum principle, first to form the Hamiltonian function. The Hamiltonian function is given by

$$\begin{aligned} &H(m_d(t), m_i(t), m_r(t), \lambda_1(t), \lambda_2(t), \lambda_3(t), w_1(t), w_2(t), w_3(t), t) \\ &= -\tilde{c}_1(t)x_1(t) - \tilde{c}_2(t)x_2(t) - \tilde{c}_3(t)x_3(t) + \lambda_1(t)x_1(t) + \lambda_2(t)x_2(t) + \lambda_3(t)x_3(t). \end{aligned}$$

The necessary conditions for an optimal solution are defined similarly. The co-state variables $\lambda_1(t)$, $\lambda_2(t)$ and $\lambda_3(t)$ is given by the following differential equation

$$\begin{aligned} \frac{d}{dt}\lambda_1(t) &= -\frac{\partial H(m_d(t), m_i(t), m_r(t), \lambda_1(t), \lambda_2(t), \lambda_3(t), w_1(t), w_2(t), w_3(t), t)}{\partial m_d(t)}, \\ \frac{d}{dt}\lambda_2(t) &= -\frac{\partial H(m_d(t), m_i(t), m_r(t), \lambda_1(t), \lambda_2(t), \lambda_3(t), w_1(t), w_2(t), w_3(t), t)}{\partial m_i(t)}, \\ \frac{d}{dt}\lambda_3(t) &= -\frac{\partial H(m_d(t), m_i(t), m_r(t), \lambda_1(t), \lambda_2(t), \lambda_3(t), w_1(t), w_2(t), w_3(t), t)}{\partial m_r(t)}, \end{aligned}$$

with terminal conditions $\lambda_1(T) = 0$, $\lambda_2(T) = 0$ and $\lambda_3(T) \leq 0 (= 0 \text{ if } m_r(T) > m_t)$ and $0 \leq t \leq T$. This results are demonstrate in the following theorem.

Theorem 1 *Consider a free final time and free final state problem with general cost function [39] which is to minimize the following performance index*

$$J(x_j(t), u_j(t), t) = S(x_j(t_f), t_f) + \int_{t_0}^{t_f} V(x_j(t), u_j(t), t)dt, \tag{10}$$

subject to the system equation

$$\dot{x}_j(t) = f(x_j(t), u_j(t), t), \tag{11}$$

along with boundary conditions as $x_j(t = t_0) = x_{j0}$, $t = t_f$ free, and $x_j(t_f)$ is free. Here $j = 1, 2, 3$. Then

(i) for fixed final time and free final state system, the boundary conditions are $x_j(t_0) = x_{j0}$ and $\lambda_j(t_f) = \left(\frac{\partial S}{\partial x_j}\right)_{t_f}$,

(ii) for fixed final time and fixed final state system, the boundary conditions are $x_j(t_0) = x_{j0}$ and $x_j(t_f) = x_{jf}$.

Proof The Hamiltonian with respect to the above problem is defined as

$$\begin{aligned} H &= H(x_j(t), u_j(t), \lambda_j(t), t) \\ &= V(x_j(t), u_j(t), t) + \lambda_j(t) f(x_j(t), u_j(t), t). \end{aligned} \tag{12}$$

For $j = 1, 2, 3$, we know that the boundary condition for the free final time, and the free final state system in terms of Hamiltonian is given by

$$\left[H + \frac{\partial S}{\partial t}\right]_{t_f} \delta t_f + \left[\frac{\partial S}{\partial x_j} - \lambda_j(t)\right]_{t_f} \delta x_{jf} = 0. \tag{13}$$

(i) For fixed final time and free final state system, $\delta t_f = 0$ and $\delta x_f \neq 0$. Then from Eq. (13) the coefficient of δx_f is zero, i.e.

$$\begin{aligned} \left(\frac{\partial S}{\partial x_j} - \lambda_j(t)\right)_{t_f} &= 0, \\ \Rightarrow \qquad \lambda_j(t_f) &= \left(\frac{\partial S}{\partial x_j}\right)_{t_f}. \end{aligned}$$

Hence, for fixed final time and free final state system, the boundary conditions are $x_j(t_0) = x_{j0}$ and $\lambda_j(t_f) = \left(\frac{\partial S}{\partial x_j}\right)_{t_f}$.

(ii) For fixed final time and fixed final state system, $\delta t_f = 0$ and $\delta x_{jf} = 0$ in general boundary condition and there is no extra boundary condition to be used other than those given in the problem formulation. But if the state system didn't acquire the desire value, then from Eq. (13), we get

$$\lambda_j(t_f) \leq 0 \ \ (= 0 \text{ if } x_j(t_f) > x_{jf}).$$

This completes the proof of the theorem. □

The co-state variables $\lambda_1(t)$, $\lambda_2(t)$ and $\lambda_3(t)$ represents per unit change in the objective function for a small change in $m_d(t)$, $m_i(t)$ and $m_r(t)$ respectively i.e. $\lambda_1(t)$, $\lambda_2(t)$ and $\lambda_3(t)$ can be interpreted as marginal cost of faults detected, isolated and removal respectively at time t. Moreover, $\lambda_1(t)$, $\lambda_2(t)$ and $\lambda_3(t)$ stands for future cost of detection, isolation and removal incurred as one more fault is detected, isolated and removed at time t respectively. So, the Hamiltonian is the sum of total current cost $\tilde{c}_1(t)x_1(t) + \tilde{c}_2(t)x_2(t) + \tilde{c}_3(t)x_3(t)$ and the total future cost $\lambda_1(t)x_1(t) + \lambda_2(t)x_2(t) + \lambda_3(t)x_3(t)$. The necessary conditions for optimality are given by:

$$\frac{\partial H(t)}{\partial w_1(t)} = 0, \quad \frac{\partial H(t)}{\partial w_2(t)} = 0, \quad \frac{\partial H(t)}{\partial w_3(t)} = 0.$$

From optimality condition, we get

$$- \tilde{c}_{1w_1}(t)x_1(t) - (\tilde{c}_1(t) - \lambda_1(t))x_{1w_1}(t) = 0, \tag{14}$$
$$-\tilde{c}_{2w_2}(t)x_2(t) - (\tilde{c}_2(t) - \lambda_2(t))x_{2w_2}(t) = 0, \tag{15}$$
$$-\tilde{c}_{3w_3}(t)x_3(t) - (\tilde{c}_3(t) - \lambda_3(t))x_{3w_3}(t) = 0. \tag{16}$$

Now solving the above Eqs. (14), (15) and (16) for the control variable $w_1(t)$, $w_2(t)$ and $w_3(t)$ respectively, we get

$$w_1(t) = \frac{-(\tilde{c}_1(t) - \lambda_1(t))}{\tilde{c}_{1w_1}(t)},$$
$$w_2(t) = \frac{-(\tilde{c}_2(t) - \lambda_2(t))}{\tilde{c}_{2w_2}(t)},$$
$$w_3(t) = \frac{-(\tilde{c}_3(t) - \lambda_3(t))}{\tilde{c}_{3w_3}(t)}.$$

Other optimality conditions are $H_{w_1w_1} \leq 0$,

$$\begin{vmatrix} H_{w_1w_1} & H_{w_1w_2} \\ H_{w_2w_1} & H_{w_2w_2} \end{vmatrix} \geq 0,$$

and

$$\begin{vmatrix} H_{w_1w_1} & H_{w_1w_2} & H_{w_1w_3} \\ H_{w_2w_1} & H_{w_2w_2} & H_{w_2w_3} \\ H_{w_3w_1} & H_{w_3w_2} & H_{w_3w_3} \end{vmatrix} \leq 0.$$

where

$$H_{w_1w_1} = -\tilde{c}_{1w_1w_1}(t)x_1(t) - 2\tilde{c}_{1w_1}(t)x_{1w_1}(t) - (\tilde{c}_1(t) - \lambda_1(t))x_{1w_1w_1}(t) \leq 0,$$
$$H_{w_2w_2} = -\tilde{c}_{2w_2w_2}(t)x_2(t) - 2\tilde{c}_{2w_2}(t)x_{2w_2}(t) - (\tilde{c}_2(t) - \lambda_2(t))x_{2w_2w_2}(t) \leq 0,$$
$$H_{w_3w_3} = -\tilde{c}_{3w_3w_3}(t)x_3(t) - 2\tilde{c}_{3w_3}(t)x_{3w_3}(t) - (\tilde{c}_3(t) - \lambda_3(t))x_{3w_3w_3}(t) \leq 0,$$
$$H_{w_1w_2} = 0,\ H_{w_1w_3} = 0,\ H_{w_2w_1} = 0,\ H_{w_2w_3} = 0,\ H_{w_3w_1} = 0,\ H_{w_3w_2} = 0.$$

Now taking derivative of Eqs. (14), (15) and (16) with respect to t

$$\dot{w}_1(t) = \frac{(\tilde{c}_{1m_d}(t)x_1(t) + \tilde{c}_1(t)x_{1m_d}(t) + \tilde{c}_2(t)x_{2m_d}(t) - \lambda_1(t)x_{1m_d}(t))x_{1w_1}(t)}{\tilde{c}_{1w_1w_1}(t)x_1(t) + 2\tilde{c}_{1w_1}(t)x_{1w_1}(t) + \tilde{c}_1(t)x_{1w_1w_1}(t) - \lambda_1(t)x_{1w_1w_1}(t)},$$
$$\dot{w}_2(t) = \frac{(\tilde{c}_{2m_i}(t)x_2(t) + \tilde{c}_2(t)x_{2m_i}(t) + \tilde{c}_3(t)x_{3m_i}(t) - \lambda_2(t)x_{2m_i}(t))x_{2w_2}(t)}{\tilde{c}_{2w_2w_2}(t)x_2(t) + 2\tilde{c}_{2w_2}(t)x_{2w_2}(t) + \tilde{c}_2(t)x_{2w_2w_2}(t) - \lambda_2(t)x_{2w_2w_2}(t)},$$
$$\dot{w}_3(t) = \frac{(\tilde{c}_{3m_r}(t)x_3(t) + \tilde{c}_3(t)x_{3m_r}(t) - \lambda_3(t)x_{3m_r}(t))x_{3w_3}(t)}{\tilde{c}_{3w_3w_3}(t)x_3(t) + 2\tilde{c}_{3w_3}(t)x_{3w_3}(t) + \tilde{c}_3(t)x_{3w_3w_3}(t) - \lambda_3(t)x_{3w_3w_3}(t)},$$

where

$$\tilde{c}_{1w_1} = \frac{\partial \tilde{c}_1}{\partial w_1}, \tilde{c}_{1w_1w_1} = \frac{\partial^2 \tilde{c}_1}{\partial w_1^2}, x_{1w_1} = \frac{\partial x_1}{\partial w_1}, x_{1w_1w_1} = \frac{\partial^2 x_1}{\partial w_1^2}, \tilde{c}_{2w_2} = \frac{\partial \tilde{c}_2}{\partial w_2}, \tilde{c}_{2w_2w_2} = \frac{\partial^2 \tilde{c}_2}{\partial w_2^2},$$

$$x_{2w_2} = \frac{\partial x_2}{\partial w_2}, x_{2w_2w_2} = \frac{\partial^2 x_2}{\partial w_2^2}, \tilde{c}_{3w_3} = \frac{\partial \tilde{c}_3}{\partial w_3}, \tilde{c}_{3w_3w_3} = \frac{\partial^2 \tilde{c}_3}{\partial w_3^2}, x_{3w_3} = \frac{\partial x_3}{\partial w_3}, x_{3w_3w_3} = \frac{\partial^2 x_3}{\partial w_3^2}.$$

4.1 Special Cases

The following scenarios are depicted to show the behaviour of the proposed model. We have taken different functional forms for detection cost, isolation cost, and removal cost to analyze the behaviour of the control model and related optimal policies.

Case-1: In this subsection, we have assumed that per unit detection cost associated with detection efforts $w_1(t)$, per unit isolation cost associated with isolation efforts $w_2(t)$ and per unit removal cost associated with removal efforts $w_3(t)$ are constant.

$$i.e \;\; \tilde{c}_1(t) = c_1, \tilde{c}_2(t) = c_2 \; and \; \tilde{c}_3(t) = c_3.$$

Then, the objective function can be written as:

$$\min \left[\int_0^T \left\{ c_1 x_1(t) + c_2 x_2(t) + c_3 x_3(t) \right\} dt \right] \tag{17}$$

subject to

$$x_1(t) = \frac{dm_d(t)}{dt} = w_1(t) b_1 \left[a - m_d(t) \right], \qquad 0 \le t \le T, \tag{18}$$

$$x_2(t) = \frac{dm_i(t)}{dt} = w_2(t) b_2 \left[m_d(t) - m_i(t) \right], \quad 0 \le t \le T, \tag{19}$$

$$x_3(t) = \frac{dm_r(t)}{dt} = w_3(t) b_3 \left[m_i(t) - m_r(t) \right], \quad 0 \le t \le T, \tag{20}$$

with the conditions $m_d(0) = 0$, $m_i(0) = 0$ and $m_r(0) = 0$. Then the Hamiltonian function is given by

$$
\begin{aligned}
H(t) &= -c_1 x_1(t) - c_2 x_2(t) - c_3 x_3(t) + \lambda_1(t) x_1(t) + \lambda_2(t) x_2(t) + \lambda_3(t) x_3(t) \\
&= -(c_1 - \lambda_1(t)) x_1(t) - (c_2 - \lambda_2(t)) x_2(t) - (c_3 - \lambda_3(t)) x_3(t) \\
&= -(c_1 - \lambda_1(t)) w_1(t) b_1 (a - m_d(t)) - (c_2 - \lambda_2(t)) w_2(t) b_2 (m_d(t) - m_i(t)) \\
&\quad - (c_3 - \lambda_3(t)) w_3(t) b_3 (m_i(t) - m_r(t)).
\end{aligned}
$$

The co-state variable $\lambda_1(t)$, $\lambda_2(t)$ and $\lambda_3(t)$ is defined as

$$\frac{d}{dt}\lambda_1(t) = \dot{\lambda}_1(t) = -b_1 w_1(t)(c_1 - \lambda_1(t)) + b_2 w_2(t)(c_2 - \lambda_2(t)), \tag{21}$$

$$\frac{d}{dt}\lambda_2(t) = \dot{\lambda}_2(t) = -b_2 w_2(t)(c_2 - \lambda_2(t)) + b_3 w_3(t)(c_3 - \lambda_3(t)), \tag{22}$$

$$\frac{d}{dt}\lambda_3(t) = \dot{\lambda}_3(t) = -b_3 w_3(t)(c_3 - \lambda_3(t)), \tag{23}$$

with the transversality conditions at $t = T^*$, $H(T^*) = 0$, and $\lambda_1(T) = 0$, $\lambda_2(T) = 0$, and $\lambda_3(T^*) \le 0 (= 0 \text{ if } m_r^*(T^*) > m_t)$. Solving Eqs. (21), (22) and (23) together with terminal conditions to get

$$
\begin{aligned}
\lambda_1(t) &= \int_0^T \Big\{ b_1 w_1(t)(c_1 - \lambda_1(t)) - b_2 w_2(t)(c_2 - \lambda_2(t)) \Big\} dt, \\
\lambda_2(t) &= \int_0^T \Big\{ b_2 w_2(t)(c_2 - \lambda_2(t)) - b_3 w_3(t)(c_3 - \lambda_3(t)) \Big\} dt, \\
\lambda_3(t) &= \lambda_3(T) + \int_0^T \Big\{ b_3 w_3(t)(c_3 - \lambda_3(t)) \Big\} dt.
\end{aligned}
$$

The necessary condition for optimality are $\frac{\partial H(t)}{\partial w_1(t)} = 0$, $\frac{\partial H(t)}{\partial w_2(t)} = 0$ and $\frac{\partial H(t)}{\partial w_3(t)} = 0$. For optimal policy, let us assume the following:

$$
\begin{aligned}
\alpha_1(t) &= (\lambda_1(t) - c_1) b_1 (a - m_d(t)), \\
\alpha_2(t) &= (\lambda_2(t) - c_2) b_2 (m_d(t) - m_i(t)), \\
\alpha_3(t) &= (\lambda_3(t) - c_3) b_3 (m_i(t) - m_r(t)).
\end{aligned}
$$

Then, the Hamiltonian can be written as

$$H(t) = \alpha_1(t) w_1(t) + \alpha_2(t) w_2(t) + \alpha_3(t) w_3(t).$$

Since Hamiltonian is linear in control parameters $w_1(t)$, $w_2(t)$ and $w_3(t)$. So, we have the following optimal policies, given in Table 1, for $w_1(t)$, $w_2(t)$ and $w_3(t)$ which maximizes the objective function.

The cases presented in Table 1, can be summarized in Fig. 2 and Table 2.

Table 1 The optimal policy for detection effort, isolation effort and removal effort for various values of α_1, α_2 and α_3

Subcase	Condition on α_1, α_2 & α_3	Optimal controls	Characterization
1	$\alpha_1 = 0, \alpha_2 = 0, \alpha_3 = 0$	w_1, w_2, w_3 not defined	Bang-Bang
2	$\alpha_1 = 0, \alpha_2 > 0, \alpha_3 > 0$	$0 \le w_1 \le 1, w_2 = 1, w_3 = 1$	Singular
3	$\alpha_1 = 0, \alpha_2 < 0, \alpha_3 < 0$	$0 \le w_1 \le 1, w_2 = 0, w_3 = 0$	Singular
4	$\alpha_1 = 0, \alpha_2 > 0, \alpha_3 < 0$	$0 \le w_1 \le 1, w_2 = 1, w_3 = 0$	Singular
5	$\alpha_1 = 0, \alpha_2 < 0, \alpha_3 > 0$	$0 \le w_1 \le 1, w_2 = 0, w_3 = 1$	Singular
6	$\alpha_1 = 0, \alpha_2 > 0, \alpha_3 = 0$	$0 \le w_1 \le 1, w_2 = 1,$ $0 \le w_3 \le 1$	Singular
7	$\alpha_1 = 0, \alpha_2 = 0, \alpha_3 > 0$	$0 \le w_1 \le 1, 0 \le w_2 \le 1,$ $w_3 = 1$	Singular
8	$\alpha_1 = 0, \alpha_2 < 0, \alpha_3 = 0$	$0 \le w_1 \le 1, w_2 = 0,$ $0 \le w_3 \le 1$	Singular
9	$\alpha_1 = 0, \alpha_2 = 0, \alpha_3 < 0$	$0 \le w_1 \le 1, 0 \le w_2 \le 1,$ $w_3 = 0$	Singular
10	$\alpha_1 > 0, \alpha_2 = 0, \alpha_3 = 0$	$w_1 = 1, 0 \le w_2 \le 1,$ $0 \le w_3 \le 1$	Singular
11	$\alpha_1 > 0, \alpha_2 > 0, \alpha_3 > 0$	$w_1 = 1, w_2 = 1, w_3 = 1$	Bang-Bang
12	$\alpha_1 > 0, \alpha_2 < 0, \alpha_3 < 0$	$w_1 = 1, w_2 = 0, w_3 = 0$	Bang-Bang
13	$\alpha_1 > 0, \alpha_2 > 0, \alpha_3 < 0$	$w_1 = 1, w_2 = 1, w_3 = 0$	Bang-Bang
14	$\alpha_1 > 0, \alpha_2 < 0, \alpha_3 > 0$	$w_1 = 1, w_2 = 0, w_3 = 1$	Bang-Bang
15	$\alpha_1 > 0, \alpha_2 > 0, \alpha_3 = 0$	$w_1 = 1, w_2 = 1, 0 \le w_3 \le 1$	Singular
16	$\alpha_1 > 0, \alpha_2 = 0, \alpha_3 > 0$	$w_1 = 1, 0 \le w_2 \le 1, w_3 = 1$	Singular
17	$\alpha_1 > 0, \alpha_2 < 0, \alpha_3 = 0$	$w_1 = 1, w_2 = 0, 0 \le w_3 \le 1$	Singular
18	$\alpha_1 > 0, \alpha_2 = 0, \alpha_3 < 0$	$w_1 = 1, 0 \le w_2 \le 1, w_3 = 0$	Singular
19	$\alpha_1 < 0, \alpha_2 = 0, \alpha_3 = 0$	$w_1 = 0, 0 \le w_2 \le 1,$ $0 \le w_3 \le 1$	Singular
20	$\alpha_1 < 0, \alpha_2 > 0, \alpha_3 > 0$	$w_1 = 0, w_2 = 1, w_3 = 1$	Bang-Bang
21	$\alpha_1 < 0, \alpha_2 < 0, \alpha_3 < 0$	$w_1 = 0, w_2 = 0, w_3 = 0$	Bang-Bang
22	$\alpha_1 < 0, \alpha_2 > 0, \alpha_3 < 0$	$w_1 = 0, w_2 = 1, w_3 = 0$	Bang-Bang
23	$\alpha_1 < 0, \alpha_2 < 0, \alpha_3 > 0$	$w_1 = 0, w_2 = 0, w_3 = 1$	Bang-Bang
24	$\alpha_1 < 0, \alpha_2 > 0, \alpha_3 = 0$	$w_1 = 0, w_2 = 1, 0 \le w_3 \le 1$	Singular
25	$\alpha_1 < 0, \alpha_2 = 0, \alpha_3 > 0$	$w_1 = 0, 0 \le w_2 \le 1, w_3 = 1$	Singular
26	$\alpha_1 < 0, \alpha_2 < 0, \alpha_3 = 0$	$w_1 = 0, w_2 = 0, 0 \le w_3 \le 1$	Singular
27	$\alpha_1 < 0, \alpha_2 = 0, \alpha_3 < 0$	$w_1 = 0, 0 \le w_2 \le 1, w_3 = 0$	Singular

Case-2: Compared with constant fault detection cost, isolation cost, and removal cost discussed in case-1, we have taken a scenario when the fault detection cost, isolation cost, and removal cost are dynamic. In this subsection, we have considered the cost per unit fault detection, isolation, and removal following the learning curve phenomenon [40]. The fault detection cost, isolation cost, and removal cost are taken

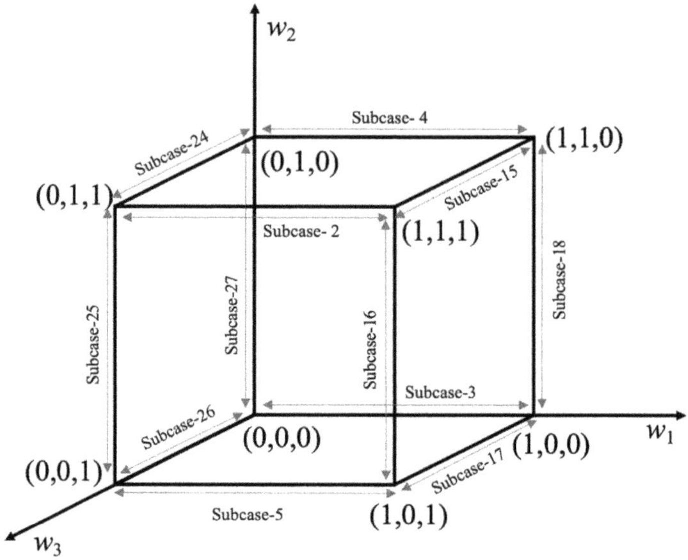

Fig. 2 Graph showing optimal policy of detection effort (w_1), isolation effort (w_2) and removal effort (w_3) at any time *t*

Table 2 Optimal policy shown by co-ordinate points

Co-ordinate points	Remarks
(0, 0, 0)	Subcase-1
(1, 0, 0)	Subcase-12
(1, 1, 0)	Subcase-13
(0, 1, 0)	Subcase-22
(0, 1, 1)	Subcase-20
(0, 0, 1)	Subcase-23
(1, 0, 1)	Subcase-14
(1, 1, 1)	Subcase-11

as a function of detection effort, isolation effort, and removal effort, respectively.

$$\tilde{c}_1(t) = b_0(w_1(t))^{m_d(t)}, \quad \tilde{c}_2(t) = c_0(w_2(t))^{m_i(t)}, \quad \tilde{c}_3(t) = d_0(w_3(t))^{m_r(t)},$$

where b_0, c_0 and d_0 are base detection cost, isolation cost and removal cost respectively. Then the Hamiltonian function can be written as

$$H(t) = -\tilde{c}_1(t)x_1(t) - \tilde{c}_2(t)x_2(t) - \tilde{c}_3(t)x_3(t) + \lambda_1(t)x_1(t) + \lambda_2(t)x_2(t) + \lambda_3(t)x_3(t).$$

The co-state variables along with terminal conditions are given below

$$\lambda_1(t) = \int_0^T \Big\{b_1 w_1(t)(\tilde{c}_1(t) - \lambda_1(t)) - b_2 w_2(t)(\tilde{c}_2(t) - \lambda_2(t)) \\ - \tilde{c}_{1m_d}(t) x_1(t) \log(w_1(t))\Big\} dt,$$

$$\lambda_2(t) = \int_0^T \Big\{b_2 w_2(t)(\tilde{c}_2(t) - \lambda_2(t)) - b_3 w_3(t)(\tilde{c}_3(t) - \lambda_3(t)) \\ - \tilde{c}_{2m_i}(t) x_2(t) \log(w_2(t))\Big\} dt,$$

$$\lambda_3(t) = \lambda_3(T) + \int_0^T \Big\{b_3 w_3(t)(\tilde{c}_3(t) - \lambda_3(t)) - \tilde{c}_{3m_r}(t) x_3(t) \log(w_3(t))\Big\} dt.$$

From the necessary condition of optimality $\frac{\partial H(t)}{\partial w_1(t)} = 0$, $\frac{\partial H(t)}{\partial w_2(t)} = 0$ and $\frac{\partial H(t)}{\partial w_3(t)} = 0$, we get

$$w_1(t) = \left\{\frac{\lambda_1(t) - \tilde{c}_1(t)}{b_0 m_d(t)}\right\}^{\frac{1}{m_d(t)}},$$
$$w_2(t) = \left\{\frac{\lambda_2(t) - \tilde{c}_2(t)}{c_0 m_i(t)}\right\}^{\frac{1}{m_i(t)}},$$
$$w_3(t) = \left\{\frac{\lambda_3(t) - \tilde{c}_3(t)}{d_0 m_r(t)}\right\}^{\frac{1}{m_r(t)}}.$$

5 Numerical Analysis

In this section, we demonstrate the behaviour of the cost minimization model numerically. This study aims to get some view into the result and study the impact of detection effort, isolation effort, and removal effort on the objective function. We have conducted several simulations by taking different values of parameters. While doing simulation, the base value of parameters is as follows.

$$a = 100, b_1 = 0.3, b_2 = 0.35, b_3 = 0.4, \quad w_1 = 0.7, w_2 = 0.75, w_3 = 0.8, b_0 = 1000, \\ c_0 = 1000, d_0 = 1000, m_t = 90.$$

In this analysis, the main aim is to check importance of efforts (w_1, w_2, w_3) on software development life cycle. During sensitive analysis, we have taken four sets of efforts (w_1, w_2, w_3) and it has been observed that, when the value of efforts (w_1, w_2, w_3) gradually increases the cumulative detected faults, isolated faults and removed faults are also increases. The four sets of efforts are: Case-1: $w_1 = 0.7$, $w_2 = 0.75$, $w_3 = 0.8$; Case-2: $w_1 = 0.75$, $w_2 = 0.8$, $w_3 = 0.85$; Case-3: $w_1 = 0.8$, $w_2 = 0.85$, $w_3 = 0.9$; Case-4: $w_1 = 0.85$, $w_2 = 0.9$, $w_3 = 0.95$. The behaviour of cumulative

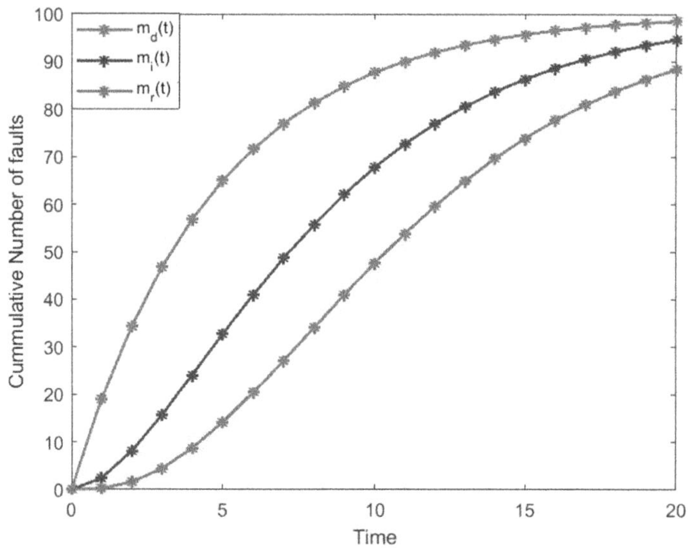

Fig. 3 Number of detected, isolated and removed faults versus time (Case-1)

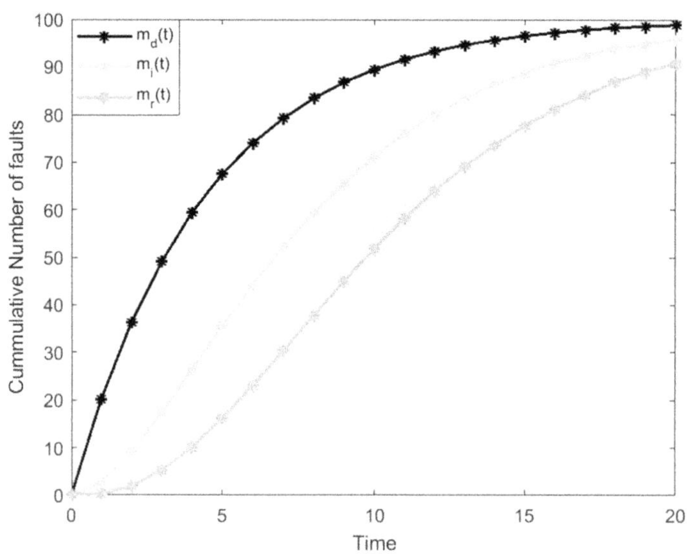

Fig. 4 Number of detected, isolated and removed faults versus time (Case-2)

number of detected faults, isolated faults and removed faults are depicted in Figs. 3, 4, 5, and 6 for Case-1, Case-2, Case-3, and Case-4 respectively.

The analysis was also done to show how the future cost of detection, isolation and removal stage for different sets of efforts (w_1, w_2, w_3). The pattern is depicted

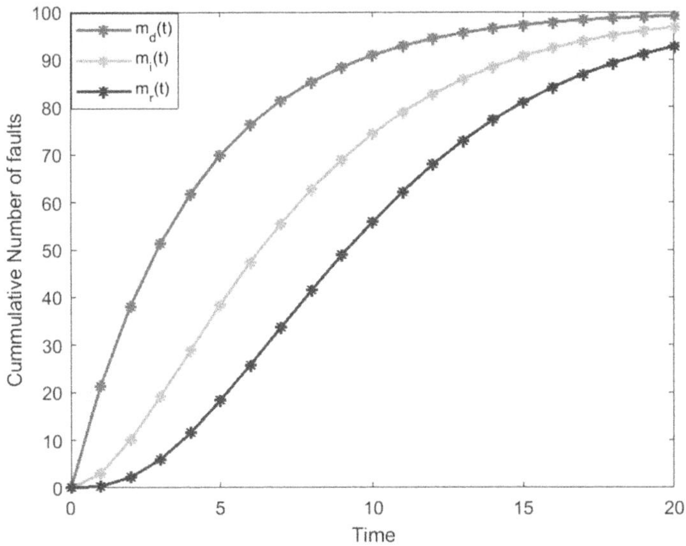

Fig. 5 Number of detected, isolated and removed faults versus time (Case-3)

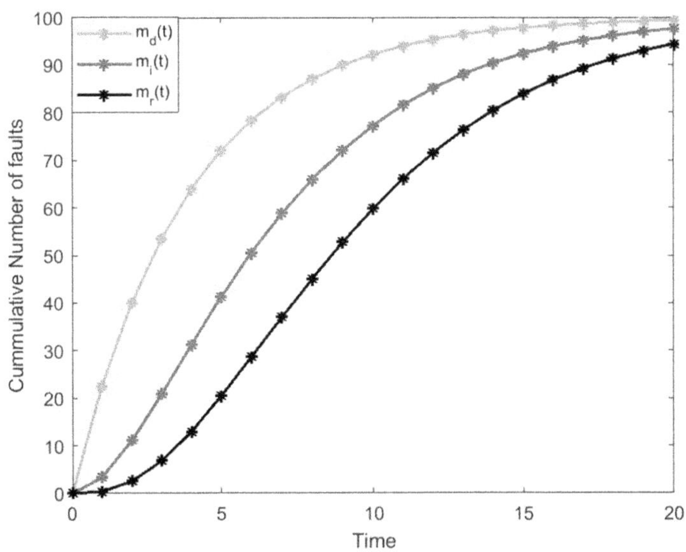

Fig. 6 Number of detected, isolated and removed faults versus time (Case-4)

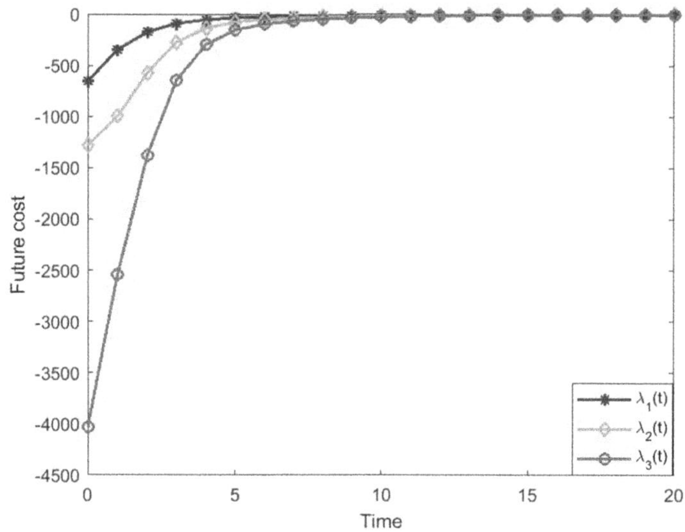

Fig. 7 Shadow cost versus time (Case-1)

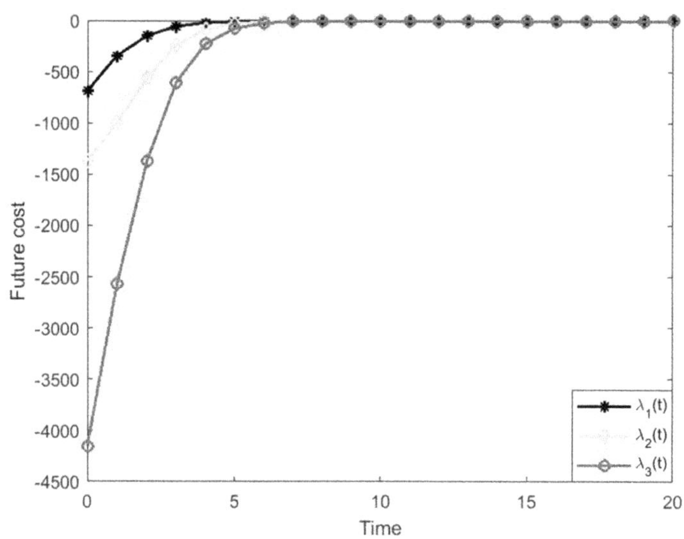

Fig. 8 Shadow cost versus time (Case-2)

in Figs. 7, 8, 9, and 10 for Case-1, Case-2, Case-3, and Case-4 respectively, which tells that the co-state variables for detection, isolation, and removal stage decrease with time and approach zero for different efforts.

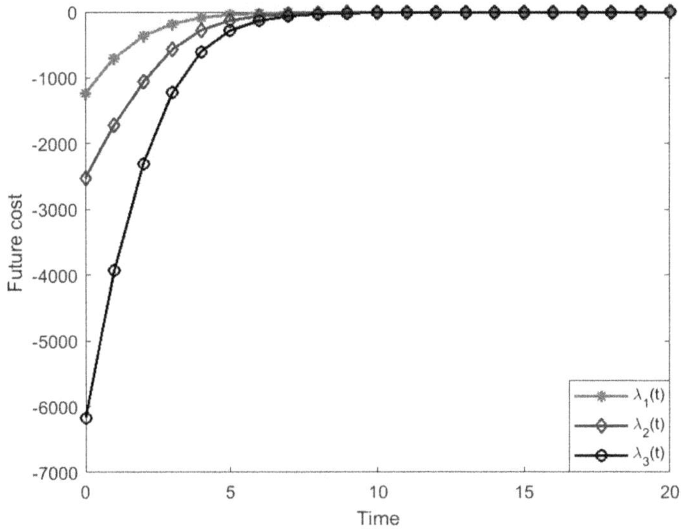

Fig. 9 Shadow cost versus time (Case-3)

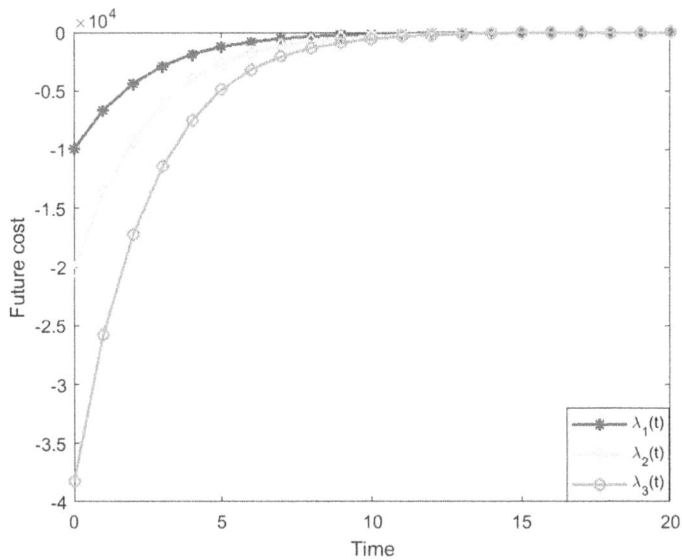

Fig. 10 Shadow cost versus time (Case-4)

6 Conclusion

In this paper, we have discussed the changing trend of software fault detection, isolation, and removal over time. Practically, software faults cannot be removed completely due to limitations of resources, operating environment, etc. Resources utilized in the detection, isolation, and removal stage are practical issues in the software development process to obtain reliable software. We have proposed the optimal policy using optimal control theory, and a control-theoretic approach is used to solve the cost minimization model. During analysis, we have observed from the graph of the future cost that, as time increases, the future cost tends to zero due to the learning curve phenomenon. And consequently, shadow cost tends to zero as time increases.

A few limitations in our control problem that suggest future research. The objective function is proposed by assuming detection, isolation, and removal costs in terms of detection, isolation, and removal. We may take a different form of detection, isolation, and removal cost functional form. To present a more realistic SRGM, a stochastic model can be considered. We can also extend the proposed model by incorporating different fault content functions. All these issues will be part of our further work [17, 18].

References

1. Goel AL, Okumoto K (1979) Time-dependent error-detection rate model for software reliability and other performance measures. IEEE Trans Reliab 28(3):206–211
2. Ohba M, Yamada S (1984) S-shaped software reliability growth models. International colloquium on reliability and maintainability, 4th edn. Tregastel, France, pp 430–436
3. Ohba M (1984) Software reliability analysis models. IBM J Res Dev 28(4):428–443
4. Ohba M, Chou X-M (1989) Does imperfect debugging affect software reliability growth?. In: Proceedings of the 11th international conference on Software engineering, pp 237–244
5. Musa JD, Iannino A, Okumoto K (1990) Software reliability. Adv Comput 30:85–170
6. Huang C-Y, Lin C-T, Kuo S-Y, Lyu MR, Sue C-C (2004) Software reliability growth models incorporating fault dependency with various debugging time lags. In: Proceedings of the 28th annual international computer software and applications conference, 2004. COMPSAC 2004. IEEE, pp 186–191
7. Ji Y, Mookerjee VS, Sethi SP (2005) Optimal software development: a control theoretic approach. Inf Syst Res 16(3):292–306
8. Kapur P, Goswami D, Gupta A (2004) A software reliability growth model with testing effort dependent learning function for distributed systems. Int J Reliab Qual Saf Eng 11(04):365–377
9. Huang C-Y, Kuo S-Y (2002) Analysis of incorporating logistic testing-effort function into software reliability modeling. IEEE Trans Reliab 51(3):261–270
10. Huang C-Y, Lyu MR, Kuo S-Y (2003) A unified scheme of some nonhomogenous poisson process models for software reliability estimation. IEEE Trans Softw Eng 29(3):261–269
11. Ohba M (1984) Inflection s-shaped software reliability growth model. In: Stochastic models in reliability theory. Springer, pp 144–162
12. Pham H (1993) Software reliability assessment: imperfect debugging and multiple fault types in software development eg&g-raam-10737. Idaho National Laboratory

13. Pham H (1996) A software cost model with imperfect debugging, random life cycle and penalty cost. Int J Syst Sci 27(5):455–463
14. Pham H, Zhang X (1997) An nhpp software reliability model and its comparison. Int J Reliab Qual Saf Eng 4(03):269–282
15. Pham H, Nordmann L, Zhang Z (1999) A general imperfect-software-debugging model with s-shaped fault-detection rate. IEEE Trans Reliab 48(2):169–175
16. Zhu M, Pham H (2018) A two-phase software reliability modeling involving with software fault dependency and imperfect fault removal. Comput Lang Syst Struct 53:27–42
17. Chang Y-P (2001) Estimation of parameters for nonhomogeneous poisson process: software reliability with change-point model. Commun Stat-Simul Comput 30(3):623–635
18. Kapur P, Goswami D, Bardhan A, Singh O (2008) Flexible software reliability growth model with testing effort dependent learning process. Appl Math Model 32(7):1298–1307
19. Li Q, Pham H (2017) Nhpp software reliability model considering the uncertainty of operating environments with imperfect debugging and testing coverage. Appl Math Model 51:68–85
20. Yamada S, Ohba M, Osaki S (1983) S-shaped reliability growth modeling for software error detection. IEEE Trans Reliab 32(5):475–484
21. Yamada S, Tokuno K, Osaki S (1992) Imperfect debugging models with fault introduction rate for software reliability assessment. Int J Syst Sci 23(12):2241–2252
22. Grottke M, Trivedi KS (2005) A classification of software faults. J Reliab Eng Assoc Jpn 27(7):425–438
23. Grottke M, Trivedi KS (2007) Fighting bugs: remove, retry, replicate, and rejuvenate. Computer 40(2):107–109
24. Laprie J-C, Arlat J, Beounes C, Kanoun K (1990) Definition and analysis of hardware-and software-fault-tolerant architectures. Computer 23(7):39–51
25. Yamada S, Hishitani J, Osaki S (1993) Software-reliability growth with a weibull test-effort: a model and application. IEEE Trans Reliab 42(1):100–106
26. Jin C, Jin S-W (2016) Parameter optimization of software reliability growth model with s-shaped testing-effort function using improved swarm intelligent optimization. Appl Soft Comput 40:283–291
27. Huang C-Y, Kuo S-Y, Lyu MR (2007) An assessment of testing-effort dependent software reliability growth models. IEEE Trans Reliab 56(2):198–211
28. Kumar V, Sahni R (2016) An effort allocation model considering different budgetary constraint on fault detection process and fault correction process. Decis Sci Lett 5(1):143–156
29. Kapur P, Pham H, Chanda U, Kumar V (2013) Optimal allocation of testing effort during testing and debugging phases: a control theoretic approach. Int J Syst Sci 44(9):1639–1650
30. Kumar V, Sahni R (2020) Dynamic testing resource allocation modeling for multi-release software using optimal control theory and genetic algorithm. Int J Qual Reliab Manage
31. Kapur P, Pham H, Kumar V, Anand A (2012) Dynamic optimal control model for profit maximization of software product under the influence of promotional effort. J High Technol Manage Res 23(2):122–129
32. Pradhan SK, Kumar A, Kumar V (2021) An optimal resource allocation model considering two-phase software reliability growth model with testing effort and imperfect debugging. Reliab Theory Appl SI 2(64):241–255
33. Kumar V, Mathur P, Sahni R, Anand M (2016) Two-dimensional multi-release software reliability modeling for fault detection and fault correction processes. Int J Reliab Qual Saf Eng 23(03):1640002
34. Ji Y, Kumar S, Mookerjee VS, Sethi SP, Yeh D (2011) Optimal enhancement and lifetime of software systems: a control theoretic analysis. Prod Oper Manage 20(6):889–904
35. Saxena P, Kumar V, Ram M (2021) Ranking of software reliability growth models: a entropy-electre hybrid approach. Reliab Theory Appl SI 2(64):95–113
36. Kumar V, Saxena P, Garg H (2021) Selection of optimal software reliability growth models using an integrated entropy–technique for order preference by similarity to an ideal solution (topsis) approach. In: Mathematical methods in the applied sciences

37. Yamada S, Osaki S (1987) Optimal software release policies with simultaneous cost and reliability requirements. Eur J Oper Res 31(1):46–51
38. Kapur P, Goswami D, Bardhan A (2007) A general software reliability growth model with testing effort dependent learning process. Int J Model Simulat 27(4):340–346
39. Naidu DS (2002) Optimal control systems. CRC Press
40. Pegels CC (1969) On startup or learning curves: an expanded view. AIIE Trans 1(3):216–222

GPSR Compliance
The European Union's (EU) General Product Safety Regulation (GPSR) is a set of rules that requires consumer products to be safe and our obligations to ensure this.

If you have any concerns about our products, you can contact us on

ProductSafety@springernature.com

In case Publisher is established outside the EU, the EU authorized representative is:

Springer Nature Customer Service Center GmbH
Europaplatz 3
69115 Heidelberg, Germany

www.ingramcontent.com/pod-product-compliance
Ingram Content Group UK Ltd.
Pitfield, Milton Keynes, MK11 3LW, UK
UKHW020245290926
15470UKWH00002B/759
* 9 7 8 3 0 3 1 0 5 3 4 6 7 *